International Conference on Applied Electronic and Engineering (ICAEE 2017)

IOP Conference Series: Materials Science and Engineering
Volume 341

Kuching, Malaysia
7 – 8 August 2017

ISBN: 978-1-5108-6117-6
ISSN: 1757-8981

Printed from e-media with permission by:

Curran Associates, Inc.
57 Morehouse Lane
Red Hook, NY 12571

Some format issues inherent in the e-media version may also appear in this print version.

Printed by Curran Associates, Inc. (2018)

For permission requests, please contact the Institute of Physics
at the address below.

Institute of Physics
Dirac House, Temple Back
Bristol BS1 6BE UK

Phone: 44 1 17 929 7481
Fax: 44 1 17 920 0979

techtracking@iop.org

Additional copies of this publication are available from:

Curran Associates, Inc.
57 Morehouse Lane
Red Hook, NY 12571 USA
Phone: 845-758-0400
Fax: 845-758-2633
Email: curran@proceedings.com
Web: www.proceedings.com

Table of contents

Volume 341

International Conference on Applied Electronic and Engineering 2017 (ICAEE2017)

7–8 August 2017, Kuching, Sarawak, Malaysia

Accepted papers received: 16 March 2018
Published online: 9 April 2018

Preface

Papers

Applied Electronic

IOP Conf. Series: Materials Science and Engineering **341** (2018) 011001 doi:10.1088/1757-899X/341/1/011001

International Conference on Applied Electronic and Engineering 2017 (ICAEE 2017)

Abdul Hadi Abdul Razak (Chair)

Centre for Electronic Engineering Studies, Faculty of Electrical Engineering, Universiti Teknologi MARA, Shah Alam, Selangor, Malaysia

Faculty of Electrical Engineering, Universiti Teknologi MARA Cawangan Sarawak, Kota Samarahan, Sarawak Malaysia

Abstract. On behalf of the Organizing Committee of ICAEE2017, and with great pleasure, we would like to welcome all delegates to Kuching, Sarawak, Malaysia. Being held from 7th to 8th August 2017 at Hilton Hotel, Kuching, the event is organized by the Centre for Electronic Engineering Studies, Faculty of Electrical Engineering, Universiti Teknologi MARA (UiTM), Malaysia. The conference creates a forum for engineers, scientists and practitioners of varied disciplines and their associated applications. Throughout this event we hope to create an opportunity for researchers, academics and students to get together and, more importantly, to welcome new peers in diverse areas of expertise. On behalf of the organizing committee we would like to take the opportunity to express our gratitude to all reviewers who have been working hard to review the papers on time and share their expertise. Special mention goes to the distinguished Keynote Speakers for their willingness to spend their time and share their valuable knowledge. My deepest gratitude goes to the Organizing Committee who have been working hard behind the scenes for the success of this event. To all participants, we would like to encourage you to fully utilize this platform for knowledge sharing and networking while enjoying your stay in Kuching, Sarawak, Malaysia.

IOP Conf. Series: Materials Science and Engineering **341** (2018) 011002 doi:10.1088/1757-899X/341/1/011002

Peer review statement

All papers published in this volume of *IOP Conference Series: Materials Science and Engineering* have been peer reviewed through processes administered by the proceedings Editors. Reviews were conducted by expert referees to the professional and scientific standards expected of a proceedings journal published by IOP Publishing.

AC to DC Bridgeless Boost Converter for Ultra Low Input Energy Harvesting

[1,2]A. H. A. Dawam, [1,2,a] M. Muhamad

[1]Faculty of Electrical Engineering, Universiti Teknologi Mara (UiTM), 40450 Shah Alam, Selangor, Malaysia, [2]Integrated Microelectronics Systems and Applications Research Group, Universiti Teknologi Mara (UiTM), 40450 Shah Alam, Selangor, Malaysia

a) Corresponding author: maizan@salam.uitm.edu.my

Abstract. This paper presents design of circuit which converts low input AC voltage to a higher output DC voltage. A buck-boost topology and boost topology are combined to condition cycle of an AC input voltage. the unique integration of a combining circuit of buck-boost and boost circuit have been proposed in order to introduce a new direct ac-dc power converter topology without conventional diode bridge rectifier. The converter achieved to convert a milli-volt scale of input AC voltage into a volt scale of output DC voltages which is from 400mV to 3.3V.

1. Introduction

In the current era of our technology, the supplies of fossil fuel are decreasing in a skyrocketing of energy generation costs, so the development of more advance methods in exploiting renewable energy sources are increasing. In a conventional way, power plant or battery are being used in supplying electricity which required electrical wiring and battery replacement. On the other hand, Energy harvesting is a technology that converts freely available energy in the environment into electrical energy which can be further used in low power electronics. The requirement for electronic devices has decreased because of semiconductor devices which lead to the development of technology like electronic devices, wireless sensors, medical implants etc. There are various sources for energy harvesting [1] such as solar, organic, vibration, thermal, near-field electromagnetic and far-field electromagnetic or Radio Frequency RF. Energy harvesting, or energy scavenging is a process where energy is captured from surrounding system's environments and extracted into useable electrical energy [2]. If there is an energy lost occur, the energy will be lost in form of heat, light, sound, vibration or movement. Energy harvesting are the greatest advancement in the power microelectronics technology which it can replace batteries for small and low power electronic devices.

This research mainly focuses on the development on the simulation of AC-DC boost converter for ultra-low input low energy harvesting to investigate it`s ability and developing the cost effective and environmentally friendly prototype. The design was simulated using power electronic simulation software (PSIM) to determine total energy converted from the power converter. The ability of the AC-DC converter was analyzed by the voltage and current output produced.

2. Methodology

2.1. Basic principle of AC-DC converter

AC-DC converters serve as a rectifier. A rectifier is an electrical devices that converts alternating current into direct current which periodically in reverse direction or polarity in a one direction [4]. It is called as rectification process. This rectifier has many applications such as component of power supplies and as the radio signal detectors. There are two basic concept of rectification which are full wave rectification and half wave rectification as shown in Figure 1. The function of diode is to clamp the negative side of the input voltage in both cases. For half wave rectification, only a single diode is used to clamp the negative side of input ac voltage to form a half positive side of waveform. Whereas, for the full bridge rectification, there is four combining diode that structured like a bridge. The diode then, will inverts the negative side of waveform and turns it into positive side and the result of the full wave is direct current.

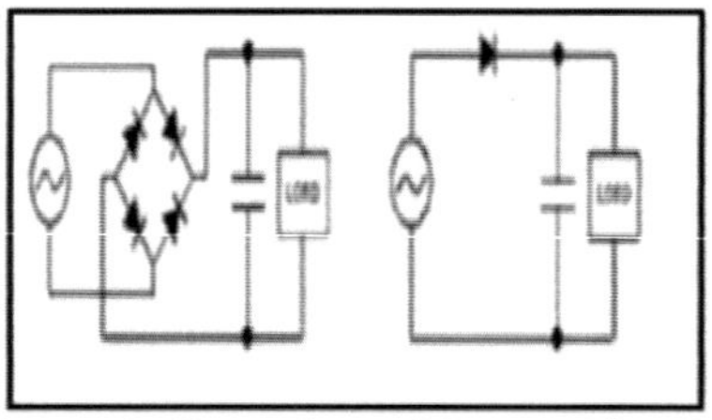

Figure 1: Operation of the rectification

After the conversion of direct current, the capacitor smoothed the waveform. The ripple of the waveform remained and as for the ripple voltage, it changed as a function of the of the capacitor and the load.

2.2. Buck-boost converter

Buck boost converter is a type of switched of mode power source that combine two type of converter which is buck and boost in a similar circuit. The circuit included in the category of non-isolated DC-DC converter which provides a regulated DC output voltage from either an AC or DC input voltage. In this case, the electronic switch used are MOSFET, which are opened and closed periodically by the converter. Furthermore, the circuit needs to operate in continuous current mode (CCM) because of the inductor never goes to zero and continuously supply the current for the output. Based on the duty cycle of the switching transistor, the output voltage was modifiable, and the polarity of the output voltage was inverse to the AC voltage input. Circuit diagram of buck-boost converter as shown in Figure 2[6].

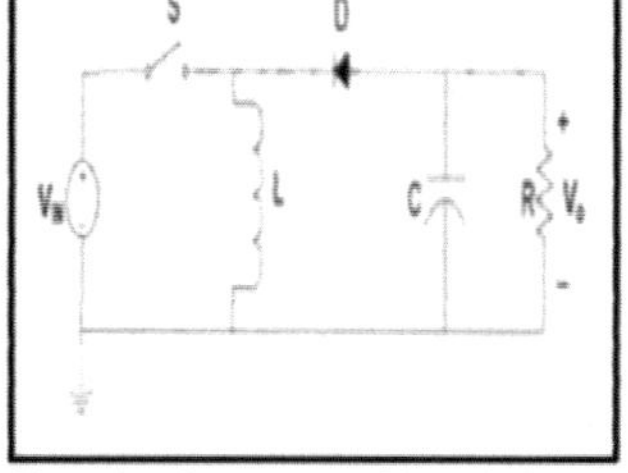

Figure 2. Circuit diagram of buck-boost converter

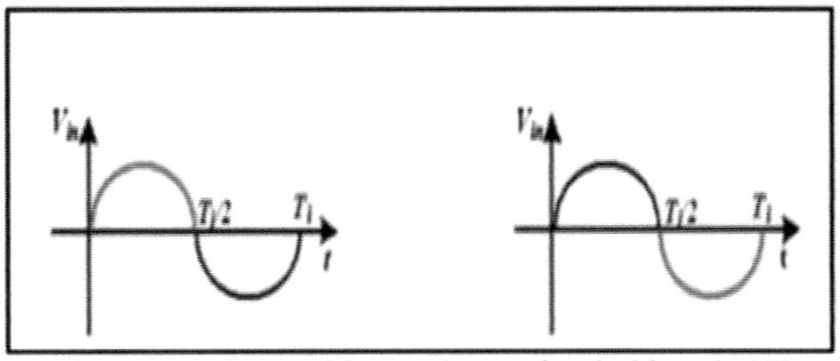

Figure 3. The positive and negative half cycle [7]

There are 2 modes of operation for buck-boost converter:
Mode 1: when switch is turned ON, the diode will be in a reverse biased state. During this state, the inductor gets charged and the inductor current increased.
Mode 2: when the switch is turned OFF, the diode be in a forward biased state as it allows the current to flow from output to input. The energy stored in inductor directed to the load side and the inductor current now will drop until the switch get turned ON again in the next cycle.

2.3. Basic principle of the proposed circuit

The proposed bridgeless boost converter topology is a unique integration of boost and buck-boost converter. The boost converter is the common power conditioning interface due to its simple structure, voltage step-up capability and its high efficiency. While for the buck-boost converter, it literally has an ability to step up the voltage input with a reverse polarity for an instance, the converter managed to condition the negative cycle of the circuit. The advantages of this proposed circuit are both of the buck-boost and boost circuit shared the same single inductor and capacitor, this may led to the reduction of circuit size where it meet the weight and size requirement for this project. Secondly, for the very low voltage micro-generators, the output voltage is so small that its rectification is feasible by the use of MOSFET in the proposed circuit compared with the use of conventional diode which are not feasible at all. 0.4 V as an input voltage with 100 Hertz of frequency sinusoidal AC voltage is adopted to simulate the output.

2.4. Modes of operations

The proposed converter containing six modes of operation which divided into two half cycles, positive half cycle and negative half cycle. The circuit operates in positive half cycle from mode 1 until mode 3 whereas for the circuit to operates in negative half cycle is from mode 4 until mode 6. When the input voltage is in the positive half cycle, S1 is turned ON while the diode D1 is reverse biased, and the circuit operated in boost modes. When the input voltages are in negative half cycle, S2 is turned ON while the diode D2 is reverse biased and the circuit operated in buck-boost modes.

Mode 1: Switch S2 is in ON state at t0, at this state, the inductor is energized by the AC voltage input. As for load, it will be powered by output filter capacitor which had stored energy. Zero Current Switching also occur at this state in order to reduce the switching losses.

Mode 2: Switch S2 is turned OFF, the energy which had been stored in the inductor earlier in mode 1 will energized the load side.

Mode 3: When there is no more energy in the inductor, the inductor current will drop to zero. As soon as this happen, diode, D2 will be in the OFF state automatically. Then, here comes the capacitor to power up the load by using stored in it. The converter will return to mode 1 as soon as switch S2 is turned ON and still in positive half cycle.

Mode 4: Switch S1 is turned ON in this mode, the inductor again will be charged by the AC voltage input. Meanwhile, the energy stored in capacitor will feed the load.

Mode 5: Switch S1 is in OFF condition while on this mode. The energy stored in inductor during mode 4 will be used to feed load side. As the inductor current drop to zero, the diode D1 will turn ON automatically. In this mode, there is an occurrence of switching losses.

Mode 6: When diode D1 is in OFF state, it means that the inductor current had already reach zero. But, the capacitor will continue to power up the load side using the stored left in it. The converter will return to mode 4 if switch S1 is turned ON again and if it still in negative cycle.

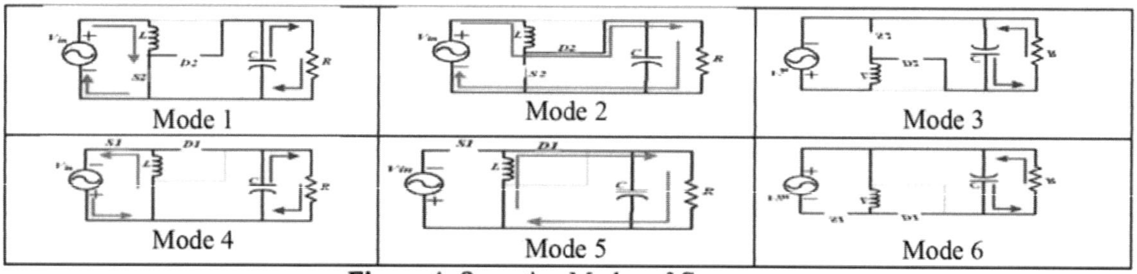

Figure 4. Operation Modes of Converter

2.5. Circuit Design

The proposed circuit contained a buck-boost circuit and boost circuit was simulated using PSIM software [5]. The proposed circuit is modelled and simulated in a closed loop operation. For the simulation in PSIM, the output of the proposed converter was around 3.4V direct current which are then is compared to a reference voltage of 3.3V and the error signal is given to a PI controller. The saturation blocks functions were to produce a hysteresis control that can limit the output of PI controller which mean, it can control the PI output to stay in the certain limits or range. Meanwhile, the saw-tooth waveform from signal generator were used to compared with the control voltage in order for gate pulses to detect the positive and negative peak in the cycle from the converter switches. The electromagnetic microgenerator represented by the sinusoidal AC voltage source in PSIM software to obtain a 3.3V DC as an output voltage. The simulation parameters are as shown below:

Table 1. Simulation parameter

Parameter	Specification
Input voltage	400mV, 100Hz
Switching Frequency	10KHz
Inductor	4.7μH
Capacitor	1000μF
Load resistance	200Ω

With the simulation parameter, the proposed circuit is now completed and successful. The circuit is simulated using PSIM software and being implemented in a closed loop control as shown in Figure 5.

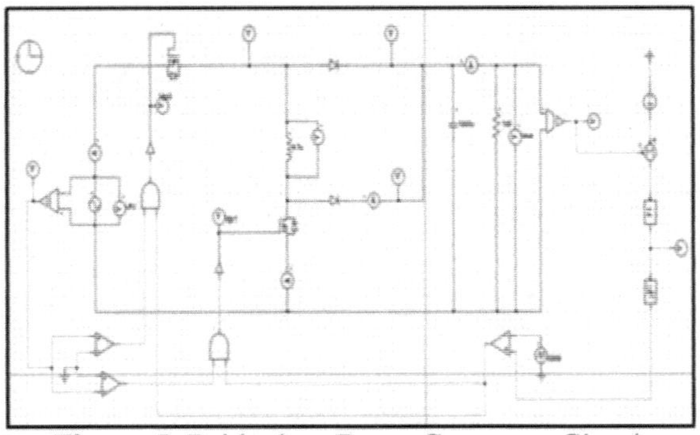

Figure 5. Bridgeless Boost Converter Circuit

3. Results and Discussion

Figure 6 shows the waveform generated for the converter circuit. The figure showed the input voltage is 400mV and the sinusoidal waveform tells that the input is AC voltage from the electromagnetic microgenerators. The figure also shows that the inductor operates during positive and negative half of cycle for the voltage input. The inductor is fully operated when the negative and positive half cycle are being conditioned by the input voltage. The AC voltage of the micro generator is being given to the proposed AC-DC step up converter which rectifies and boosts the AC voltage to a DC voltage of constant magnitude which is found to settle at 3.35V and 0.0166A. The waveform of a sawtooth voltage and PI controller act as a control voltage. The function of both of this voltage are to provide a gate pulses to be given to both switches where the PI controller voltage will cut the sawtooth voltage.

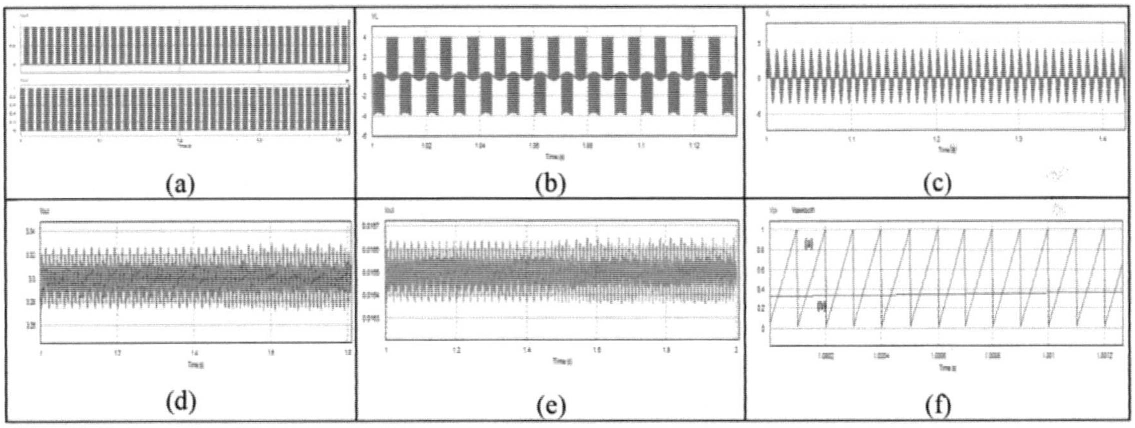

Figure 6. Waveform for the Buck-Boost Converter. (a) Positive & negative half cycle of both switches, (b) Inductor`s voltage, (c) the inductor`s current, (d) Output voltage, (e) Output current, (f) sawtooth voltage and voltage of PI controller

The simulation had been conducted to determine the DC output voltage and current waveform for low energy harvesting devices. Here, to get the desired of DC output voltage, the actual voltage was being compared to 3.3V which are the reference voltage. This is because the input voltage of the proposed circuit is connected to a transducer which getting its energy by converting mechanical energy to electrical energy. The converting process is done by electromagnetic coupling and the transducer used for this experiment are electromagnetic micro-generators which build from mass-spring damper based arrangement. As for the method used in this project, the buck-boost topology was combined with the boost topology. The reason why both topologies were combined together are because buck-boost converter can operate the input voltage in a reverse polarity. So, it is the perfect candidate for the proposed converter to condition the negative half cycle of the switch. A performance`s comparison between the proposed bridgeless converter with the other low energy harvesting converter are depicted in Table 2. For a feedforward and feedback DC-DC PWM boost converter topology is proposed as in table. The system is designed to produce 35mW, but the system does not work well for low input voltages. As for the dual polarity boost converter system, the system works well for low input voltage, but the output power produced is only 23mW and the output voltages is 4.16V. It is not the desired output voltage for the low power converter energy harvesting project.

Table 2. The performance`s comparison of energy harvesting converters

Type of energy harvesting converters	Input voltage (V_{in})	Output voltage (V_{out})	Output power (P_o)
Feed-forward and feedback DC-DC PWM boost converter [9]	0.4V – 1.4V	3.3V	35mW
Dual polarity boost converter [10]	0.4V	4.16V	23mW
This work (bridgeless boost converter)	0.4V	3.3V	55mW

4. Conclusion

In this paper, the unique integration of a combining circuit of buck-boost and boost circuit have been proposed to introduce a new direct ac-dc power converter topology without conventional diode bridge rectifier. The feasibility of a new ultra-low input voltage power management circuit that act as the proposed circuit for micro-energy harvesting system has been demonstrated and simulated in the PSIM software. The proposed power converter circuit supplied by an AC sinusoidal voltage input are about 400mV had generated a boosted output voltage such as 3.35V will be successfully drives the low electronic load such as wireless sensors, low electronic devices and medical implants.

5. Acknowledgement

The authors wish to thank Faculty of Electrical Engineering, Universiti Teknologi MARA (UiTM) for the funding of this research paper. Also, special thanks to the Ministry of Higher Education (MOHE) for the funding of this work under the Research Acculturation Collaborative Effort (RACE) grant No. 600-RMI/RACE 16/6/2(2/2015).

References

[1] J. M. Conrad, "A survey of energy harvesting sources for embedded systems," in *IEEE SoutheastCon* 2008, 2008, pp. 442-447.

[2] Wischke, M., Masur, M., Kroener, M., & Woias, P. Vibrating Harvesting In Traffic Tunnels To Power Wireless Sensors. *Smart Material and Structures*, pp. 1-8. 2011

[3] A. Sriramulu, B. Madhu. and A. Nachiappan, "Implementation of an efficient AC-DC converter for low voltage energy harvesting," *Engineering Education: Innovative Practices and Future Trends* (AICERA), 2012 IEEE International Conference on, Kottayam, 2012, pp. 1-6.

[4] M.N.M.Hussain, N. Ismail, "*Power Electronics* power semiconductors,conveters and applications", pp.51, 2010.

[5] Hemant Mehar, "The Case Study of Simulation of Power Converter Circuits Using PSIM Software in Teaching", pp.1-6. 2013

[6] F. Saleemi, "Analysis and Design of Multiphase Multi-Interleave DC-DC Converter with Input-Output Bypass Capacitor," no. September, p. 119, 2008.

[7] H. Wang, Y. Tang, and A. Khaligh, "A bridgeless boost rectifier for low-voltage energy harvesting applications," IEEE Trans. Power Electron., vol. 28, no. 11, pp. 5206–5214, 2013.

[8] V. J. Abu, P. E. Sebastian, and P. R. P. Thomas, "A Comparative Study of Various AC-DC Converters for Low Voltage Energy Harvesting," vol. 3, no. 3, pp. 50–55, 2016.

[9] X. Cao, W. J. Chiang, Y. C. King, and Y. K. Lee, "Electromagnetic energy harvesting circuit with feedforward and feedback dc-dc PWM boost converter for vibration power generator system," IEEE Trans. Power Electron., vol. 22, no. 2, pp. 679–685, 2007.

[10] M. M. Rahman, A. Riduan, M. Foisal, and A. Gafur, "AC to DC Converter for Low Voltage Small Scale Generator," *IOSR J. Electron. Commun. Eng. Ver. I*, vol. 10, no. 5, pp. 2278–2834, 2015.

Finite-Difference Time-Domain Analysis of Tapered Photonic Crystal Fiber

M.I. Md Ali[1], S. N.Sanusidin[1] and M. H. M. Yusof[2]

[1]*Faculty of Electrical Engineering*
[2]*Faculty of Applied Science*
University Teknologi MARA
40450 Shah Alam, Selangor

Email: masiz135@salam.uitm.edu.my

Abstract— This paper brief about the simulation of tapered photonic crystal fiber (PCF) LMA-8 single-mode type based on correlation of scattering pattern at wavelength of 1.55 μm, analyzation of transmission spectrum at wavelength over the range of 1.0 until 2.5 μm and correlation of transmission spectrum with the refractive index change in photonic crystal holes with respect to taper size of 0.1 until 1.0 using Optiwave simulation software. The main objective is to simulate using Finite-Difference Time-Domain (FDTD) technique of tapered LMA-8 PCF and for sensing application by improving the capabilities of PCF without collapsing the crystal holes. The types of FDTD techniques used are scattering pattern and transverse transmission and principal component analysis (PCA) used as a mathematical tool to model the data obtained by MathCad software. The simulation results showed that there is no obvious correlation of scattering pattern at a wavelength of 1.55 μm, a correlation obtained between taper sizes with a transverse transmission and there is a parabolic relationship between the refractive index changes inside the crystal structure.

Keywords— tapered, photonic crystal fiber, scattering pattern, transmission spectrum, optiwave

1. Introduction

Photonic crystals (PCs) have attracted much attention in the past two decades [1], [2]. It has become a key issue in material engineering to control the optical properties of materials. PCs are materials which have a periodic dielectric constant in three- orthogonal spatial directions. Based on variation in refractive index, PCs are classified as one-dimensional (1D), two- dimensional (2D) and three-dimensional (3D). Fig. 1(a-c) show the schematic illustrations for PCs where yellow colour indicate permittivity, ε at high region and pink colour indicate ε at low region.

A complete band gap along all dimensions in space can be best realized in a 3D system in principle. However, 3D crystals with band gaps in the optical regime prohibit the progression of many applications which lead to the difficulty in fabricating. On the other hand, 2D PCs have been extensively studied because they provide the possibility to control the propagation of light yet remain comparatively easy to fabricate [3-5].

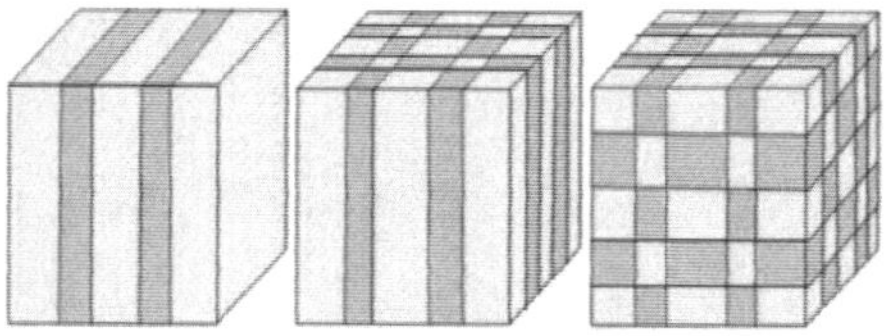

(a) (b) (c)

Figure 1(a-c): From left: 1D-periodic in one direction, 2D-periodic in two directions and 3D-periodic in three directions

FDTD method has been developed and applied to the analysis and investigation of PCs [6-8]. By using an adaptive finite element method in time domain, 2D PC in-plane light propagation is investigated in this work. In the simulation model, both the transverse electric (TE) and transverse magnetic (TM) modes are considered as the polarization characteristics. The PCs' dielectric function profile is discretizing by employed a finite-element method and calculating the in-plane band structures by applying eigenvalue equations with proper periodic boundary conditions following the Bloch theorem [9], [10]. A 2D PC is periodic in two directions (x,y) and homogeneous in the third (z). For light propagating in the xy-plane, we can separate the modes into two independent polarizations, TM and TE modes, and consider the band structure and photon density of states of each. The propagation properties of TM and TE modes can be characterized by the field components parallel to the rods, $Ez(x,y)$ and $Hz(x,y)$, respectively as show in Fig. 2.

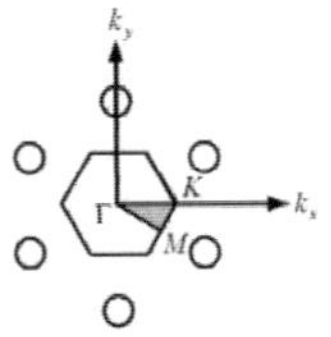

Figure 2: The corresponding reciprocal lattice with irreducible Brillouin zone

PCF is a new dielectric structure with a refractive index that varies periodically in the transverse plane, with a period of the order of an optical wavelength [11]. PCFs or known as microstructured optical fibers (MOFs) are dielectric optical waveguides having a complex air-silica cross-section. Due to the novel optical properties, silica PCFs, also known as holey fibers (HFs) have received considerable attention. By varying the size of air-holes, their number and position, PCF structure can be designed with desirable anomalous group velocity dispersion and modal properties [13][14]. All-silica fiber consist of a solid core with the absence of an air-hole at the lattice site forms a region of raised refractive index, which is surrounded by microstructured cladding, that are typically hexagonally packed, with pitch, Λ (natural length of separation between the two nearest air-holes) and d, the air-hole diameter [15], [16] as can be observed in Fig. 4.

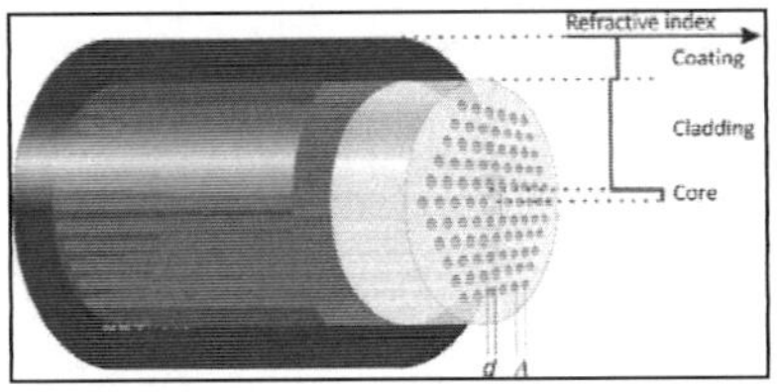

Figure 3: Schematic of an index guided PCF [16]

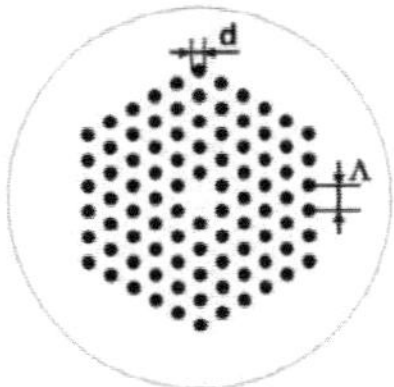

Figure 4: Cross section of triangular cladding PCF [15] [16]

Index guiding (IG) PCFs are made of undoped silica that provides very low losses, sustains high powers and temperature levels, and may withstand nuclear radiation [16]. Fig. 3 shows the IG of a waveguide consists of a solid core and a cladding with an array of air holes. The guided modes may be trapped in a core with a higher refractive index than an averaged index of the cladding. IG is guiding light by total internal reflection between a solid core and a cladding region with multiple air- holes. IG possess the especially attractive property of great controllability in chromatic dispersion by varying the hole diameter and hole-to-hole spacing [17], [18].

In this project, LMA-8 is chosen due to its commercially available. LMA-8-ultraviolet (UV) from NKT Photonics has a hexagonal lattice of air holes surrounding a solid core. LMA-8- UV is a custom fiber made from high-OH fused silica (Heraeus Quarzglas F110), with a core size of 8.6±0.5 *mm* and a cladding diameter of 240±2 *mm*. The cladding diameter of this fiber is larger than that of typical single-mode fibers; this reduces micro-bending, which can increase propagation losses at short wavelengths [19]. An optical micrograph of the fiber cross- section for LMA-8-UV is shown in Fig. 5 [20].

From manufacturer's data, LMA-8 having Λ=5.6 μm and d/Λ=0.49 [21], [22], [23]. The results of LMA-8 PCF tapering are presented in this paper. The focus of the paper is on describing fundamental properties of tapered PCFs, characterization techniques that probe the air-hole collapse looking at the different index.

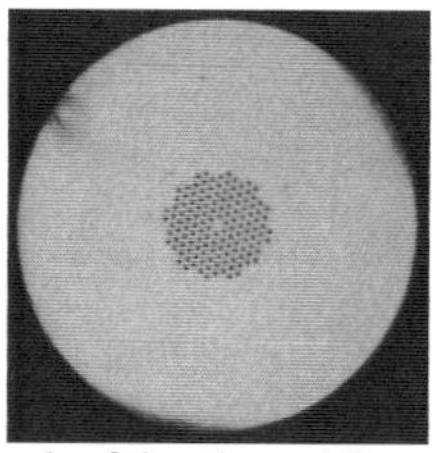

Figure 5: An optical micrograph of the cleaved facet of an uncollapsed LMA-8- UV fiber. The pattern of holes which forms the PC can be seen in the centre. The guided mode propagates through the solid core in the centre of the array of holes [20]

The paper is structured as follows: In Sect. II, the principle of tapering optical fibers are reviewed, and in particular, tapering of PCFs. The air-hole structure of PCF can be well maintained during tapering under certain conditions is demonstrated and a non-destructive method of probing and profiling the PCF microstructure along the taper are described. In Sec. III, the results and discussions of three main objectives are presented and discussed to correlate the scattering pattern of tapered LMA-8 at wavelength of 1.55 μm, to analyse the transmission spectrum of tapered LMA-8 at wavelength over the range of 1.0 until 2.5 μm and to correlate the transmission spectrum with the refractive index change in PC holes with respect to taper size of 0.1 until 1.0. In Sec IV, an overall conclusion of the objectives are concluded and suggested recommendations are listed to achieve the objectives.

2. Methodology

The fiber taper geometry schematic is shown in Fig. 6. A standard flame brushing technique [24], in which the fiber is brushed with a butane flame while simultaneously stretched by motorized stages was applied. In the heated section of the fiber the viscosity decreases and the glass is allowed to flow during this process [25].

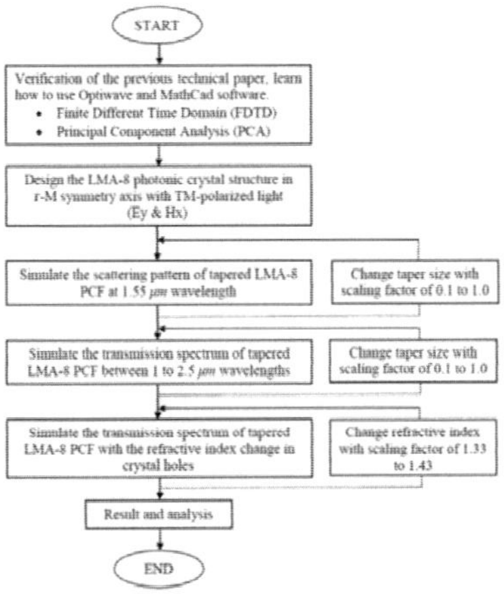

Figure 6: The methodology flow chart

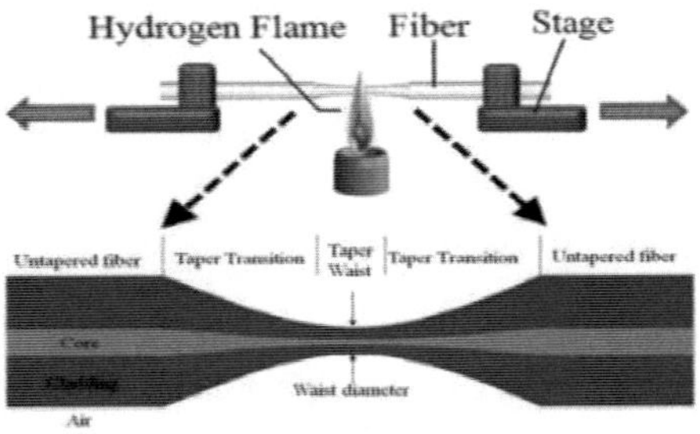

Figure 7: Schematic of a fiber tapering rig

The examination of the PCF microstructure has required the fiber to be cleaved for scanning electron microscope (SEM) analyses in the past. This process is time-consuming, destroys the taper and is most inappropriate for examination of the longitudinal variation in the microstructure although it yields accurate results [21]. Therefore, a novel transverse probing technique is used to analyze the internal structure along the PCF taper, non-destructively [25], [26], [27].

The periodic nature of the PCF microstructure is exploited and the transverse probing technique is illustrated in Fig. 7. The PCF is sandwiched between two butt-coupled SMFs, coupled to a broadband source and an optical spectrum analyzer (OSA), and a transmission spectrum is measured transversely across the PCF. The PCF is sandwiched between two butt-coupled SMFs, coupled to a broadband source and an optical spectrum analyzer (OSA), and a transmission spectrum is measured transversely across the PCF.

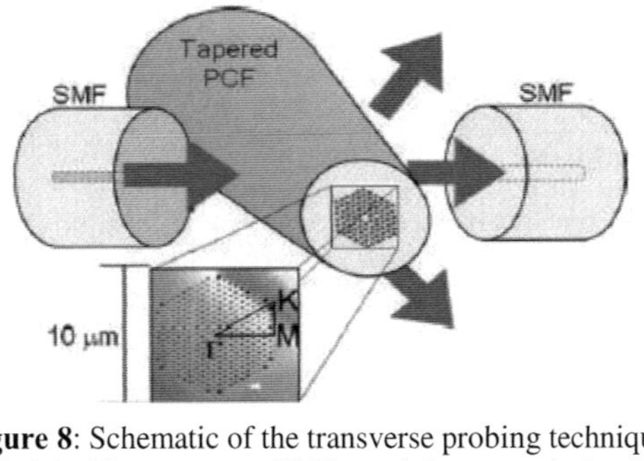

Figure 8: Schematic of the transverse probing technique. The tapered PCF is sandwiched between two SMFs, and the transmission is measured transversely across the taper [25]

Along the taper, the transverse transmission spectra are measured across different positions then in the pitch, the longitudinal variation can be profiled. In the transmission spectrum, the spectral features of PC microstructure are directly linked to the pitch, diameter and regularity of the holes which gives rise to partial photonic bandgaps [28]. By relating the measured bandgap wavelengths to the photonic band structure of the uniform array of holes, the local pitch can be inferred [29] obtained with plane wave expansion (PWE) calculations [25]. By measuring the transverse transmission spectra across different positions along the taper, the longitudinal variation in the pitch can be profiled.

The purpose of simulating and collecting data using Optiwave simulation software over the taper size of 0.1 to 1.0 is to characterize the non-destructive technique. In this case, for an initial experiment, it is crucial to creating a taper size chart to categorize the spectrum pattern but first, we need to cut the PCF as the starting point or reference to correlate between the spectrum and its actual size. The actual size can be determined by using scanning electron microscope (SEM).

This technique is used to ensure the simulation is consistent with the results reported in [25]. The next step is to do PCA and smoothing method using MathCad software. PCA is a technique used to emphasize variation and bring out strong patterns in a dataset and often used to make data easy to explore and visualize [30] meanwhile smoothing method is used to eliminate the noise so that the pattern will not be cluttered

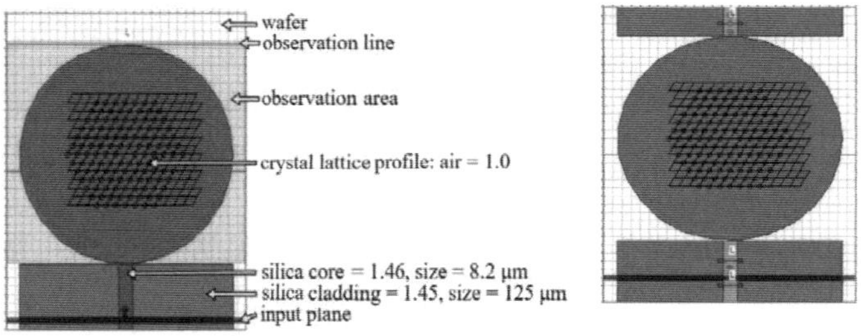

Indices	Material
1.33	Pure Water
1.34	Seawater
1.35	Milk
1.36	Ethyl Alcohol
1.37	Acetic Acid
1.38	Sugar Solution (30%)
1.39	Kerosene
1.40	Octane
1.41	Decane
1.42	Sugar Solution (50%)
1.43	Ethylene Glycol

(a) (b) (c)

Figure 9: (a) The scattering pattern design, (b) the transmission spectrum design of 2D photonic crystal structure measured along the Γ-M direction using Optiwave simulation software, (c) the material of refractive index

As to achieve the second objective on the simulation of transmission spectrum, Bragg's Law equation is used due to constructive interference occurred. The equation is shown in Fig. 9(d) where the calculation is in μm, where λ is wavelength, n is an integer, d is a lattice spacing, and we used maximum sin value of 1 respectively. When x-rays are scattered from a crystal lattice, peaks of scattered intensity are observed which correspond to the following conditions:

- The angle of incidence = angle of scattering.
- The path length difference is equal to an integer number of wavelengths.

The condition for maximum intensity contained in Bragg's law above allow us to calculate details about the crystal structure, or if the crystal structure is known, to determine the wavelength of the x-rays incident upon the crystal.

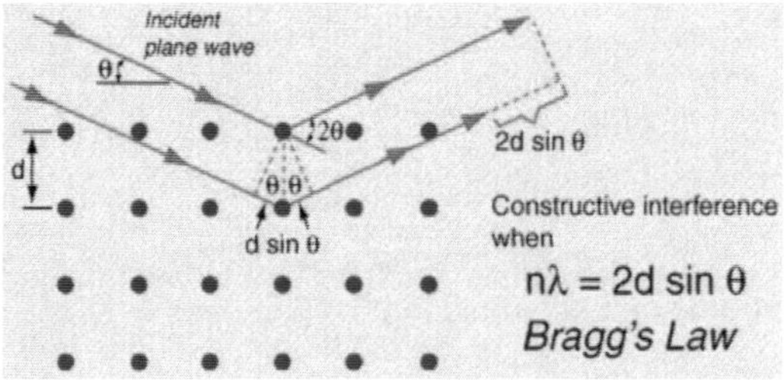

Figure 9: (d) The Bragg's Law illustration and formula

IOP Conf. Series: Materials Science and Engineering **341** (2018) 012002 doi:10.1088/1757-899X/341/1/012002

3. Results and Discussions

Scattering is a technique used to observe the quality of the crystal inside LMA-8, meanwhile scattering pattern is pattern observed at a single wavelength. The first objective is to investigate the correlation of scattering pattern of tapered LMA-8 between taper ratios of 0.1 to 1.0 at a wavelength of 1.55 μm. The wavelength of 1.55 μm is chosen due to telecommunication wavelength.

Fig. 10(a-b) shows the scattering pattern measured along r-M symmetry axis (Fig. 8) with TM-polarised light at a wavelength of 1.55 μm where the x-axis is a spatial distance (μm) and the y-axis is the scattering (dB) respectively.

Fig. 10(a-b) shows there is no obvious correlation pattern between the scattering of tapered LMA-8 and taper ratio and Fig. 10(c-d) illustrated the observation area of Ey and Hx scattering pattern respectively from the Optiwave simulation software. PCA used to look at the correlation between the different pattern and Fig. 10(e-f) shows there is no trend and uncluttered PCA of both Ey and Hx respectively. All in all, it can be concluded that when LMA-8 is tapered, there is no relationship in the pattern, no bandgap and uncluttered PCA with respect to taper ratio. Hence, this scattering technique is not suitable to do at the internal structure of LMA-8.

Table 1. Taper ratio indication from 0.1 to 1.0

Taper Ratio	Legend
0.1	
0.2	
0.3	
0.4	
0.5	
0.6	
0.7	
0.8	
0.9	
1.0	

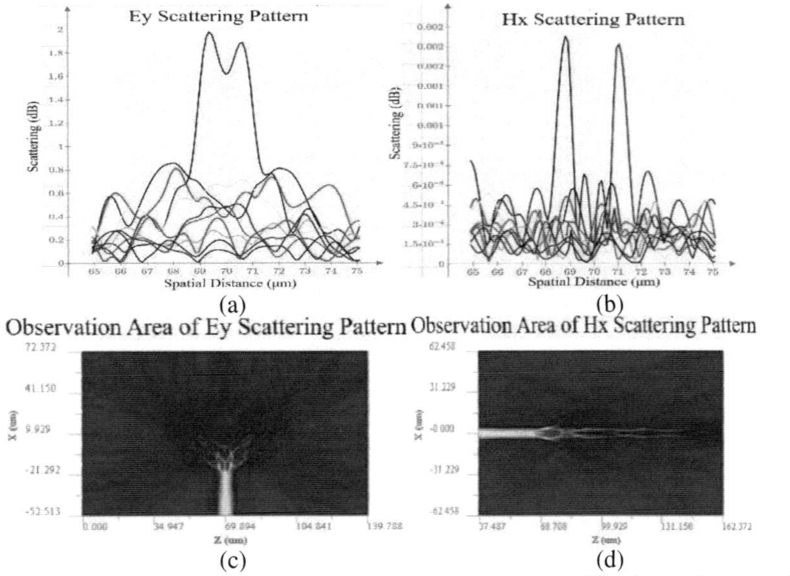

Figure 10 (a) Ey scattering pattern (b) Hx scattering pattern (c) (c) The observation area of Ey scattering pattern, (d) the observation area of Hx scattering pattern from the Optiwave simulation software

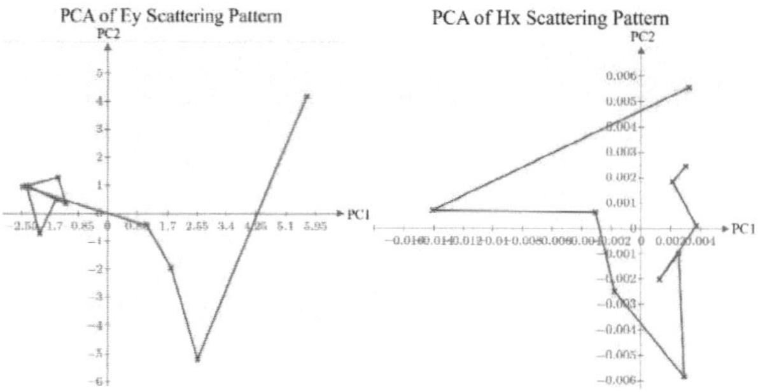

(e) (f)

Figure 10: (e) The PCA of Ey pattern, (f) the PCA of Hx pattern

The second objective is to analyze the transmission spectrum of tapered LMA-8 at a wavelength over the range of 1.0 until 2.5 μm. This can be done by first verification of previous technical paper [25] as shown in Fig. 11(a-b) to ensure the simulation is correct, then we proceed to analyze whether this transmission technique can be applied to LMA-8 or not. Fig. 11(a-b) shows the correct verification of the transverse transmission spectra of the LMA-8 PCF measured along the r- M symmetry axis (as in Fig. 8) with TM-polarised light with taper ratio over the range of 0.40 to 0.75 where x-axis represent the wavelength (μm) and the y-axis represents the transmission (dB).

The PCF has a hole diameter-to-pitch ratio d/Λ of 0.70 and an initial pitch Λ of 1.28 μm. The bandgap shifts to shorter wavelength with decreasing taper size which is similar to the result of previous technical paper [25]. In this case, the bandgap shows that there is no transmission over the wavelength of 1 to 2.5 μm. This is because the holes size, period and pitch are exceptionally small. The purpose of showing the transverse transmission spectra is to prove the fact that we can characterize the taper ratio through transverse transmission using the non- destructive technique without the need to cut the fiber by transversing the light across the LMA-8 PCF, the percentage and position of the point where the light can be transverse can be acquired. Fig 11 (b) illustrates that there is an obvious correlation between the transmission spectrum of previous technical paper with taper size. The similar result was obtained in LMA 8.

Table 2. The indication of taper ratio of 0.4 to 0.75 respectively

Taper Ratio	Legend
0.40	
0.45	
0.50	
0.55	
0.60	
0.65	
0.70	
0.75	

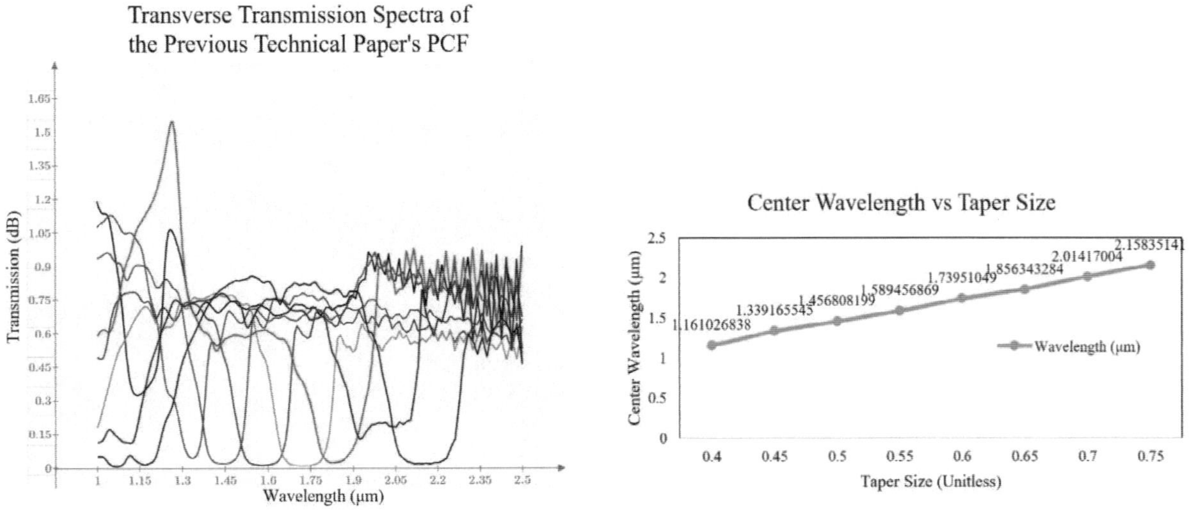

(a) (b)

Figure 11: (a) The transverse transmission spectra of previous technical paper with 0.03 smoothing (b) The graph of center wavelength vs taper size

Fig. 12(a-c) illustrates the transverse transmission spectra of tapered LMA-8 measured along the Γ-M symmetry axis (refer to Fig. 7) with TM-polarised light at wavelength over the range of 1.0 until 2.5 μm with respect to taper size of 0.2, 0.3 and 0.5 respectively where the x-axis is the wavelength (μm) and y-axis represent the transmission (dB). The selection of wavelength range between 1.0 to 2.5 μm is due to the wavelength range of optical spectrum analyzer (OSA) inside the laboratory.

By applying the Bragg's Law, there is a correlation between the transmission spectrum of tapered LMA-8 with taper size. The bandgap wavelength was observed at a longer wavelength. Bandgap transmission occurred when there is no transmission and a reflection at the wavelength. It can be discussed that as the taper ratio increased, the bandgap moves towards a longer wavelength which is known as a red or infrared shift in terms of sensor. But if the bandgap move towards shorter wavelength, it is known as blue shift. This is because the crystal size and pitch of LMA-8 is double than the previous technical paper which is 5.6 μm pitch value and 0.49 d/λ value. To be concluded, the transmission spectrum technique is possible to characterize and correlate LMA-8 crystal with taper size.

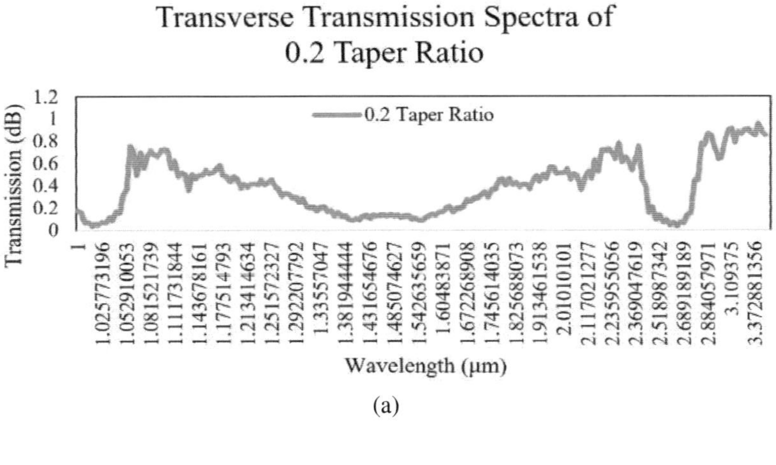

(a)

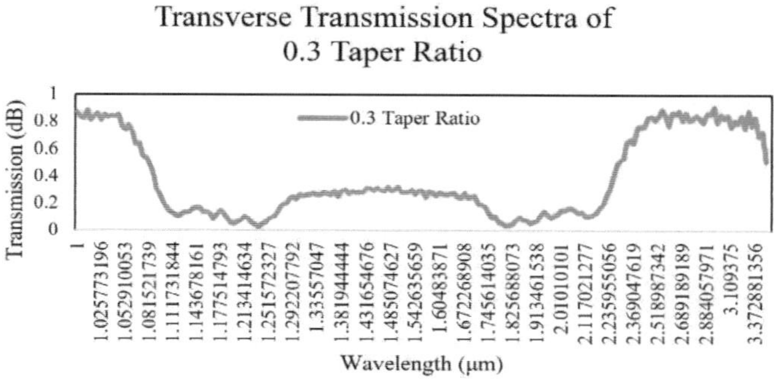

(b)

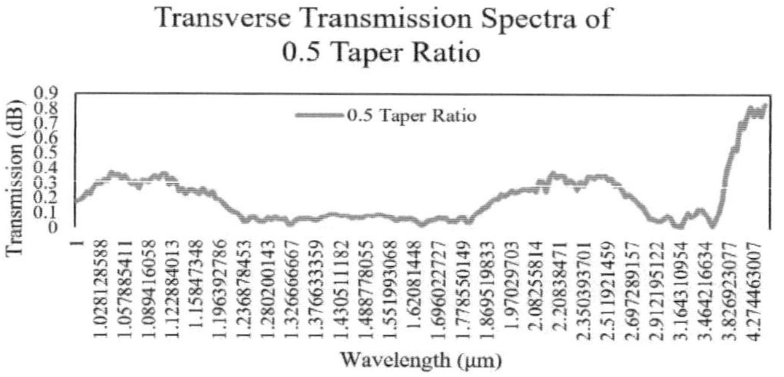

(c)

Figure 12: (a) The transverse transmission spectra of LMA-8 tapered at (a) 0.2 taper ratio observed a total of 3 bandgap(b) 0.3 taper ratio observed a total of 2 bandgap (c) 0.5 taper ratio observed a total of 2 bandgap

The third objective is to correlate the transmission spectrum of tapered LMA-8 with the refractive index change from 1.33 to 1.43 in photonic crystal holes with respect to taper size of 0.1 until 1.0. Fig. 13(a-b) illustrates the correlation of transmission spectra of LMA-8 measured along the г-M symmetry axis (as shown in Fig. 2) with TM-polarised light where x-axis represent the wavelength (*μm*) and the y-axis represents the transmission (dB) tapered at 0.5 and 1.0 respectively. Normally in sensing, the refractive index of distilled water which is 1.33 being used as a standard. We want to investigate on how much power being transferred at different taper size and refractive index from 1.33 to 1.43. The reason on the selection range of refractive index is because we do not want the LMA-8 to be a high-power rod-type PCF and used for correlating material inside the crystal.

To be concluded, there is a relationship between the transmission spectra with respect to taper ratio and there is a pattern on PCA (Fig. 13(c-d)). The two ways to analyze the relationship between the transmission spectrum with refractive index change in LMA-8 PCF holes are multispectral analysis and single spectra at 1.0, 1.5 and 1.9 *μm* wavelength. The multispectral analysis is analyzing and comparing the whole spectrum by using PCA. We observed that there is a trend between the index of the crystal holes with the value of principal component (Fig. 13(a-d)). Furthermore, principal component is a vector that is perpendicular to one another which represent in for X, Y and Z-axis.

Meanwhile, Fig. 14 shows the correlation of a single spectrum at 1.0, 1.5 and 1.9 *μm* wavelengths with respect to the refractive index in the crystal holes. Based on the range of investigation, the gradient demonstrates the fact that a steeper curve at a wavelength of 1.0 *μm* will give the most sensitive sensor compared to 1.5 and 1.9 *μm* for un-tapered LMA-8. At 1.0 *μm*, the index change of 0.01 causing a large change in normalized power meanwhile at 1.9 *μm*, the index change causes a small change in normalized power where sensitivity is ΔP over ΔN. ΔP denote as the rate of change of normalized power and ΔN is denoted as the refractive index change of LMA-8 crystal holes.

It is advisable to use the transmission spectrum with respect to 1.0 tapered size. This proved that the LMA-8 PCF works equally well in tapered or un-tapered condition. Fig. 13(d) shows the parabolic relationship obtained which act as the standard reference of how LMA-8 being as a sensor by first the

LMA-8 being transverse propagate, then we analyzed the crystal structure by using PCA, afterward the spectrum reading obtained is 1.355 which falls in between 1.35 and 1.36 line of parabolic relationship.

Table 3. The color indication and the value of refractive index change

Refractive Index	Legend
1.33	
1.34	
1.35	
1.36	
1.37	
1.38	
1.39	
1.40	
1.41	
1.42	
1.43	

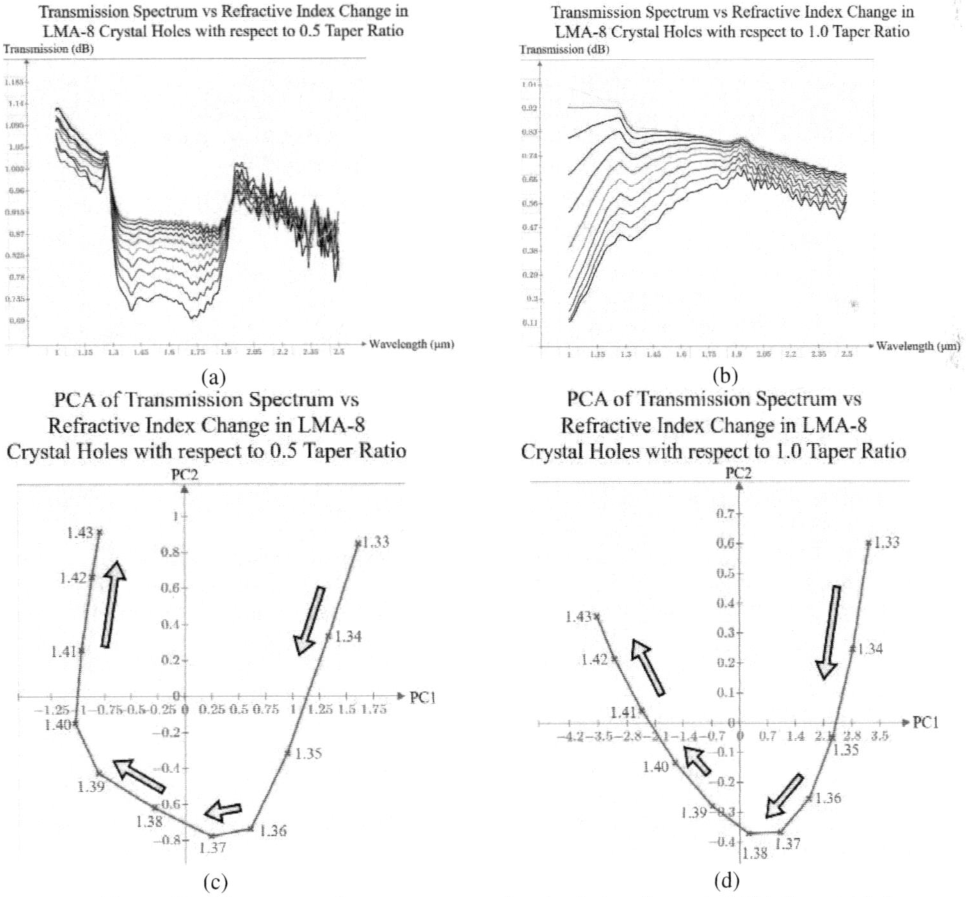

Figure 13: The transmission spectrum vs refractive index change in LMA-8 crystal holes with respect to (a) 0.5 taper ratio (b) 1.0 taper ratio and PCA of transmission spectrum for (c) 0.5 tapered (d) 1.0 tapered

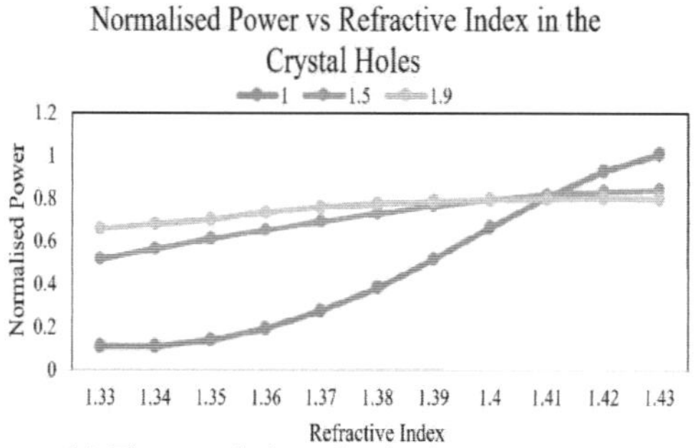

Figure 14: The correlation of a single spectrum at 1.0, 1.5 and 1.9 μm wavelength with respect to the refractive index in the crystal holes.

4. Conclusion and Recommendation

In summary, the advantage of scattering pattern and transverse transmission techniques is the capability to observe the diffraction pattern with light transverse and without having to cut the LMA-8 to destroy the sample. We have demonstrated that the scattering pattern is not a suitable technique to model the LMA-8 PCF as there is no relationship in pattern and uncluttered PCA with respect to taper ratio of 0.1 to 1.0 at 1.55 μm wavelength. This is because the LMA-8 fiber has bigger crystal size and bigger pitch size. Thus, this technique is more suitable to be applied when the crystal size is very small. As for the analysis of the transmission spectrum, there is a relationship between the transmission spectra and taper ratio of 0.1 and 1.0 at a range of 1.0 to 2.5 μm wavelength. Hence, this technique can be used to observe the difference in taper size. We proved that the LMA-8 PCF works equally well in tapered or un- tapered condition for the correlation of the transmission spectrum against the refractive index change from 1.33 to 1.43 in LMA-8 crystal holes. As a conclusion, this is a totally new research to be investigated and would produce interesting findings.

Acknowledgement

This work was partly supported by the Faculty of Electrical Engineering, Universiti Teknologi Mara Malaysia, and Ministry of Higher Education, Malaysia (research grant #FRGS/1/2015/TK04/UITM/03/5).

References

[1] J. D. Joannopoulos, R. D. Meade, and J. N.Winn, *Photonic Crystals* (Princeton University Press, Princeton, New Jersey 1995).
[2] J. D. Joannopoulos, P. R. Villeneuve, and S. Fan, "Photonic crystals: putting a new twist on light," Nature (London) 386, 143-149 (1997).
[3] K. Sakoda, *Optical Properties of Photonic Crystals* (Springer, Berlin 2001).
[4] M. M. Sigalas, R. Biswas, K. M. Ho, and C. M. Soukoulis, "Theoretical investigation of off-plane propagation of electromagnetic waves in two- dimensional photonic crystals," Phys. Rev. B **58**, 6791-6794 (1998).
[5] Z. Y. Li and Y. Xia, "Omnidirectional absolute band gaps in two- dimensional photonic crystals," Phys. Rev. B **64**, 153108 (2001).
[6] A. J. Ward and J. B. Pendry, "Calculating photonic Green's functions using a nonorthogonal finite-difference time-domain method," Phys. Rev. B **58**, 7252-7259 (1998).
[7] H. Y. D. Yang, "Finite difference analysis of 2-D photonic crystals," IEEE Trans. Microwave Theory Tech. **44**, 2688-2695 (1996).
[8] C. T. Chan, Q. L. Yu, and K. M. Ho, "Order-N spectral method for electromagnetic waves," Phys. Rev. B **51**, 16635-16642 (1995).
[9] A. Yariv and P. Yeh, *Optical Waves in Crystals* (Wiley, New York 1984).
[10] C. Kittel, Introduction to Solid State Physics (Wiley, New York 1976).
[11] T. A. Birks et al., "Full 2-D photonic bandgaps in silica/air structures", Electron. Lett., vol. 31, pp. 1941–1945, 1995.
[12] T.A. Birks, J.C. Knight, and P.St.J. Russell, "Endlessely single mode photonic crystal fiber," Opt. Lett.,

vol. 22, pp. 961-963, (1997).

[13] Philip St J. Russell, "Photonic-Crystal Fibers," J. Lightwave Technol.,vol. 24, pp. 4729-4749, (2006).

[14] J.C. Knight, T.A. Birks, P.St.J. Russell, and D.M. Atkin, "All silica single-mode optical fiber with photonic crystal cladding," Opt. Lett., vol.21, pp. 1547-1549, (1996).

[15] J.C. Knight, T.A. Birks, P.St.J. Russell, and J.P. Sandro, "Properties of photonic crystal fiber and the effective index model," J. Opt. Soc. Am. A, vol. 15, pp. 748-752, (1998).

[16] "Tutorial: Fiber Optic Basics." *Fiber Optic Basics*. Web. 27 May 2016.

[17] J. Arriaga, J.C. Knight, and P.St. Russell, "Modeling Photonic Crystal Fibers", Physica E 17, Elsevier Science, pp. 440-442, 2003.

[18] S. Soussi, "Modeling Photonic Crystal Fibers, Applied Mathematics", Elsevier Science, vol.36,Issue 3,pp.288- 317, 2006.

[19] M. D. Nielsen, N. A. Mortensen, and J. R. Folkenberg, "Reduced microdeformation attenuation in large-mode area photonic crystal fibers for visible applications," Opt. Lett. 28, 1645–1647 (2003).

[20] Colombe, Yves, Daniel H. Slichter, Andrew C. Wilson, Dietrich Leibfried, and David J. Wineland. "Single-mode Optical Fiber for High- power, Low-loss UV Transmission." *Opt. Express Optics Express* 22.16 (2014): 19783. Web.

[21] Szustakowski, M., N. Palka, and W. Grabiec. "Simple Method for Determination of Photonic Crystal Fibers Geometry." *Photonic Crystal Fibers* (2007).

[22] Sharma, Dinesh Kumar, and Anurag Sharma. "Spot Size of Photonic Crystal Fibers: An Analytical Method of Calculation." *2012 International Conference on Optical Engineering (ICOE)* (2012).

[23] Nielsen, M. D., J. R. Folkenberg, N. A. Mortensen, and A. Bjarklev. "Bandwidth Comparison of Photonic Crystal Fibers and Conventional Single-mode Fibers." *Opt. Express Optics Express* 12.3 (2004): 430.

[24] T.A. Birks, Y.W. Li, J. Lightwave Technol. **10**, 432 (1992).

[25] Nguyen, Hong C., Boris T. Kuhlmey, Eric C. Magi, Michael J. Steel, Cameron L. Smith, and Benjamin J. Eggleton. "Tapered Photonic Crystal Fibres: Properties, Characterisation, and Applications (Invited Paper)." *Photonic Materials, Devices, and Applications* (2005).

[26] E.C. Magi, P. Steinvurzel, B.J. Eggleton, J. Appl. Phys. **96**, 3976 (2004).

[27] J.C. Jasapara, Opt. Express **13**, 1228 (2005).

[28] H.C. Nguyen, P. Domachuk, B.J. Eggleton, M.J. Steel, M. Straub,M. Gu, M. Sumetsky, Opt. Express **12**, 1528 (2004).

[29] E.C. Magi, P. Steinvurzel, B.J. Eggleton, J. Appl. Phys. **96**, 3976 (2004). [30] "Principal Component Analysis Explained Visually." *Explained Visually*. Web. 28 May 2016.

3D Surface Reconstruction for Lower Limb Prosthetic Model using Radon Transform

S S Mohd Sobani[1], N H Mahmood[2], N A Zakaria[2], M A Abdul Razak[2]

[1] Faculty of Biosciences & Medical Engineering, Universiti Teknologi Malaysia, 81310 UTM Johor Bahru, Johor, Malaysia.
[2] Faculty of Electrical Engineering, Universiti Teknologi Malaysia, 81310 UTM Johor Bahru, Johor, Malaysia.

nasrulhumaimi@utm.my

Abstract. This paper describes the idea to realize three-dimensional surfaces of objects with cylinder-based shapes where the techniques adopted and the strategy developed for a non-rigid three-dimensional surface reconstruction of an object from uncalibrated two-dimensional image sequences using multiple-view digital camera and turntable setup. The surface of an object is reconstructed based on the concept of tomography with the aid of performing several digital image processing algorithms on the two-dimensional images captured by a digital camera in thirty-six different projections and the three-dimensional structure of the surface is analysed. Four different objects are used as experimental models in the reconstructions and each object is placed on a manually rotated turntable. The results shown that the proposed method has successfully reconstruct the three-dimensional surface of the objects and practicable. The shape and size of the reconstructed three-dimensional objects are recognizable and distinguishable. The reconstructions of objects involved in the test are strengthened with the analysis where the maximum percent error obtained from the computation is approximately 1.4 % for the height whilst 4.0%, 4.79% and 4.7% for the diameters at three specific heights of the objects.

Keywords. three-dimensional reconstruction, uncalibrated image, multiple views, tomography

1. Introduction

Modern computer technology has given a rise in the research on prosthetic design to overcome the limitations of the existing methods. The prostheses have to be designed and customized specific to one's anatomical shape. It cannot be simply designed or manufactured through mass production since it is specific for an individual. However, the demand on the prostheses has increased as the number of amputees are also increasing and it is necessary to realize ad-hoc with high precision design methodologies [1, 2]. In addition, optimal design of prosthesis is very important for every single amputee. This is to restore and encourages functional muscular activity, pressure-sensitive areas can be relieved, and proper attachment of the prosthesis during ambulation able to be maintained [3]. The data can be acquired in several forms, such as a set of image sequences or a video, views from multiple cameras set up, or multi-dimensional data from a scanner. It has been seen that the computer-aided design and computer-aided manufacturing (CAD/CAM) accompanied with medical imaging technology has been introduced in prosthetic practice [4] where the image of a certain part of a patient is captured using X-ray computed tomography scan (CT-scan), magnetic resonance imaging (MRI), or ultrasound imaging [5]. However, capturing images using such devices are very expensive and the processing task requirements are enormous. The medical device also required to be operated by an expertise in a specifically designated room. It is also bulky and taking up a large space to be assembled.

The basic flow of this research is referred to the software operational framework by [6] that used CT cross-sectional scanning for the data acquisition. Utilizing CT-scan is very advantageous to detailed reconstruction and accuracy which is hardly can be denied. However, it involves a very high operational cost that also requires help by an individual who is expert in radiology known as radiographer or radiologic technologist. Thus, in contrast to an expensive method this research use uncalibrated digital camera is used to capture images of a residual limb model in multiple views for the data acquisition. The outcome from this research is expected to be able to compute and overcome possible constraints in the design of prosthetic part for lower limb amputees fitting to their body size, where the input data is 2D image sequences of amputated lower limb model captured by multiple views camera and a turntable setup in specified projections as the first step in adapting the research field. The advantage of this research is that it can be done with simple experimental setup, as the devices used are usually owned by most people like phone camera and affordable. In addition, reconstruction of three-dimensional surface in this research neither involves expensive equipment nor require any service by an expert to handle sophisticated mechanical scanning.

2. Methodology

2.1. Data Acquisition

Data acquisition for this research involves a very simple setup that consists of a digital camera mounted on a mini tripod stand, and a simple turntable together with black or dark color background that provides better contrast between object and background in the image going to be shot. The material used must not be able to reflect light that save the time and trouble computing silhouette image. An object is placed on the center of a wooden turntable. The object is approximately 700*mm* away from the lens of the digital camera as described in figure 1. The object is static and will not easily move or overturn when turntable is rotated. A mini tripod stand keeps the digital camera static whilst capturing images of the object on its target. A piece of metal is attached under the round turntable, thus helping it to turn smoothly to any desired angle of rotation.

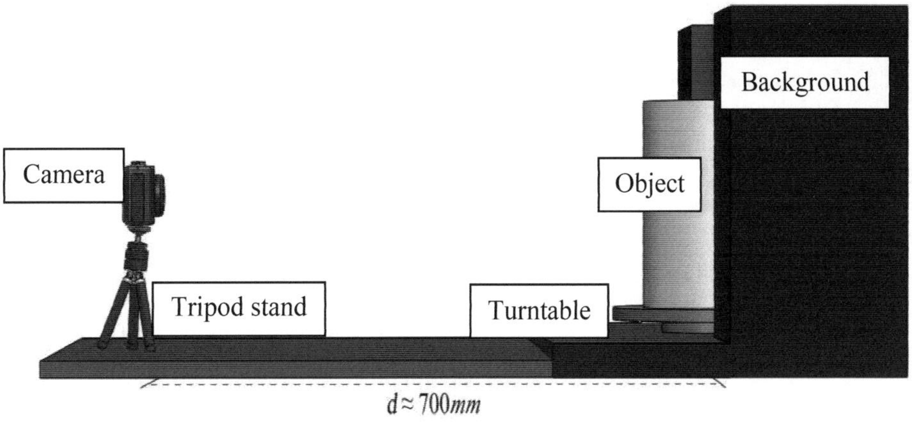

Figure 1. Illustration of the experimental setup at the side view.

It is important to know the field of view (*FOV*) of a camera in order to determine the optimal distance between camera and object to be placed in the scene. It is to make sure that the object is well-fitted into the image especially the height since the image is captured in landscape orientation of an image. For an example, figure 2 illustrates the relationship between the intrinsic parameters of a camera lens and the optimal distance. Hence, the field of view of a camera specifically for y direction, FOV_y, can be calculated by knowing the focal length of the camera lens, f , and the dimension of the sensor which either the height or the width, S_m , as expressed in (1).

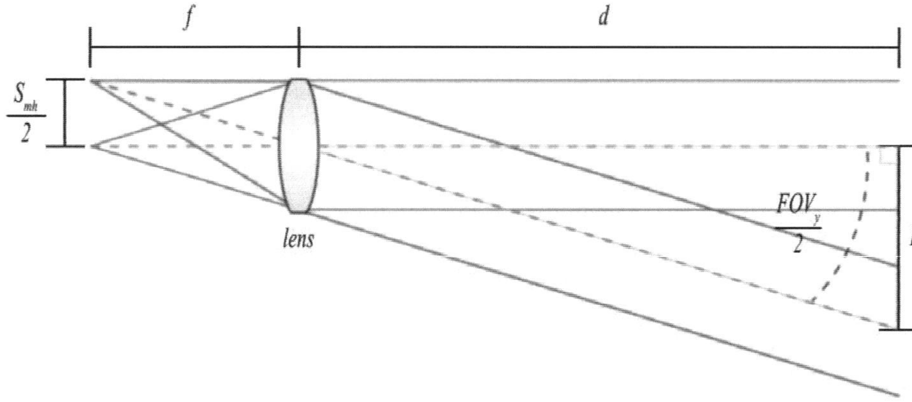

Figure 2. The relationship between the intrinsic parameters of a lens and the optimal distance.

$$FOV_y(°) = 2\tan^{-1}(\frac{S_{mh}}{2f}) \quad (1)$$

The turntable is rotated manually by hand in a fixed sequence which is practically challenging therefore it is limited to a total of thirty-six different views of the object with 10° of difference between each projection to keep the consistency of the input data. The turntable is rotated in a clockwise manner, pausing at every 10° to allow the digital camera to capture the image around the object on target at different projections to obtain thirty-six different views. The images of the object are then transferred to the computer as input for image processing tasks.

2.2. Image Processing

Image-based 3D reconstruction is computationally intensive with a full 8-*bit* colour and high resolution image and with such amount of data in each image, it is impossible to do calculation manually without any computed program. The whole process of reconstructing 3D surface of an object involved at least thirty-six images to be manipulated sequentially, thus iterative and looping in the programming are necessary to simplify the computation and reduce time consumption.

With modern computers, it is possible to perform pixel-by-pixel image processing on the images but the task consumes longer time considering the huge amount of data involves in each image. In addition, RGB color might not works well with some of the image processing where any additional colors information is not helpful and working with grey-scale, the possibility to solve image segmentation problems is higher. The computation of converting the RGB image to grey-scale image is done based on the weighted average [7] as expressed in (2) with specific value of integer for red, green and blue colors respectively while figure 3 shows an example of the result obtained.

$$a = 0.299R + 0.587G + 0.114B \quad (2)$$

(a) (b)

Figure 3. Image of an object (a) before and (b) after grey-scaling.

Reconstructing 3D surface of an object involves thirty-six images and there are millions of data obligated to undergoes the serial computation in achieving a desired result. This number of data requires take a very long processing time, therefore it might not be an efficient system aiming for a fast computation. As an alternative, every single image is resized by reducing the resolution of the images using image resampling method [8]. All images of each object at 36 different projections are down-sampled from 5152×3864pixel by 0.16 scale factor to 309×232pixel.

Otsu thresholding eliminates the hue and saturation information of a color from the image while at the same time it retains the chromaticity or the quality of a color regardless of its luminance. This method works by selecting optimal local threshold value adapted to the image content which generate within satisfactory level results [9, 10]. An image $f(x, y)$ is defined by selecting a threshold value, T then the pixel within the range will be selected as the foreground, while the rest of the pixel will be rejected to the background forming binary image $g(x, y)$ [11, 12] and this can be summarized as in (3) followed with figure 4 displays the silhouette image obtained.

$$g(x, y) = \begin{cases} 1 & for \quad f(x, y) > T \\ 0 & for \quad f(x, y) \leq T \end{cases} \tag{3}$$

(a) (b)

Figure 4. Image of an object (a) before and (b) after Otsu thresholing.

The concept of Radon transform is used to compute multiple silhouette images, $g(x, y)$ to generate a sinogram where the silhouette images obtained are equivalent to indication of electron absorbed into the object as in x-ray procedure [13]. Based on the experimental set up explain in previous section, projection from different view angles, $\rho_\theta(s)$ can be obtained by rotating the object on the turntable at a constant rotation. The task is done based on the idea of following line expression (4) where s is the shortest distance to the origin and theta, θ is the angle of the projection.

$$s = x\cos\theta + y\sin\theta \tag{4}$$

2D cross-section of an object can be reconstruct with the knowledge of integrals computed along the straight line [14]. Radon Transform of an image represented by the function $R(s,\theta)$ such that perpendicular to s and it can be defined as a series of straight line integrals through $h(x, y)$ at different offsets from the origin which is parallel to y -axis with a projection line of $R(s,\theta)$, at angle θ, of an ideal situation [15] and it can be defined mathematically as in (5).

$$R(s,\theta) = \int_{-\infty}^{\infty}\int_{-\infty}^{\infty} h(x, y)\delta(s - x\cos\theta - y\sin\theta)dxdy \tag{5}$$

The sinogram, $R(s,\theta)$ is generated layer by layer corresponding to the number of rows equivalent to height of image $g(x, y)$. Inverse Radon transform reconstruct $h(x, y)$ at a specific layer utilizing. This is performed by projecting each of the layers and aligned the data to its actual rotation with back-projection, B that helps to form the cross-section of the object [16].

2.3. 3D Reconstruction

Edge detection is used for data extraction to find set of points in a digital image which represents an abrupt change of the image intensity [17, 18]. This method is applied on the images generated from inverse Radon transform to identify the contour of the cross-section (if present). The method for edge detection used in this computation is the Canny operator which is the optimal edge detector for image segmentation [19] and also extract border of a shape or an object which is less susceptible to the interference noise [20]. Figure 5 shows cross-section images of an object at a specific layer before and after Canny edge detection is applied. The extracted boundary of the cross-section is the key attribute in obtaining the *xy*-coordinates to reconstruct the surface of the object from the contour lines.

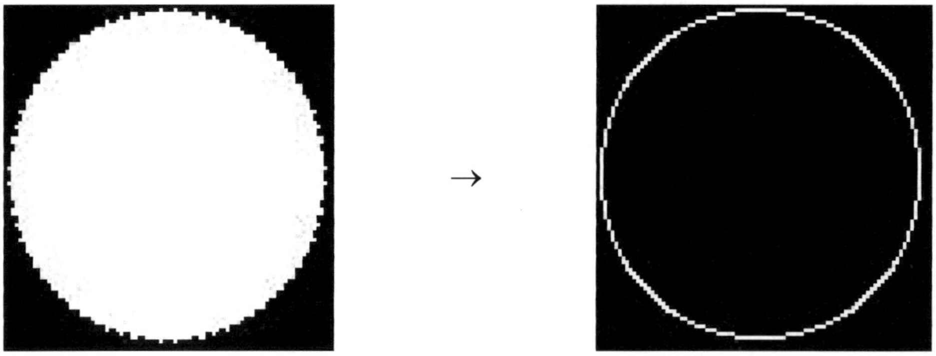

Figure 5. Canny edge detection on cross-section of an object.

The point extracted from the layers of the cross-section images forms a point cloud where each point registered is comprises of *x*, *y*, and *z* coordinates information. Then, Delaunay triangulation is used for mapping and connect all the points to define 3D surface of the object where the result is shown in figure 6. Some parts at the side of the body of the 3D surfaces are reconstructed as a straight curve although the physical structure of the object is actually curvy. Delaunay triangulation provides sets of triangles to be used as polygons in the surface by creating meshes from the 3D point cloud. The meshes are refined to smooth up the surface and minimize distortion between a point and another neighboring point. Thus, the rough edge generated from the thresholding of the low-resolution images has caused the reconstructed 3D surface to be straight curve at the sides of the body of the object. But why Delaunay triangulation? This is because this method has several advantages where the triangles can construct stable structure and ensures that any point on the surface is as close as possible to a node [21-23].

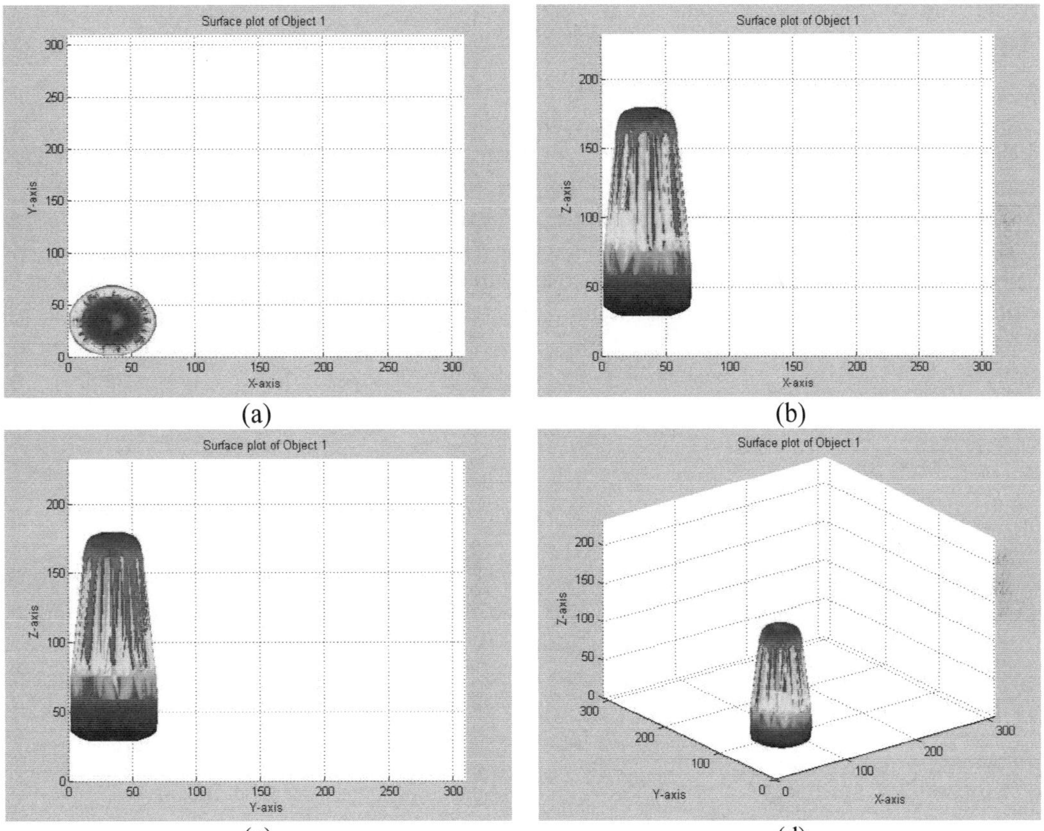

Figure 6. Image of Object 1 at (a) xy-view, (b) xz-view, (c) yz-view, and (d) 3D-view.

3. Analysis and Discussions

The proposed method has been evaluated on four different objects which are models of amputated lower limbs at the knee with average height of 400*mm* and 200*mm* diameter. Four different objects used are labelled with number one to four. The objects are placed upside down where the flat part is at the bottom serving as the base of the object while the amputated part is at the top. Height for the reconstructed 3D surface of the object is calculated and labelled as Position 1, *P1*. The diameter is also measured at three different height of the object which are the Position 2, *P2* is the diameter at the base, Position 3, *P3* Position 4, *P4* are the diameters at the one-third and two-third of the total height of the object respectively as shown in figure 7.

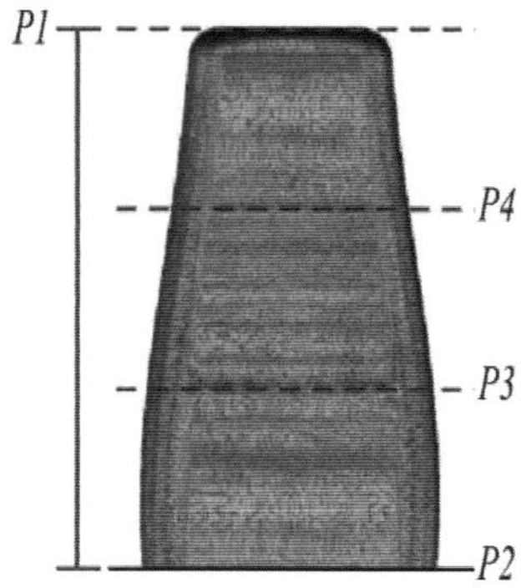

Figure 7. Position measured for model involved in the experiment.

A laptop with is used to provide visual display and also for the computation of the 3D reconstruction process including the error analysis which computes the percent error for experimental measurement over the actual measurement as summarized in figure 8. Reconstruction of 3D surface is highly dependent to the acquired images. High contrast 2D images contribute to higher accuracy of the reconstruction. Overall, the height measurements have better accuracy then the diameters of the objects because the reconstruction is done layer-by-layer by referring to the silhouette generated. The side curve of the object appears rough in the silhouette images, thus affect the accuracy of the diameters. It can be concluded from the analysis that the maximum percent error of the computation is approximate 1.4% for the height and also approximate 4.0%, 4.7%, and 4.7% for the diameters at three specific heights of the objects.

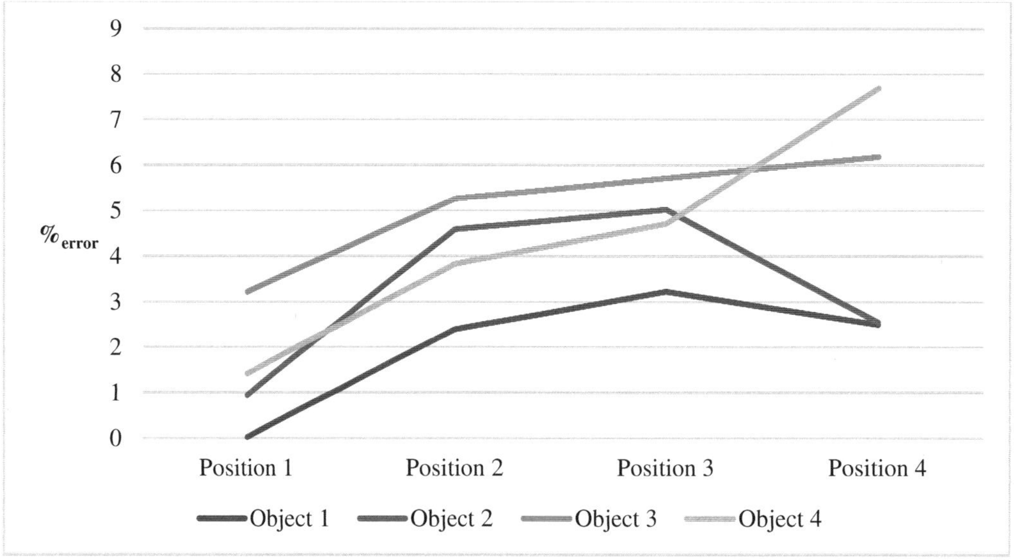

Figure 8. Percent error of four different position measured for four objects.

Shape-from-silhouette method is particularly sensitive to either the errors from silhouette extraction, or noise. Consequently, even small errors in extracted silhouette contributes large effects on the accuracy of a reconstruction, especially down-sampled image because the outline of the object probably only a few pixels wide in the original image with larger dimension has been compressed thus it also causes the errors in the measurement of the objects.

4. Conclusion

The main purpose of this research is to develop 3D surface from uncalibrated 2D image sequences of an object acquired from multiple views imaging system with digital camera as data acquisition device. The advantage of this research is that it can be done with simple experimental setup. The devices used are usually owned by most people and affordable. Therefore, digital camera with detailed specification is used but it does not require any sophisticated mechanical scanning system or expensive devices. However, the results do not support the idea that the study is practicable for the prosthetic design which demand high accuracy in the measurement. Future works should find a better solution to overcome the limitations and weaknesses of the developed method in order to obtain more accurate result and measurement as it is vital for prosthetic design especially at the irregular side curve shape of the model towards the top. This is because the 2D image sequences used are resampled to low resolution to simplify and reduce the time consumption for the whole processing. Low resolution image provides less information on the curve of the object which means this is one of the reason for the slight errors in the measurements of the reconstructed 3D surface that causes the inconsistent and the precision of the height and width of the objects. However, the images are resampled in order for a low specification laptop is able to reconstruct 3D image without any problem in computing all the data involve. Furthermore, it is possible to improve the accuracy by using an automatic image acquisition system to avoid any unintended movement on the object or the camera while capturing the images that will cause inaccuracy of the input data and it can also simplify the entire process.

Acknowledgement

The authors are grateful for a research grant from eScienceFund (Vote Number: 4S027) which has been provided by the Ministry of Science Technology and Innovation (MOSTI), as well as a scholarship from the Ministry of Higher Education of Malaysia.

References

[1] Senin N and Groppetti R 2005 Surface Microtopography Design and Manufacturing through Topography Descriptors: An Application to Prosthetic Implant Surfaces *Computer-Aided Design* **37** 1163-1175.

[2] Colombo G, Filippi S, Rizzi C, and Rotini F 2010 A New Design Paradigm for the Development of Custom-Fit Soft Sockets for Lower Limb Prostheses *Computers in Industry* **61** 513-523.

[3] Frieden R A, Brar A K, Esquenazi A, and Watanabe T 2012 Fitting an Older Patient with Medical Comorbidities with a Lower-Limb Prosthesis **4** 59-64.

[4] Knopf G K and Al-Naji R 2001 Adaptive Reconstruction of Bone Geometry from Serial Cross-Sections *Artificial Intelligence in Engineering* **15** 227-239.

[5] Shah P, Mahajan S, Nageswaran S, and Paul S K 2016 *A Novel Way to Acquire Foot Contour Measurements of Remotely Located Patients Having Foot Deformities IEEE 1st Int. Conf. on Control, Measurement and Instrumentation* pp 211-214.

[6] Shuxian Z, Wanhua Z, and Bingheng L 2005 3D Reconstruction of the Structure of a Residual Limb for Customising the Design of a Prosthetic Socket *Medical Engineering & Physics* **27** 67-74.

[7] Huang D S, Zhang X, Garcia C A R, and Zhang L 2010 *Advanced Intelligent Computing Theories and Applications: With Aspects of Artificial Intelligence: 6th Int. Conf. on Intelligent Computing Proc.* (Changsha: Springer) p 685.

[8] Mohan P G, Prakash C, and Gangashetty S V 2011 *Bessel Transform for Image Resizing 18th Int. Conf. on Systems, Signals and Image Processing* pp 1-4.

[9] Sthitpattanapongsa P and Srinark T 2011 *A Two-Stage Otsu's Thresholding Based Method on a 2D Histogram IEEE 7th Int. Conf. on Intelligent Computer Communication and Processing* pp 345-348.

[10] Belhedi A and Marcotegui B 2016 Adaptive Scene-Text Binarisation on Images Captured by Smartphones *IET Image Processing* **10** 515-523.
[11] Nagabhushana S 2005 *Computer Vision and Image Processing* (New Age International) p 242.
[12] Choras R S 2013 *Image Processing and Communications Challenges 5* (Springer International Publishing) p 420.
[13] Chandra S and Svalbe I 2009 *A Fast Number Theoretic Finite Radon Transform Digital Image Computing: Techniques and Applications* pp 361-368.
[14] Venkatraghavan V, Rekha S, Chatterjee J, and Ray A K 2011 *Modified Radon Transform for Texture Analysis Annual IEEE India Conf.* pp 1-4.
[15] Zakaria Z, Jaafar N H, Yazid N A M, Mansor M S B, Rahiman M H F, and Rahim R A 2010 *Sinogram Concept Approach in Image Reconstruction Algorithm of a Computed Tomography System Using Matlab Int. Conf. on Computer Applications and Industrial Electronics* pp 500-505.
[16] Pintavirooj C and Sangworasil M 2002 *3D Shape Recovery Based on Radon Transform Digital Image Computing Techniques and Applications* (Melbourne).
[17] Punarselvam E and Suresh P 2011 *Edge Detection of CT Scan Spine Disc Image Using Canny Edge Detection Algorithm Based on Magnitude and Edge Length 3rd Int. Conf. on Trendz in Information Sciences and Computing* pp 136-140.
[18] Zhao H, Qin G, and Wang X 2010 *Improvement of Canny Algorithm Based on Pavement Edge Detection 3rd Int. Congress on Image and Signal Processing* vol 2 pp 964-967.
[19] Tang Z and Shen D 2012 *Canny Edge Detection Codec Using Vlib on Davinci Series DSP Int. Conf. on Computer Science and Service System* pp 221-224.
[20] Agarwal A and Goel K 2016 *Comparative Analysis of Digital Image for Edge Detection by using Bacterial Foraging and Canny Edge Detector 2nd Int. Conf. on Computational Intelligence and Communication Technology* pp 125-129.
[21] Wencheng Y, Hu J, and Stojmenovic M 2012 *NDTC: A Novel Topology-Based Fingerprint Matching Algorithm using N-Layer Delaunay Triangulation Net Check 7th IEEE Conf. on Industrial Electronics and Applications* pp 866-870.
[22] Nguyen V S, Bac A, and Daniel M 2014 *Triangulation of an Elevation Surface Structured by a Sparse 3D Grid IEEE 5th Int. Conf. on Communications and Electronics* pp 464-469.
[23] Melkemi M and Elbaz M 2014 *A Simple and Efficient Algorithm for Dot Patterns Reconstruction IEEE Int. Conf. on Image Processing* pp 4727-4731.

Design and Development of Smart Medicine Box

Ekbal Rosli[1], Yusnira Husaini[1,2]

[1]Faculty of Electrical Engineering, Universiti Teknologi Mara Shah Alam 40450 Selangor, Malaysia
[2]IMSaA, Universiti Teknologi Mara Shah Alam 40450 Selangor, Malaysia

E-mail: yusni458@salam.uitm.edu.my

Abstract
The Smart Medicine Box is successfully designed in helping the introvert patients taking their medicine without help of others. This project is to develop a robotic device that can assist patient to take medicine alone by implementing an IOT apps system for controlling the Smart Medicine Box where it will overcome an emotional disturbance experience by the introvert patients. There are four sensors such as PIR, IR, temperature and ultrasonic sensors use for the project. The purpose of PIR sensor is to detect hand movement near the device, while IR sensor is to detect the line follower on the floor. The LM 35 acts as the detection of the temperature inside the box and the ultrasonic acts as the detection of the obstacle in front of the device. The MIT Apps Invention 2 is used to develop an apps and collect the data from sensors through Arduino microcontroller. A proof of concept design has implemented and demonstrated successfully.

Keywords— Arduino, MIT Apps Invention 2, IOT system, PIR sensor

1. INTRODUCTION

According to Health Facts 2015 in Hospital Kuala Lumpur (HKL), the number admission of the patients for 2015 are 131, 639 while 128, 836 is the number admission of the patients in 2014 [1]. The increasing number of patients will increase the number of ward needed to cater all. Therefore, the nurses will be busy to handle the patients at the same time.

Nowadays, the nurses still used the manual method to give the medicine to the patients. They will go to the patient's bed for three times per day and give the medicine to the patients. By using this method, the nurses will waste the time because they should go to the patients' ward one by one. Besides, they also cannot monitor the patients properly because they do not have enough time because they will busy to monitor the other patients. So, the patients will skip the medicine by throwing them in the dustbin. Some patients who are introverts prefer to be alone. Introverts are known for gaining energy from spending time alone [2]. Usually, they will stay in the private ward and do not like to share that ward with another person. Introversion is the common ways of defining personality types that describe how individuals derive their energy, process information, and respond to stimulation.

For this problem, there is no specific solution that can help the nurse to solve this problem. They just can give the advice to take the medicine and monitor them if they have enough time to do that. This project is developed to help the problem occur with introvert patients. The smart medicine box is to solve and help the nurse to handle the introvert patients specifically.

2. METHODOLOGY

The design process of our project started with the decision making of the project title. The group members brainstormed together to come up with a suitable project. After deciding on one specific title, consultation with the lecturer was done, and the lecturer approved our title. We proceed to the proposal part of the project. The literature review was done, and all requirements for the proposal was analyzed. After completing the literature review, the proposal was complete. Therefore, after completing the proposal, we make an analysis on the system that we are going to use; software and hardware designing process can take place. We then test the prototype for any problem and the presentation will take place with a full report to submit.

2.1 Functional block diagram

From the block diagram, the input of power supply, 12V is needed to activate the system. Then the Arduino will act as a controller and the detection will be work by sensors which are PIR motion sensor, temperature sensor, ultrasonic sensor, infrared sensor to gives the output to apps controller. next, the servo motor will be used in the box to drop the medicine, DC motor will move from one place to another place when receiving the command from the microcontroller. Lastly, buzzer and LED will act as an output to give a warning if it faces the obstacle.

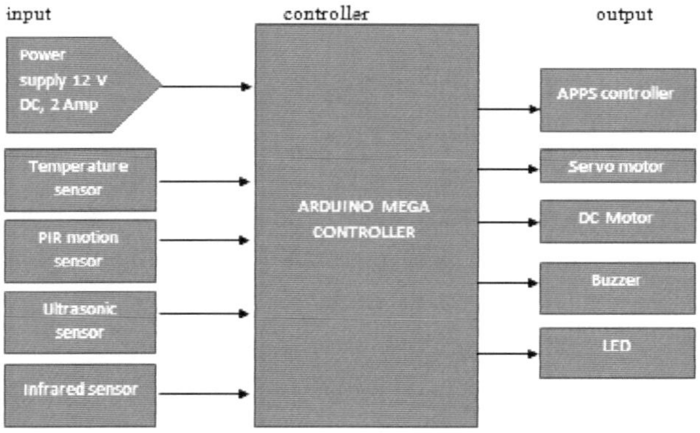

Figure 1 : The Block of Smart Medicine Box

2.2 Theory and calculations

In this project, Proteus 8 professional has been used to build the simulation of the project. It is the best software for simulation because its' contains a lot of electronic components that can be run virtually on the computer. This software can calculate and show the currents and voltages needed by the electronic components on the board.

Line follower system is used for the smart medicine box. So that it can move the box from one place to another. On another side, the photodiode is used with the photodetector for tracking line based on the light reflection by using infrared sensor module. If it is no light detection on the photodetector, the microprocessor will give the command to the DC motor to reroute the track. From this module, the uniform movement of the product can be maintained. In this system, 3V-6V DC motor is used to make use of DC supply of the system.

In addition, the project that had been built, it is related to robotic technology that may reduce nurse burden in handling their patient in the hospital. The title for this project is smart medicine box really related to the problem faced by hospital staff nowadays. To build the robotic system some sensor needs to consider, especially ultrasonic sensor, an infrared sensor, PIR sensor and temperature sensor.

Firstly, the ultrasonic sensor module is used to detect the obstacle of the surroundings, and automatically stop the DC motor. In fact, by doing this the project can avoid any collision with other things especially human. The sensor consists of 4 pins, which VDD pin, GND pin, Trigger pin, ECHO pin and connect to the high-performance 8-bit microcontroller. VDD pin act as an input voltage to the sensor, and GND pin to stabilize the circuit. The most importance pins are Trigger pin and Echo pin to control the range for the obstacle to be detected.

The voltage needed for this sensor to operate are around 4.95 V to 5.05 V with 15mA working current. This ultrasonic sensor used for safety purposes to avoid the project harming the surrounding. Ultrasonic functioning based on the speed of sound around 340 ms^{-1} or 29 microseconds per centimeter. The trigger will cause ping to travel out of the ultrasonic sensor and the distance can be altered by setting the speed of sound range from 0 ms^{-1} until 340 ms^{-1} in the c programming of the Arduino command.

Next, the infrared sensor uses to differentiate the amount of light absorb after it is reflected from the obstacle. This project used IR sensor for line follower purposes. The threshold used for the infrared sensor is 300 to distinguish the black and white color of the track. The IR sensor operated at 5 V supply with input output pin. It has one IR LED to send the light signal from the system and one photodiode that converts light into current. This current will send through the signal pin to the Arduino board.

When the IR receiver does not receive a signal, the potential at the inverting input goes higher than that non-inverting input of the comparator IC (LM339). Thus, the output of the comparator goes low, but the LED does not glow. When the IR receiver module receives signal to the potential at the inverting input goes low. Thus, the output of the comparator (LM 339) goes high and the LED starts glowing. Resistor R1 (100), R2 (10k) and R3 (330) are used to ensure that minimum 10 mA current passes through the IR LED Devices like Photodiode and normal LEDs respectively. Resistor VR2 (preset=5k) is used to adjust the output terminals. Resistor VR1 (preset=10k) is used to set the sensitivity of the circuit Diagram.

Otherwise, for the temperature sensor, LM 35 is used to detect the temperature changes, especially in medicine storage container in the smart medicine box. This component used analog signal to transmit the signal to the Arduino board. It has 3 pins which VCC pin for voltage input, analog out pin for signal output and ground pin. The temperature can be calculated from the signal value as follows:

$$\text{Temperature (°C)} = \left(\frac{signal\ value}{1024.0}\right) \times 500 \qquad (1)$$

This formula (1) is to get the Celsius value from 10-bit analog value. Lastly, we are using voltage divider rule to get an output 3.3 V for Bluetooth input.

In the design, the ultrasonic sensor is used to detect any obstacle for the product to move and it will control the DC motor to rotate or stop. When the ultrasonic sensor detects the obstacle, the LED and buzzer will switch on and alert the person or living obstacle to moving away from the track and the product will continue moving by following the track.

Next, LM35 component is used to detect temperature changes in the medicine box for protection. If the temperature is too high for the medicine, it will cause a chemical reaction in the medicine and change the properties. This may cause more problems to the patient.

In this design, we used two PIR sensors for two conditions which one for morning purpose and another for evening purpose. First PIR sensor, will control the servo motor to rotate 90° angle to drop the medicine from the box when it detects patients' hand. This condition is used in the morning for patients to consume the medicine. Next PIR sensor, is used to control the servo motor to rotate at 180° angle in the evening.

Lastly, we used the Wi-Fi module to connect the device and the smartphone for communication or sending the medical condition to the nurse on duty. In this simulation, RX module and TX module is used to show the Wi-Fi module function in protest simulation because there is no Wi-Fi library in the software. On the smartphone, the temperature of the medicine box will be displayed and the number of the medicine consumption by the patient will be recorded. Virtual terminal shown in Figure 16, is the purpose output that will be shown.

3. RESULT AND DISCUSSION

3.1 Proteus simulation results

The reason used this sensor is to detect the changes of the temperature in medicine box. The virtual terminal on the proteus simulation will show the detected temperature and the output. Thumbs touch was used to stimulate the sensor of LM 35 and the output reappears on the virtual terminal. Next, the motion stated as shown in the figure above show two motions. The motion belongs to two PIR sensor. The first one for PIR sensor one used to drop the medicine in the morning and another one used to drop the medicine in the evening. The distance at the fourth line is set in range or distance of the ultrasonic sensor to detect an obstacle. If the distance below 150 cm, the LED and buzzer will turn on to warn the user or obstacle. During this moment, the DC motor will stop the product movement due to the detection of an obstacle.

Servo motors are used to rotate the medicine output from the medicine box where we set two conditions which are 90° and 180°. When running the simulation, first rotation shows 90° then turn back to zero and continue to 180° for the second condition and back to 0° and stop the rotation of servo motor. The reason why it rotates because of the PIR sensor detect the hand or obstacle that trigger the sensor to send the input to the microcontroller and cause the servo motor to rotate. The rotation will halt about 20 seconds because the delay time is 20 seconds in the programming coded. Medicine will drop for two section which is during morning and evening only.

3.2 Mit Apps Inventor 2 Display & Programming

The MIT apps inventor 2 that used in this project is to monitor the temperature of the medicine and the medicine's consumptions. The data can be collected directly from the data storage and sending to and can be displayed directly as well as accurately on the smartphone that running the Android operating system. This is due to the APK file that cannot be installed on the smartphone that running the Microsoft operating system and IOS. The MIT apps inventor 2 that use in this project are consists from screen 1 until screen 4.

MIT apps inventor is programmed to send the data from the Bluetooth data storage to Arduino Mega. Lastly, we want to use the Wi-Fi module to connect the device and the smartphone for communication or sending the medical condition to the nurse on duty. The problem faced during simulation is we cannot know the exact IP address for Wi-Fi connection except try the Wi-Fi module hardware exactly on the board. Based on the real design the voltage divider is used to get an input of 3.3 V for the input of the Tx Wi-Fi module. The reason for voltage divider is used because the Wi-Fi module only operated on a 3.3V supply only more than that it will damage the system. After a lot try and error we move to use Bluetooth to get the output because we have a lack of time to complete the project. Luckily, we managed to get the output from the device and appeared on the applications that we used.

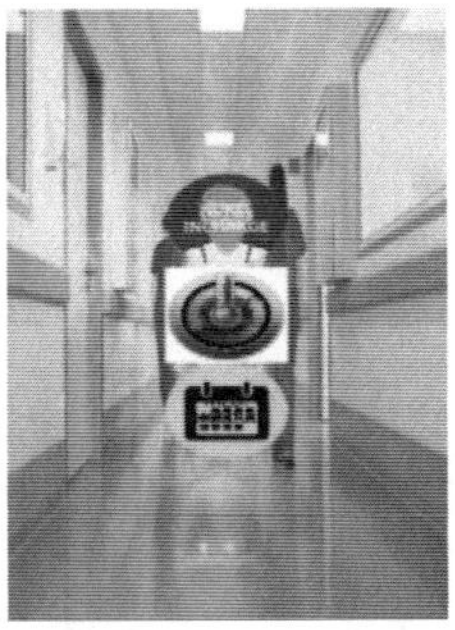

Figure 2: The home page of MIT Apps Inventor 2 & its programming (Screen 1)

Figure 3: MIT apps inventor page & programming at screen 2

Figure 4: MIT Apps Inventor Page & Programming at Screen 3

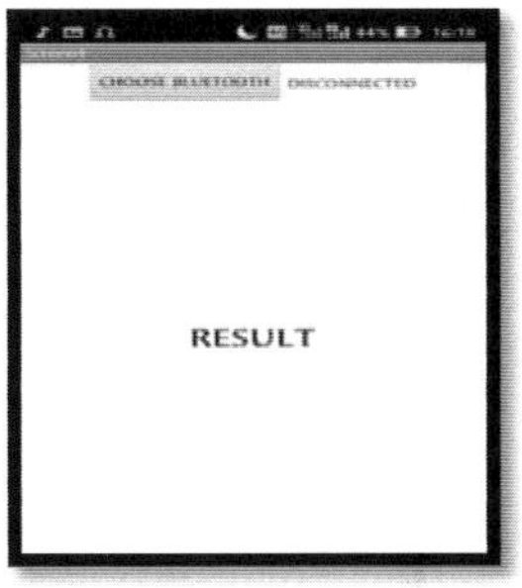

Figure 5 : MIT Apps Inventor Page & Programming at Screen 4

i. **Screen 1**
Displays the home page that contain the button for nurse's schedule, nurses in-charge and a button for temperature running.

ii. **Screen 2**
The detail that needs to enter by a nurse in order proceed by their schedule. After entering their name, duty period and date, the data will collect and save. This data is need as it's important to know whose nurse that responsible at the time need by the patient to take their medicine either in morning or evening So, this page is to ensure nurse has responsible to their duty.

iii. **Screen 3**
Displays the Bluetooth functions as a connector to Arduino Mega after receives the data from MIT Apps Inventor 2.

iv. **Screen 4**
Displays the result of medicine's temperature as the temperature is detected by the Temperature's Sensor (LM35). After the result was showed, the nurse can choose to disconnect the Bluetooth by the button that contains in this screen.

3.3 Application on the device.

The experimental setup of our project is to test all sensors functioning, safe moderate testing for the balancing and the movement of the base part. Then, for the combination of all circuit to the product, or integration as a complete system. The results are satisfactory considering that all the sensors are working as required.

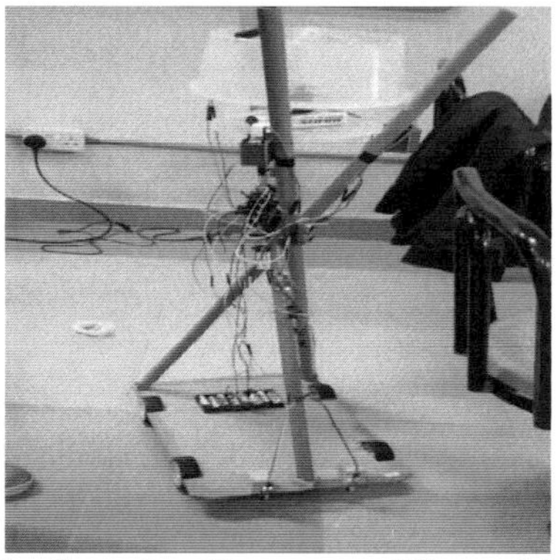

Figure 6: The Complete Prototype of Smart Medicine Box

Figure 6 above shows the complete prototype of Smart Medicine Box for final presentation by using four type of sensor that operated with their functions. This device is successfully achieved as the result that we want.

4. CONCLUSION

In conclusion, this device can help and give advantage to the nurses. The main objective for this innovation is to monitor the consumption of medicine intake for intrinsic patients. It is practical in the morning and evening but also can be used at night. This device is controlled by using Bluetooth system, so the nurse does not need go to the personal ward to give the medicine. This system is a very good to apply in the hospital because it can make the nurse job easier besides making the patients more comfortable to stay at the hospital.

5. REFERENCES

[1] H. K. Lumpur, "Health facts 2015," pp. 1–10, 2015.
[2] P. Types and P. Strengths, "Strengths of Introvert and Extrovert Personality Types Research Bytes," pp. 1–7, 1976.
[3] "Medicines Optimisation : Helping patients to make the most of the medicines," no. May 2013.
[4] H. Lin and N. W. Bergmann, "IoT Privacy and Security Challenges for Smart Home Environments," 2016.
[5] O. F. Science and I. I. N. Accordance, "Malaysian Standards Approved On 04 / 06 / 2013 By The Ministry Of Science , Technology And Innovation In Accordance With The STANDARDS OF MALAYSIA ACT (ACT 549) New Malaysian Standard," no. Act 549, 2013

6. ACKNOWLEDGMENT

The authors would like to appreciate the support of the ARAS grant (600-RMI/DANA 5/3/ARAS (51/2015)) in supporting the extension of the project to be published.

IOP Conf. Series: Materials Science and Engineering **341** (2018) 012005 doi:10.1088/1757-899X/341/1/012005

Third-Order Elliptic Lowpass Filter for Multi-Standard Baseband Chain Using Highly Linear Digitally Programmable OTA

Mohamed B. Elamien and Soliman A. Mahmoud

Department of Electrical and Computer Engineering, University of Sharjah, UAE

E-mail: {melamien, solimanm}@sharjah.ac.ae

Abstract. In this paper, a third-order elliptic lowpass filter is designed using highly linear digital programmable balanced OTA. The filter exhibits a cutoff frequency tuning range from 2.2 MHz to 7.1 MHz, thus, it covers W-CDMA, UMTS, and DVB-H standards. The programmability concept in the filter is achieved by using digitally programmable operational transconductors amplifier (DPOTA). The DPOTA employs three linearization techniques which are the source degeneration, double differential pair and the adaptive biasing. Two current division networks (CDNs) are used to control the value of the transconductance. For the DPOTA, the third-order harmonic distortion (HD3) remains below -65 dB up to 0.4 V differential input voltage at 1.2 V supply voltage. The DPOTA and the filter are designed and simulated in 90 nm CMOS technology with LTspice simulator.

1. Introduction

Multi-standard systems have become indispensable in today wireless and wireline communication. Through these systems, it is possible to connect to networks using a variety of communication standards inducing Bluetooth, W-CDMA, UMTS, DVB-H and IEEE 802.11a/g standards. The multi-mode filter is a fundamental component in these systems. It serves as a channel selector by adjusting its bandwidth with regards to the targeted standard specifications[1], [2].

The main adopted structures for implementing the multi-mode filter are the Gm-C, active-RC and MOSFET-C. For high frequency and low power application, the Gm-C structure is preferable due to the open loop operation of the operational transconductor amplifier (OTA). On the other hand, the active-RC and MOSFET-C structures consume high power in the high-frequency applications because they use local feedback around the active elements [3]. The main disadvantage of the Gm-C filter is the high nonlinearity due to the open loop operation of the used OTA. Hence, many techniques have been reported in the literature in order to enhance the linearity of the OTA [4].

The paper is organized as follow: Sec. 2 presents the design of the digitally programmable OTA and the current division network. Sec. 3 illustrates the proposed third-order elliptic lowpass filter. The simulation results of the DPOTA and the complete filter are given in Sec. 4. In [5], the same proposed DPOTA is used to implement a 4th-order Butterworth lowpass filter for multi-standard receiver baseband chain. However, the elliptic transfer function has the best selectivity among all other types.

2. Digitally Programmable OTA

The presented DPOTA circuit employs three different linearization techniques [6], [5]. The first technique is the source degeneration using MOS transistors [7] in which a resistance is connected between the two sources of the differential pair's transistors. The resistance can be implemented by a resistor or by MOS transistors working in the linear region. In the case of using a resistor of value (R), to improve the linearity significantly a large R is needed. However, the overall transconductance will be limited ($G_m \approx (1/R)$). The second technique is the adaptive biasing [8], the concept of adaptive biasing allows the input voltage to contribute in the biasing current. Hence, the non-linear term is canceled by the input voltage. The third technique is the double differential pair, which enhances the linearity performance by canceling the third-order distortion[9]. A common mode feedback circuit (CMFB) is added to the balanced OTA. The CMOS realization of the balanced OTA is shown in Fig. 1. Each of the following pairs of transistors are identical (M_1, M_2), (M_3, M_4), (M_5, M_6) and (M_7, M_8). The transistors M_5, M_6, M_7, M_8 operate in linear region. Then the output differential current of the first source degenerated differential pairs is given by

$$I_o = I_1 - I_2 = \frac{\sqrt{2K_{n1}I_{T1}}}{a_1} V_d \sqrt{1 - \frac{K_{n1}V_d^2}{8a_1^2 I_{T1}}} \tag{1}$$

where $a_1 = 1 + (K_{n1}/4K_{n5})$. From Taylor expanding, the third-order distortion is canceled if the following condition is satisfied

$$\frac{I_{T1}}{I_{T2}} = (\frac{K_{n1}}{K_{n3}})^3 (\frac{a_2}{a_1})^6 \tag{2}$$

By proper transistors sizing and choosing proper biasing currents according to (2), the third-order harmonic distortion can be eliminated. However, due to mobility degradation effect, the third-order harmonic can be reduced only. The biasing voltage V_b is used to bias the squaring circuits in order to achieve the maximum linearity.

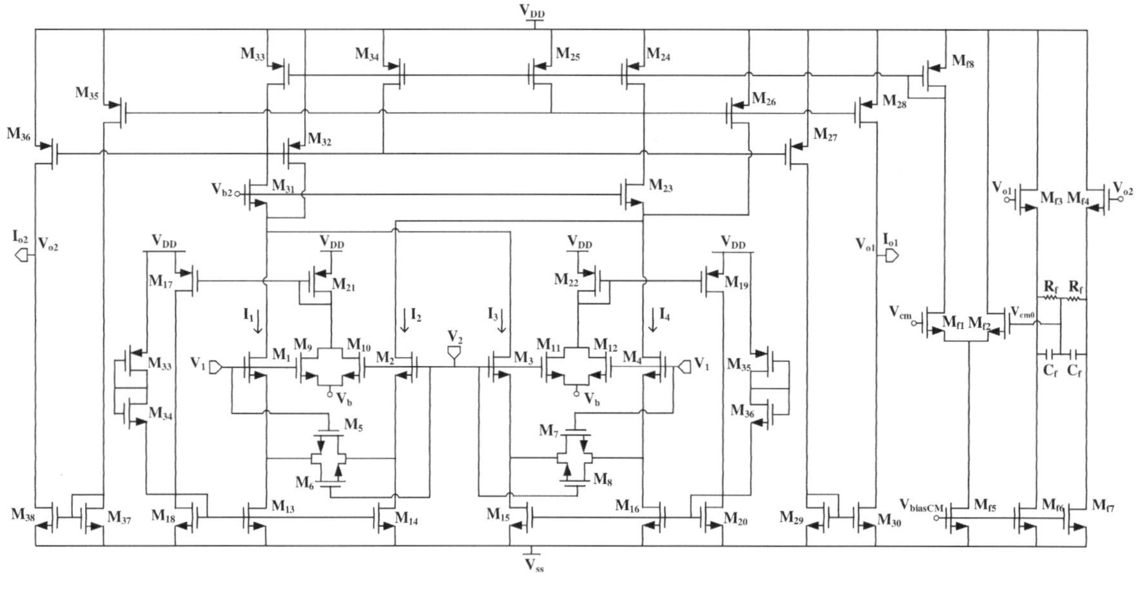

Figure 1. The CMOS realization of the proposed OTA.

The squaring circuit [8] is added to the proposed OTA to make the biasing currents adaptive to the input voltage. Then the non linear terms in each differential pair is reduced.

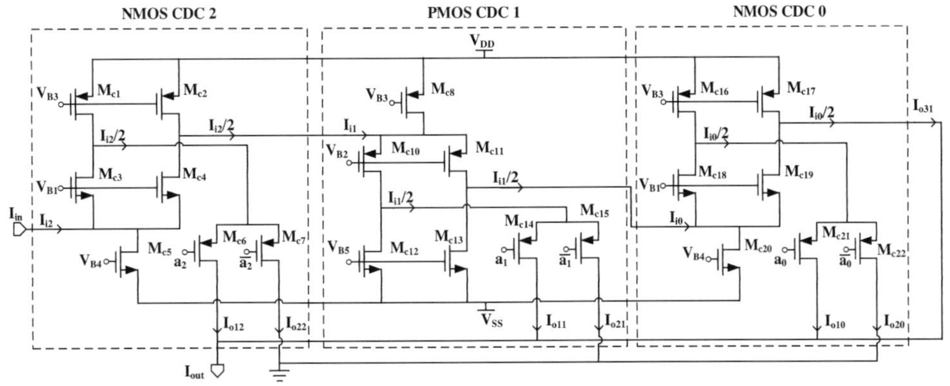

Figure 2. Three-bit CDN CMOS realization [10].

2.1. The current division network

The programmability concept in the presented DPOTA is introduced by using two current division network circuits [10] as shown in Fig. 2. The adopted CDN based on cascaded NMOS current division cell (CDC) by PMOS CDC. It has no restriction on the number of the controlling bits due to the supply voltage limitation where the voltage levels are generated from one cell to another. Moreover, the structure of the CDC has the advantages of the low input impedance and high output impedance. Hence, no need for virtual ground nodes [10]. Also, this CDN can divide bidirectional input current and the transistors matching requirements are relaxed since only the transistors within each CDC required to be matched.

Each of the two output currents (I_{o1},I_{o2}) of the proposed OTA in Fig. 1 is connected as input for one CDN. The two CDNs are controlled by 3-bit word (a_0,a_1,a_2). The 3-bit CDN circuit divides the input current with controlling factor (α), that can be expressed as

$$\alpha = \frac{1}{2^n}\Big(1 + \sum_{i=0}^{i=n-1} 2^i a_i\Big) \tag{3}$$

where n is the number of the digital control bits. The adopted tuning technique preserve the overall linearity performance of the proposed OTA. The overall output currents of the proposed digital programmable OTA is given by

$$I_{out1} = -I_{out2} = \alpha(a_0, a_1, a_2)G_m(V_1 - V_2) \tag{4}$$

where G_m is the transconductance value. The current gain $\alpha(a_0, a_1, a_2)$ is a function of the digital control bits as explained in (3).

3. Third-Order Elliptic Lowpass Filter Design

A third-order elliptic lowpass filter is designed using the presented DPOTA. The elliptic transfer function has the best selectivity, however it suffers from the ripples in pass and stop bands. The passive implementation of the filter is shown in Fig. 3. F ully differential implementation of the filter using seven identical DPOTA and five capacitors is shown in Fig. 4, where the passive inductor is implemented by four DPOTA ($G_{m3} - G_{m6}$) based on gyrator-C structure. G_{m2} and G_{m7} implement active resistors and G_{m1} acts as buffer transconductance. The values of the capacitors are extracted such that to make the tuning over WCDMA (2.2 MHz), UMTS (5 MHz) and DVB-H (7 MHz) standards. The capacitance of C_1, C_2, C_3 and C_i are 0.2 pF, 0.2 pF, 50 fF and 0.2 pF, respectively. The cutoff frequency of filter can be tuned digitally from

2.2 kHz to 7.1 MHz by changing the digital input of the DPOTA. More pass-band gain can be achieved by increasing the transconductance of G_{m1}.

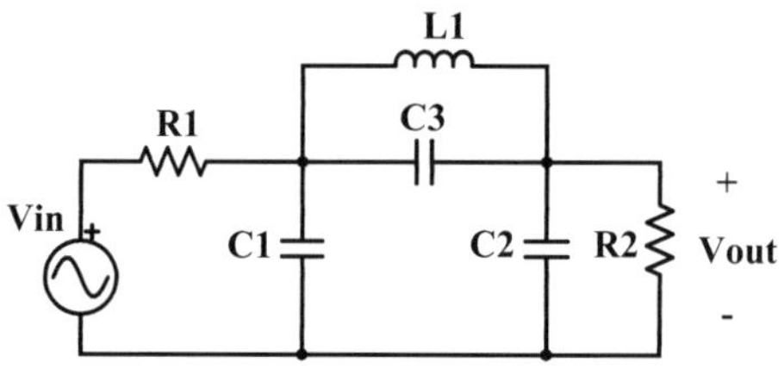

Figure 3. The passive prototype of the third-order elliptic lowpass filter.

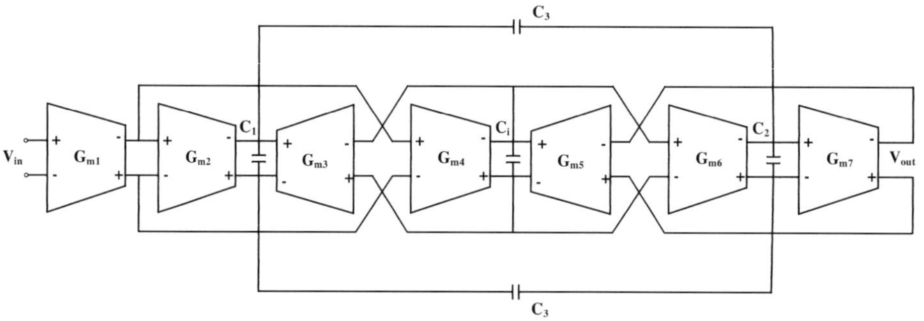

Figure 4. The active implementation of the third-order elliptic lowpass filter.

4. The Simulation Results

4.1. DPOTA simulation

The proposed circuit is validated through simulation tests. LTspice simulator is used with 90 nm BSIM4 (level 54) technology under ±0.6 V supply voltage. Fig. 5 shows the transconductance of the proposed programmable OTA while α is varied from 0.125 to 1 with 0.125 steps. The values of the transconductance can be controlled from 11.65 μA/V to 93.5 μA/V. The output differential current versus the input differential voltage is shown in Fig. 6. The linearity performance of the proposed OTA is evaluated by measuring the third harmonic distortion (HD3) of the differential output current while the input voltage frequency is 10 MHz and peak to peak amplitude is varied from 0.1 V to 0.5 V. Fig. 9 shows that the HD3 remains below -60 dB. The bandwidth of the OTA is varying from 68 MHz to 74.2 MHz as shown in Fig. 10. Fig. 7 shows the input referred noise spectral density which is inversely proportional to α, at 1 MHz and for ($\alpha = 0.125$) it is 268 nV/$\sqrt{Hz}$ while for ($\alpha = 1$) it is 139.6 nV/$\sqrt{Hz}$. The overall power consumption of the OTA is 669 μW as shown in Fig. 8.

4.2. The filter simulation

The proposed filter in Fig. 4 is simulated in 90 nm CMOS technology with LTspice simulator. The filter consumes 4.68 mW under 1.2 V supply voltage. The magnitude response of the filter in Fig .11 shows the tuning range from 2.2 MHz to 7.1 MHz. The filter simulation results are summarized in table 2. The DC gain, cutoff frequency, the input referred noise and the third-order inter-modulation distortion at an input voltage of 200 mV_{pp} amplitude with frequencies 0.8 MHz and 0.9 MHz for LPF are given.

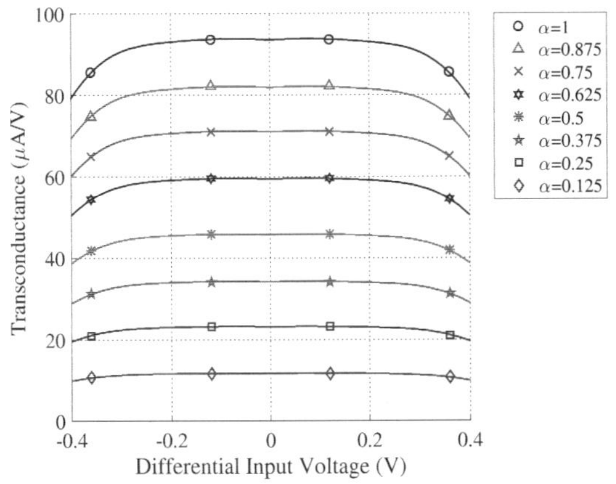

Figure 5. The transconductance of the proposed programmable OTA while α is varied from 0.125 to 1 with 0.125 steps.

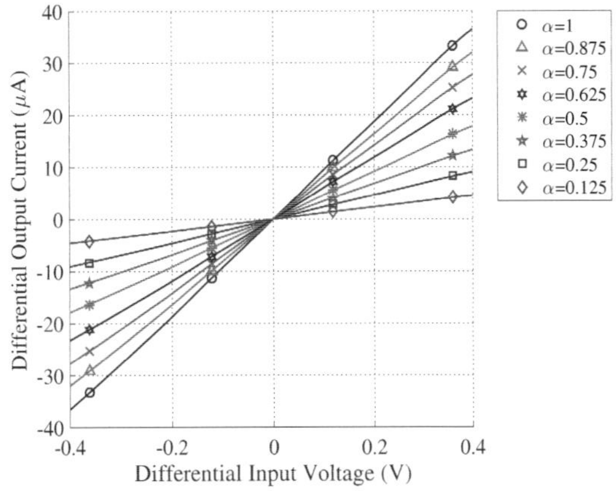

Figure 6. The output differential current of the proposed programmable OTA while α is varied from 0.125 to 1 with 0.125 steps.

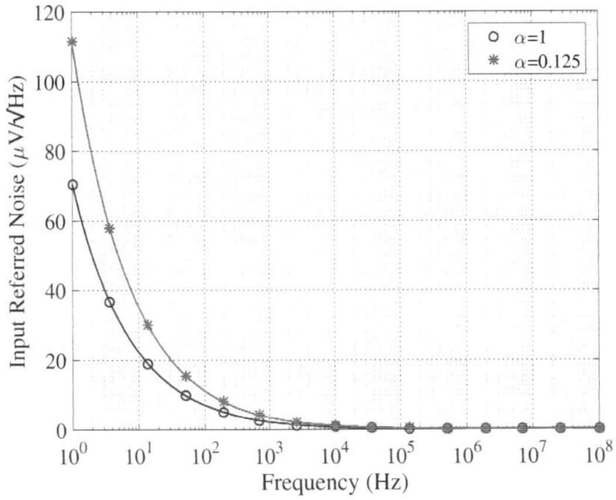

Figure 7. The input referred noise for minimum and maximum α.

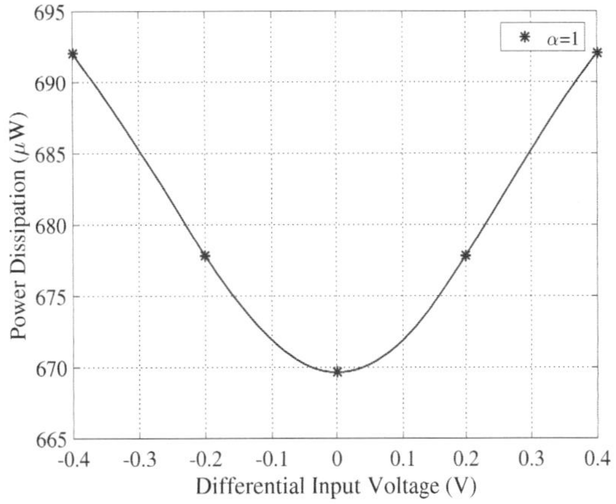

Figure 8. The total power consumption.

Table 1. Main specifications of the proposed DPOTA

Specifications	This work
Supply volyage (V)	$\pm$ 0.6
Power dissioation (μW)	669
Transconductance (μA/V)	11.65 - 93.5
Controlling bits	3
Input referred noise (nV/$\sqrt{Hz}$)	268.5 - 139.6
Technology	90 nm

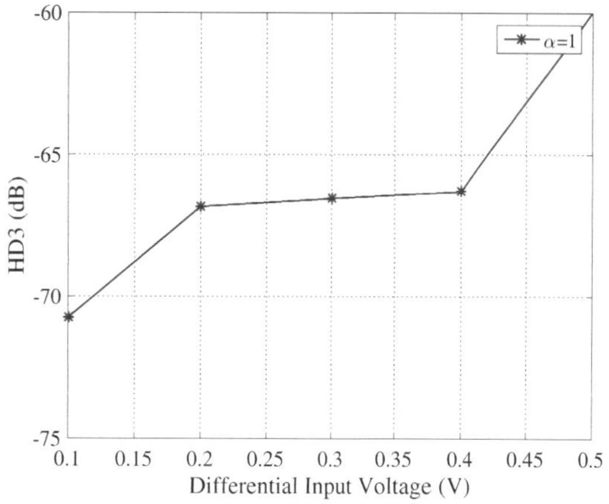

Figure 9. The third-order harmonic distortion of the differential output current at the maximum transconductance value.

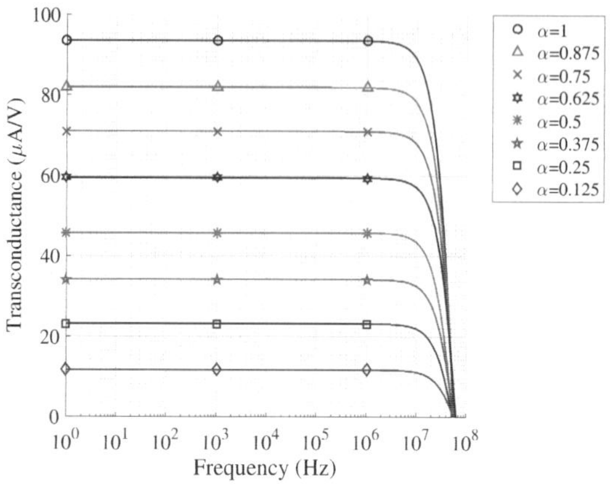

Figure 10. The frequency response of the proposed programmable OTA while α is varied from 0.125 to 1 with 0.125 steps.

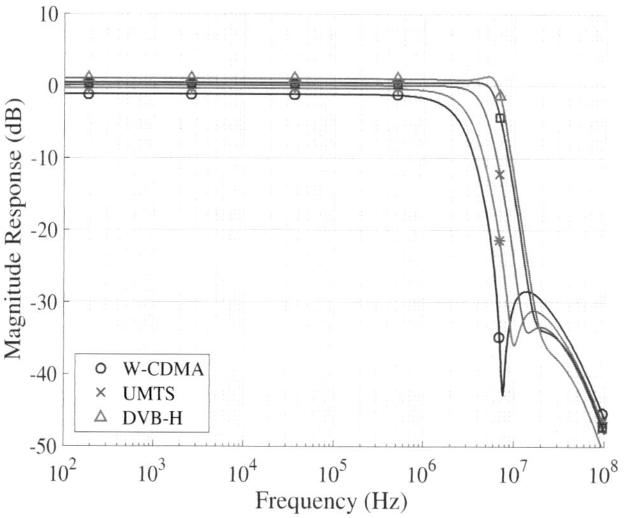

Figure 11. The magnitude response of the filter.

Table 2. Third-order elliptic lowpass simulation results for different standards.

Parameter	W-CDMA	UMTS	DVB-H
Cutoff Frequency (MHz)	2.2	4.5	7.1
DC Gain (dB)	-1.15	0.09	0.99
IM3 (dB)	-27	-36.54	-47.3
Input referred noise (nV/$\sqrt{Hz}$)	451.6	370.7	447.3

5. Conclusion

A digital programmable OTA with linearity enhancement has been proposed. The adopted linearity techniques are the source degenerated using MOS transistor, double differential pair and the adaptive biasing. The tuning capability is introduced by using 3-bit current division network. The adopted tuning technique preserves the overall DPOTA linearity performance. The DPOTA consumes 669 μW under 1.2 V supply voltage. The proposed DPOTA performance has been evaluated through LTspice simulator using 90 nm technology. The simulation results shows HD3 of the DPOTA is below -60 dB up to 0.5 V differential input voltage. A third-order elliptic lowpass filter is designed using the proposed DPOTA. The filter cutoff frequency can be tuned from 2.2 MHz to 7.1 MHz which makes it suitable for multi-standard systems.

References

[1] S. Hori, T. Maeda, N. Matsuno, and H. Hida, "Low-power widely tunable gm-c filter with an adaptive dc-blocking, triode-biased mosfet transconductor," in *Solid-State Circuits Conference, 2004. ESSCIRC 2004. Proceeding of the 30th European*, pp. 99–102, IEEE, 2004.

[2] S. A. Mahmoud and E. A. Soliman, "Multi-standard receiver baseband chain using digitally programmable ota based on ccii and current division networks," *Journal of Circuits, Systems and Computers*, vol. 22, no. 04, p. 1350019, 2013.

[3] J. Chen, E. Sanchez-Sinencio, and J. Silva-Martinez, "Frequency-dependent harmonic-distortion analysis of a linearized cross-coupled cmos ota and its application to ota-c filters," *IEEE Transactions on Circuits and Systems I: Regular Papers*, vol. 53, no. 3, pp. 499–510, 2006.

[4] D. A. Johns and K. Martin, *Analog integrated circuit design.* John Wiley & Sons, 2008.

[5] M. B. Elamien and S. A. Mahmoud, "Multi-standard lowpass filter for baseband chain using highly linear digitally programmable ota," in *40th International Conference on Telecommunications and Signal Processing*, (Barcelona, Spain), 2017 (unpublised yet).

[6] M. B. Elamien and S. A. Mahmoud, "A linear cmos balanced output transconductor using double differential pair with source degeneration and adaptive biasing," in *IEEE 59th MWSCAS*, pp. 253–256, 2016.

[7] F. Krummenacher and N. Joehl, "A 4-mhz cmos continuous-time filter with on-chip automatic tuning," *Solid-State Circuits, IEEE Journal of*, vol. 23, no. 3, pp. 750–758, 1988.

[8] K.-C. Kuo and A. Leuciuc, "A linear mos transconductor using source degeneration and adaptive biasing," *Circuits and Systems II: Analog and Digital Signal Processing, IEEE Transactions on*, vol. 48, no. 10, pp. 937–943, 2001.

[9] C. Popa and O. Mitrea, "Constant g m rail-to-rail cmos input stage with improved linearity," in *Image and Signal Processing and Analysis, 2001. ISPA 2001. Proceedings of the 2nd International Symposium on*, pp. 511–515, IEEE, 2001.

[10] S. A. Mahmoud, M. A. Hashiesh, and A. M. Soliman, "Low-voltage digitally controlled fully differential current conveyor," *IEEE Transactions on Circuits and Systems I: Regular Papers*, vol. 52, no. 10, pp. 2055–2064, 2005.

IOP Conf. Series: Materials Science and Engineering **341** (2018) 012006 doi:10.1088/1757-899X/341/1/012006

Soluble Solid Content Determination of Limau Madu Using Microwave Sensing Technique at 2.0 -2.6 GHz

Hudaliyana Ghazali[1], Rafidah Rosman[2] and Zaiton Sharif[3]

Faculty of Electrical Engineering
Universiti Teknologi MARA
40450 Shah Alam Malaysia
[1]huda.ghaz@gmail.com
[2]rafidah504@salam.uitm.edu.my
[3]zaiton623@salam.uitm.edu.my

Abstract. Orange is a fruit that essentially made up of a lot of nutrients. It is rich in vitamin C and folate. These are the qualities that must be maintained. Maintaining the qualities of this fruit is necessary. One of the prominent techniques is by measuring the solid soluble content of these local oranges non-destructively. In this study, the monopole sensor has been selected as the medium to find out the sweetness. The relationship of reflection coefficient, S_{11} and dielectric constant is observed. It is given by coefficient of determination, $R^2 = 0.9103$ at 2.4 GHz. The statistics show a promising result of frequency at 2.4GHz

1. Introduction

The prominent type of orange in Malaysia is citrus suhuiensis. Or well known to the local as *limau madu*. Sometimes they are being called *limau manis* due to their sweetness taste. Orange generally is known as important sources of vitamin C and folate [1]. Besides these nutrients, it also has significant amount of antioxidants, such as beta-carotene and flavonoid compounds [2]. These nutrients are simplified in the tables below.

Table 1. Nutrition facts of 1 orange [2]

Nutrients	Content	Nutrients	Percentage
Total fat	0.13 g	Vitamin A	1%
Saturated fat	0.015 g	Vitamin C	138%
Cholesterol	0 mg	Calcium	6%
Sodium	1 mg	Iron	2%
Total Carbohydrate	16.28 g	Thiamine	6%
Dietary Fiber	3.4 g	Folate	12%
Sugars	11.9 g		
Protein	1.4 g		

2. Literature review

Dielectric studies gained interest in learning the penetration of electric fields can be utilized in product quality determination non-destructively [3]. Microwave techniques also have been widely used in various studies due to its penetration [4]. The internal quality that is most important for all type of fruits is solid soluble content (mostly sugar), which can be determined only by destructive method. Hence, in order to preserve the condition of the fruit several non-destructive methods are proposed. Microwave measurement is potential in providing information related to permittivity.

Permittivity is the magnitude of dielectric properties which stimulus electromagnetic waves reflection and attenuation within materials[5] . Relative permittivity of a material can be expressed as:

$$\varepsilon_r = \varepsilon_r' - j\varepsilon_r'' \quad (1)$$

An optimal monopole probe is designed to measure the moisture content of mangosteen [4]. This study used the Misra-Blackham model for permittivity calculation. The magnitude of reflection coefficient is highest at 2.45 GHz. As the primary results, this study is able to prove that microwave techniques by open-ended coaxial probe can be effectively used in grading mangosteen non-destructively.

R. Rosman et. al observed the effect on reflect coefficient, Γ when an electromagnetic wave is emitted from the coaxial line port through a microwave monopole antenna [6]. On top of it, the idea of antenna monopole is to investigate the sweetness level of Malaysian local oranges non- destructively. The significance of this study is the author designed a unique holder for sample that is made up of nylon.

3. Materials and Methods

This experiment will be conducted into two separate forms. The first experiment is non-destructive, where the reflectance coefficient of each sample is measured at frequencies of 85MHz to 5 GHz. The second experiment is on destructive which will measure pH, brix, acidity and moisture content.

3.1. Sample

There are 40 samples of oranges that are bought at local market. These samples are being kept at temperature of 26̊C. They are weighed by an electronic balance. Diameter of the oranges is measured using a digital caliper. These oranges were numbered accordingly and their weight is recorded every day.

3.2. Non-destructive method

3.2.1. Microwave. This method is used to determine the reflection coefficient S_{11} of local oranges fruit. It is done by connecting the cabanR54 reflectometer to a personal computer. The sample is then touched at the end of the pole which connected to the CABAN R54 reflectometer.

IOP Conf. Series: Materials Science and Engineering **341** (2018) 012006 doi:10.1088/1757-899X/341/1/012006

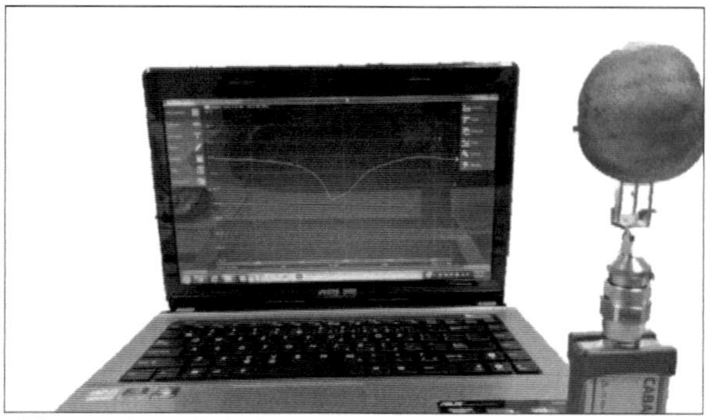

Figure 1. S_{11} is being measured using CABANR54

3.3. *Destructive method.* During this method, the brix, acidity, pH and moisture content is identified. Brix and acidity can be obtained by using digital hand-held pocket refractometer (Hybrid PAL-BX|ACID F5 ATAGO). While pH is measured with a Combo pH/EC/TDS/Temperature with Only One Tester (Model HI 98129). The fleshes of oranges were placed back into their skins respectively.

Figure 2. (Left) pH measurement of orange juice (Right) Dried oranges after 24 hours

After 24 hours, the dried oranges are weighted. The moisture content (mc%) is then obtained by [7];

$$mc\% = \frac{m_{initial} - m_{dried}}{m_{initial}} \text{ x } 100\% \quad (2)$$

4. Results and Discussion

Initially, it was necessary to investigate the linearity of the relationship between the permittivity of the material and the resulting microwave attenuation. During the process it was essential to ensure the consistency of the results. Figure 3 shows the reflection coefficient of local oranges at different days. As the day increases, the magnitude of S_{11} also increases.

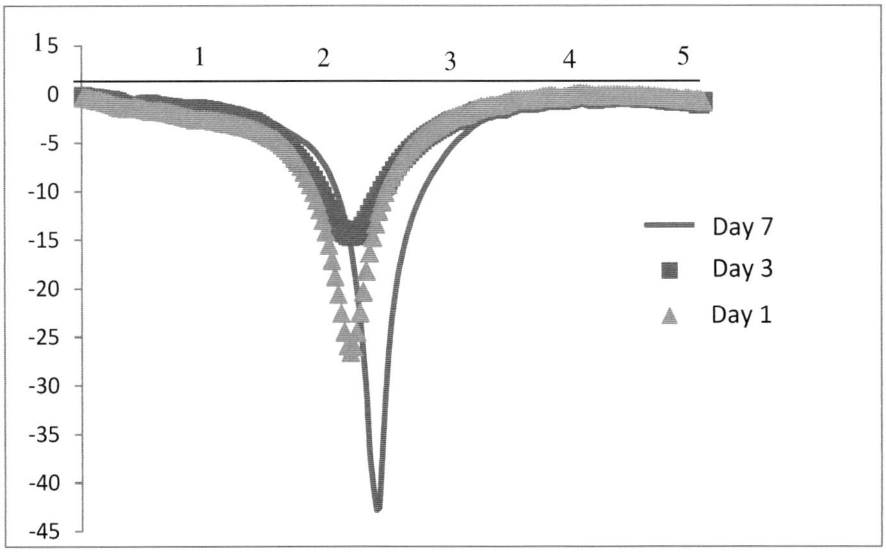

Figure 3. S_{11} is being measured using CABANR54

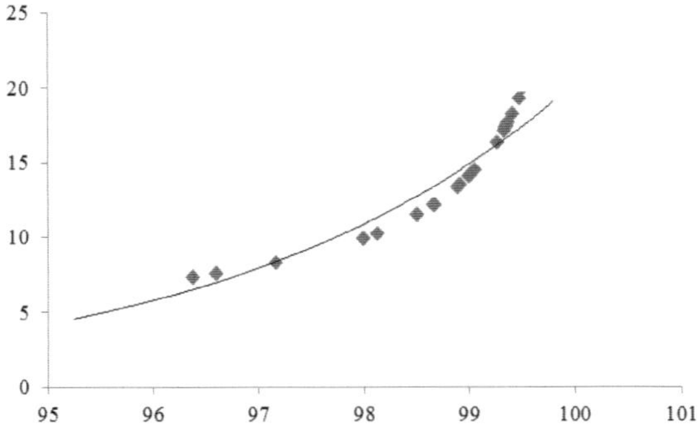

Figure 4. ε_r and S_{11} at 2.4 GHz

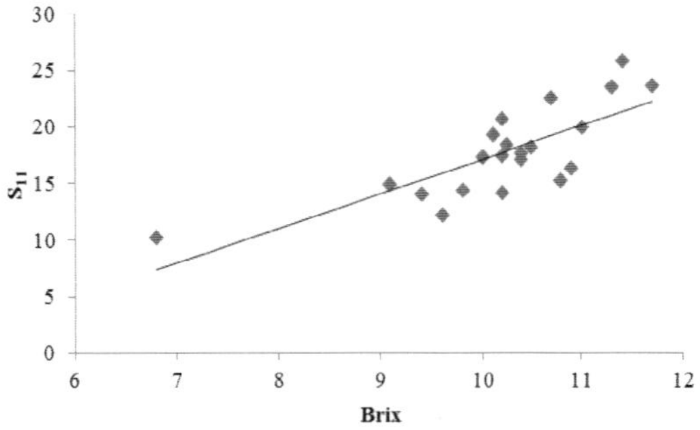

Figure 5. SSC and S_{11} at 2.4 GHz

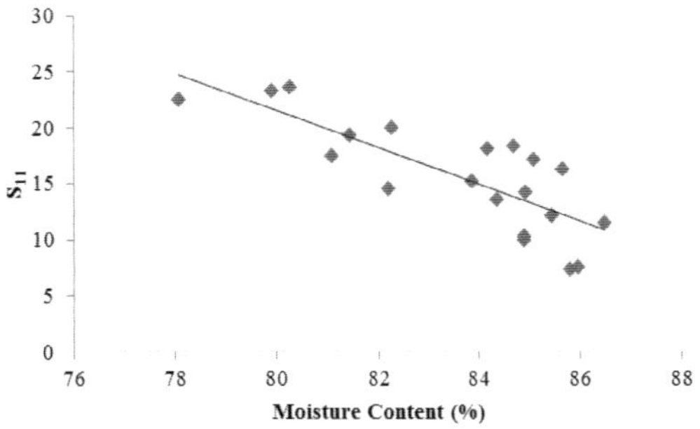

Figure 6. Moisture content and S_{11} at 2.4GHz

Figure 4, 5 and 6 shows the relationship of ε_r, SSC, moisture content (MC) at 2.4 GHz. It is consider a good operating frequency. The linearity and sensitivity of the antenna is determined at 2.4GHz to test performance of the antenna [8].

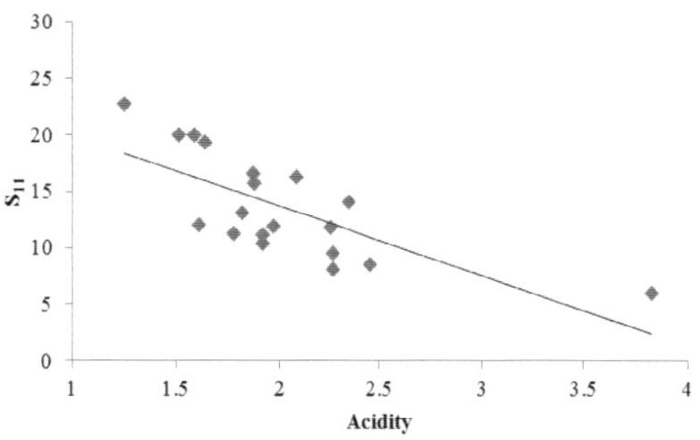

Figure 7. Acidity and S_{11} at 2.6 GHz

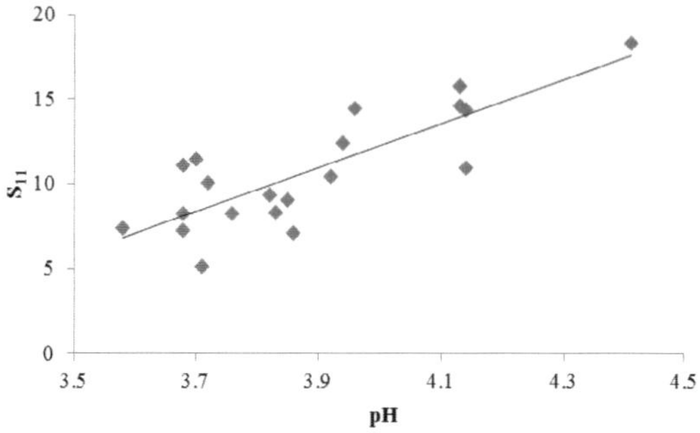

Figure 8. pH and S_{11} at 2.0 GHz

Figure 7 and 8 shows the percentage of acidity and pH value at different frequency. The frequency is chosen according to the highest value of coefficient determination. Percentage of acidity decreases while pH value is increasing. This can relate to the condition of *limau madu* which in general as it is riper, the fruit is less acidic [2].

Table 2. Multiple linear regression equations at 2.0 GHz, 2.4 GHz and 2.6 GHz

	Function	R^2	Frequency
ε_r	$= 5E\text{-}13e^{0.3143x}$	0.9279	2.4 GHz
SSC	= 3.0288x - 13.18	0.6016	2.4 GHz
Moisture Content	= -1.6572x + 154.11	0.6387	2.4 GHz
Acidity	= -6.1766x + 26.065	0.5246	2.6 GHz
pH	= 13.007x - 39.803	0.6701	2.0 GHz

Table 2 shows the equation and value of R^2 at 2.0 GHz, 2.4 GHz and 2.6 GHz. Dielectric constant shows the maximum value of R^2 at 0.9279.

5. Conclusion

The paper focuses on reflection coefficient, S_{11}. This measurement is obtained open-ended coaxial line probe technology and a network analyzer over the frequency range from 85 MHz to 5 GHZ. At increasing S_{11}, permittivity, brix or SSC(mostly contain sugars) and pH also increases. Only percentage of acidity decreases. This proves that as the fruit gain maturity it is less acidic. During the decline in S_{11}, the percentage of moisture content increases. It is observed that at 2.6 GHz, S_{11} is at the highest magnitude of -42.84 dB. However, it does not comply with other measurements. Averagely, frequency of 2.4 GHz shows an interesting relationship with dielectric constant, solid soluble content and also moisture content. Hence, it can be said that 2.4GHz is the best operating frequency.

References

[1] O. I. A. Oluremi, J. Ngi, and I. A. Andrew, "Phytonutrients in citrus fruit peel meal and nutritional implication for livestock production," *Livest. Res. Rural Dev.*, vol. 19, no. 7, 2007.

[2] J. E. Snart, M. L. Arpaia, and L. J. Harris, "Oranges : Safe Methods to Store , Preserve , and Enjoy," 2006.

[3] R. B. Russell, S. O. Nelson, W. Guo, and S. Trabelsi, "Investigation of Dielectric Sensing for Fruit Quality Determination," *Sensors Appl. Symp. 2008. SAS 2008. IEEE*, pp. 54–56, 2008.

[4] T. Tongleam, N. Jittiwaranghl, P. Kumhoml, and K. Chamnongthai, "Non-Destructive Grading of Mangosteen by Using Microwave Moisture Sensing Ilhn- '," vol. 2004, pp. 650–653, 2004.

[5] S. Wang *et al.*, "Dielectric properties of fruits and insect pests as related to radio frequency and microwave treatments," *Biosyst. Eng.*, vol. 85, no. 2, pp. 201–212, 2003.

[6] R. Rosman, M. N. Dimon, and Y. K. Yeow, "CST simulation on monopole sensor for determination of malaysian local oranges sweetness quality non-destructively," *Proc. - UKSim-AMSS 16th Int. Conf. Comput. Model. Simulation, UKSim 2014*, 2014.
[7] C. V. Kandala and J. Sundaram, "Nondestructive measurement of moisture content using a parallel-plate capacitance sensor for grain and nuts," *IEEE Sens. J.*, vol. 10, no. 7, pp. 1282–1287, 2010.
[8] H. H. Ali, Z. Bin Abbas, J. B. Hassan, A. Bin Jusoh, and R. B. M. Zahari, "Microwave antenna sensing technique for determination of moisture content in hevea latex from Hevea brasiliensis tree," *Aust. J. Crop Sci.*, vol. 5, no. 11, pp. 1326–1333, 2011.

Acknowledgments

This research was supported by Ministry of Higher Education in funding for Fundamental Research Grant Scheme (FRGS) under sponsorship file no. FRGS/1/2015/TK04/UITM/03/9 and Research Management Centre (RMC), Universiti Teknologi MARA (UiTM) Shah Alam. Authors are thankful to the Faculty of Electrical Engineering for providing space to conduct the experiment.

IOP Conf. Series: Materials Science and Engineering **341** (2018) 012007 doi:10.1088/1757-899X/341/1/012007

Interactive Design and Development of Real Arm Movements for Application in Rehabilitation

Rafidah Rosman[1], Muhammad Zaidan Abdul Hadi[2], and Nurulliyana Abu Bakar[3]

Faculty of Electrical Engineering
Universiti Teknologi MARA
40450 Shah Alam Malaysia
[1]rafidah504@salam.uitm.edu.my
[2]zaidanhadi92@yahoo.com
[3]nurulliyanaabubakar@yahoo.com

Abstract - An interactive real arm movements for application in rehabilitation is designed and developed. The aim is to encourage hand paralysis patients performing their physical therapy by introducing games application in replacing conventional hand therapy module and methods. In this project, the accelerometer is used for tracking the orientation of the arm. As the arm moves, the values from x, y and z axis from the accelerometer changes and are being read by the Analog Inputs of the Arduino Board. After being read by the Analog Inputs of the Arduino Board, the 3D model moves as well. Solidworks software was used to modeled the hand in which the data is then transferred to Matlab/Simulink using SimMechanicalLink from Mathworks. Lastly, the sensor glove was programmed to work as a controller of games application in hand rehabilitation thus makes it an enjoyable therapy process.

1. Introduction

Recently, advanced technology has contributed significantly in the development of medical field. Physical rehabilitation that are gives disadvantages to hand paralysis patients that require long-term repetitive movement of hand is awaited to be performed by sensor glove to encourage hand paralysis patients enthusiastic in physical therapy instead of human helped. For example, hand rehabilitation is an important case because hand movement is one of the simplest actions in daily life. Generally, hand paralysis is treated with the assistance of a physical therapist. Physical therapist holds and repeatedly moves the paralysis's fingers affected through the maximum range of its joint angle, as shown in Fig.1. In this case, it takes a few months to improve the range of joint angle to recovery through which finger can move because of uncontrollable muscle tightness and stiffness which make difficulty in movement of finger. Thus, in general, hand rehabilitation is expensive and time consuming. Furthermore, the unavailability of physical therapist underscores the requirement for engineering solutions for physical rehabilitation. Therefore, we aim to develop a hand rehabilitation application that can encourage hand paralysis patients enthusiastic in physical therapy that can act as a substitute for traditional physical therapy.

The new invention of computer technology is developing in the human environment. The ubiquitous systems are more common and controlling in computing that made to appear anytime and everywhere. The technology of graphical user interfaces (GUI) has changed to gesture interface, capturing the motion of the hands and controlling the devices in more natural and practical. Hand gesture may include multi-touch screen interface, surface computing, camera in gesture recognition and also in gaming industries [1]. Gesture recognition is general to Virtual Augmented Reality (VAR) as the main input system and now become popular just in films like Iron Man and Minority Report [2].

Human Machine Interaction is keeps moving more closely towards the interaction between people and computers from time to time. Human have a good clench and controlling ability with their hands and thereby interfaces like keyboard and mouse are more popular in nowadays. The gesture accordance with sensor glove is been used in sign language operating and training [3], but nowadays it is also used in robotics field to control robot's arms just wearing a glove [4]. Nowadays, the development of the new technology in interfacing the devices with the computer such as Nintendo Wii, the Apple iPhone and the iPad, SixthSense device developed at MIT Media Lab and the Kinect system for the Xbox [5] in the movement of the system using the devices that act as controller.

In this project, we mainly focus on the real-time input and output of the data from the sensor glove and successfully and accurately grasping the actions. Sensor glove is an electronic device equipped with accelerometer sensor and linear slide potentiometer that senses the movements of hand and finger's individually, and passes those movements to computer in analog signal continuously. Thus, nowadays sensor gloves are used in many research fields including virtual reality, gaming [1], robotics [4], character recognition and verification [6].

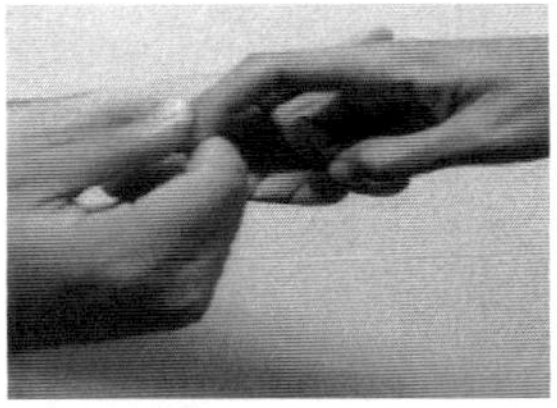

Figure 1. Rehabilitation Therapy

2. Methodology

The development of this project includes the buildings of the devices which include 3-axis accelerometer, linear slide potentiometers, HC-05 Bluetooth module and Arduino Mega 2560 that applied on a Sensor Glove development. In addition, this topic also covered on how the experimental testing was conducted using MATLAB and Simulink, Solidworks and Arduino (Mega 2560) software. The flow chart below (Fig. 2) shows how it works in the process of the system application.

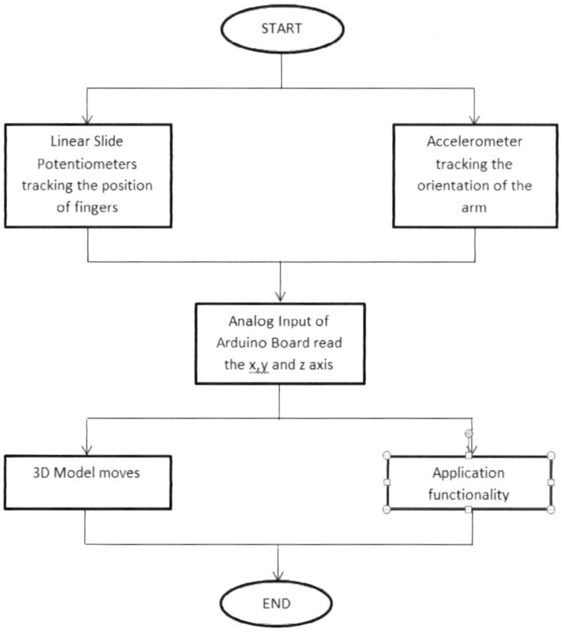

Figure 2. Flow Chart of the System Application

2.1 Building the Device

As shown in Fig. 3, the linear slide potentiometers are used for tracking the position of each finger. Furthermore, the accelerometer is used for tracking the orientation of the arm. As the arm moves, the values from the x, y and z axis from the accelerometer changes and are being read in the Analog Inputs of the Arduino Board. After being read by the Analog Inputs of the Arduino Board, the 3D model moves as well and application for gaming is functionality (Fig. 4). From Fig. 5, the block diagram of ADXL335 Accelerometer is taken from the datasheet. The ADXL335 is a small, thin, low power, complete 3-axis accelerometer with signal conditioned voltage outputs. The product measures acceleration with a minimum full-scale range of ±3 g. It can measure the static acceleration of gravity in tilt-sensing applications, as well as dynamic acceleration resulting from motion, shock, or vibration. The user selects the bandwidth of the accelerometer using the CX, CY, and CZ capacitors at the XOUT, YOUT, and ZOUT pins. Bandwidths can be selected to suit the application, with a range of 0.5 Hz to 1600 Hz for the X and Y axes, and a range of 0.5 Hz to 550 Hz for the Z axis. Lastly, the circuit schematic of the sensor glove was constructed as shown in Fig. 6.

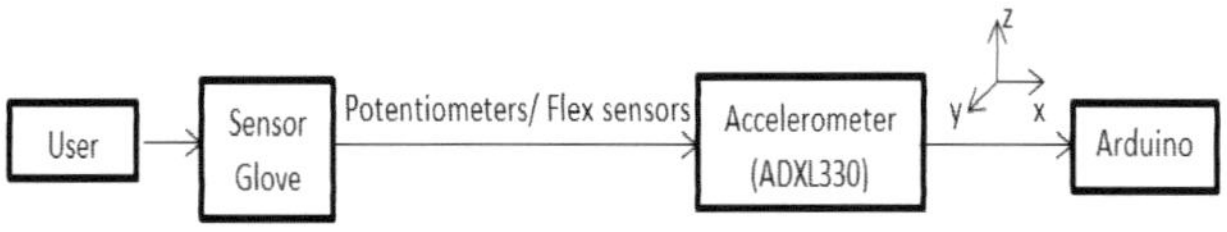

Figure 3. Functional Block Diagram in Hardware

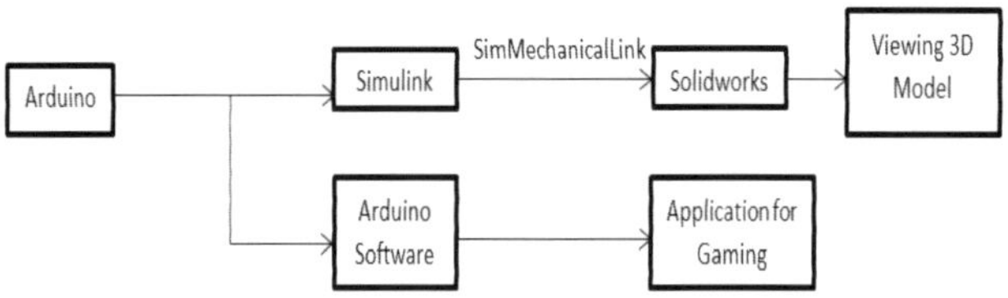

Figure 4. Functional Block Diagram in Software

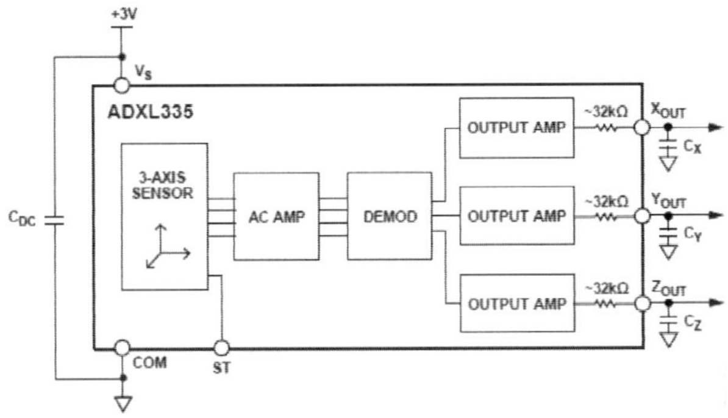

Figure 5. Block Diagram of ADXL335 Accelerometer
(from datasheet)

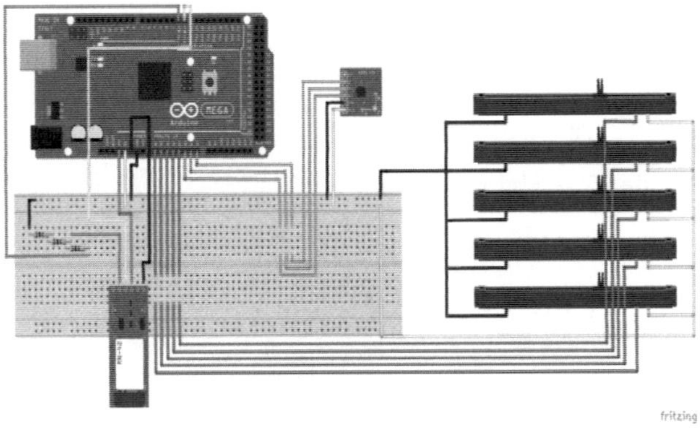

Figure 6. Circuit Schematic of the Sensor Glove

2.2 The Matlab/Simulink Model

Before building the Simulink model, the Arduino IO Package is installed which included the Simulink library for communication with the Arduino board. Furthermore, by using Arduino IDE, the Arduino board will be uploaded the code that comes with the package to enable the Simulink library. The Simulink model was created and it was organized into six subsystems in each of the block as illustrated in Fig. 7.

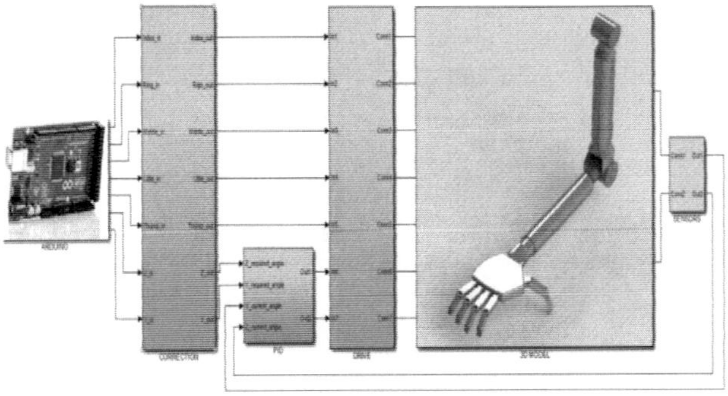

Figure 7. The Simulink Model

The Simulink model is organized into six subsystems:

IOP Conf. Series: Materials Science and Engineering **341** (2018) 012007 doi:10.1088/1757-899X/341/1/012007

- Arduino analog inputs block for the values from the accelerometer and the potentiometers sensor.
- Correction block which the values from Arduino analog inputs.
- PID control block for getting smoother movements.
- Driving block for the joints of the 3D model.
- The 3D Model that is automatically generated using the SimMechanicsLink.
- Sensor block for tracking the position of the arm in 3D model.

Based on the Fig. 7, the Simulink model will be organized into six subsystems.

i. Arduino analog inputs block for the values from the accelerometer and the potentiometers sensor
ii. Correction block which the values from Arduino analog inputs
iii. PID control block for getting smoother movements
iv. Driving block for the joints of the 3D model
v. The 3D Model that is automatically generated using the SimMechanicsLink
vi. Sensor block for tracking the position of the arm in 3D model

2.3 The 3D Model

The 3D Model as illustrated in Fig.8 is a representation of a human arm. The 3D Model was designed using Solidworks and then transferred it into Matlab/Simulink using the SimMechanicsLink from Mathworks as illustrated in Fig. 15. The purpose of transferring it to the Simulink is it will automatically generate an equivalent multibody model in the Simulink.

Figure 8. The 3D Model Design using Solidworks

3. Results and Discussion

This topic covered on the result obtained from the project analysis. All the results were displayed and illustrated in the figure respectively. A part of hardware assembly is shown in Fig.9

Figure 9. Hardware assembly of the Sensor Glove

The linear slide potentiometers are used for tracking the position of the fingers. After that, it will be attached to a spring (Fig. 18) to each of the potentiometer. The spring holds the potentiometer slider at a certain position and as the fingers move the slider is being pulled and the resistance of the potentiometer is changed.

Figure 10. Linear Slide Potentiometer attached with a spring

The accelerometer on the other hand is used for tracking the orientation of the arm. As the arm moves, the values from the x, y and z axis from the accelerometer change and are being read by the Analog Inputs of the Arduino Board. After being read by the Analog Inputs of the Arduino Board, the 3D model moves as well.

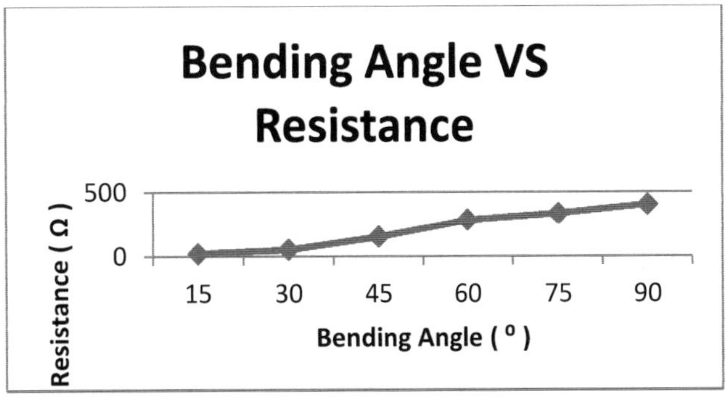

Fig. 11. Graph of Bending Angle Vs Resistance Measured at Index Finger

Based on Fig.11, the analysis had been made at index finger to measure the resistance of linear slide potentiometer based on the angles of the finger actuated. As a result, the resistance will be increase if the angles of the finger actuated increase.

For the monitoring of the 3D model in the Simulink, when the simulation is running the 3D model will automatically appear in the Simulink as shown in Fig. 12. As the result, the 3D model will move as same with the hand of user that is wear the sensor glove.

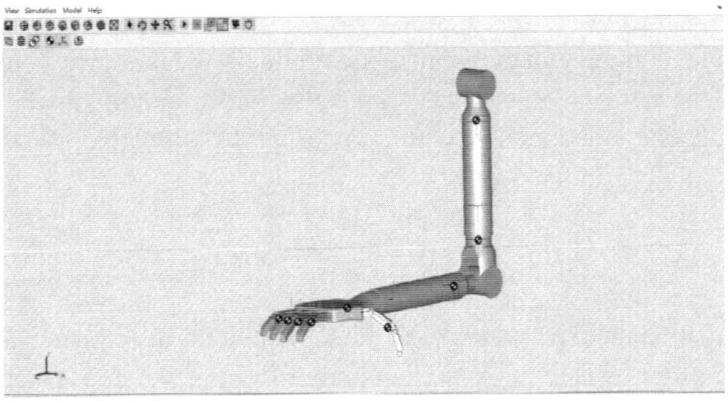

Fig. 12. The View of the 3D Model After Run the Simulation in the Simulink

For the application of Arduino game controller, firstly that it will working when the sensors of the glove capture the movements of the arm. After that, the captured values are sent into the Arduino board of analog inputs. Then, from the Arduino they are sent to the Processing IDE via Serial Communication.

Lastly, from the Processing IDE they are sent to the video game. For applications in video games, analysis has been made through the easy to difficult levels based on the bending of each finger, the finger movement frequency and speed of movement of the each finger in the application of video games. The example for the easy level of the video game that are used in the application which is Super Mario gaming as illustrated in Fig. 13. In this Super Mario gaming, each of the fingers is setting to switch the keyboard key. In this application is used up, down, left and right arrow key are switched into each of the fingers according to the fingers move the slider is being pulled and the resistance of the potentiometer is changed. Besides that, the sensor glove also are working in Need for Speed gaming that is used accelerometer which is to moves the hand to control to turn left and to turn right as shown in Fig. 14 that will give more difficulty to the user.

Fig. 13. Arduino Game Controller work in Super Mario Gaming

Fig. 14. Arduino Game Controller work in Need for Speed Gaming

In addition, this project has been analysed to the respondent who like to play video games and do not like playing video games. Based on the analysis made, the sensor glove is act more easily to the respondent who likes to play video games because too common for them to accept and learn new video game controller compared to the respondent who do not like playing video game. However, the respondent did not like playing video games require a short process only to learn control of sensor gloves in a video game if the frequency of use many times.

4. Conclusion

This project has found that generally hand exercises are beneficial in improving strength and dexterity regardless of whether the stroke patient is just beginning to get hand movement or already has good hand range of motion. In addition, there are exercises that can help improve hand functionality that have deteriorated after a stroke. Therefore, the purpose of the current study was to develop a device that enable real arm interaction movements with 3D computer models and for hand rehabilitation monitory and controlling purposely to serve the elderly and the handicapped. Hence, the sensor glove had been able to monitor and controlling to serve the elderly and the handicapped.

Furthermore, the purpose of creating this project is to rehabilitate patients who have problems with the movement of the fingers and hands more enjoyable through the application in the computer games without the need from human aid. In this case, the project has been successful in providing better patient satisfaction and eager to regain their fingers and hands.

Besides that, the sensor glove had been able to recognize hand gestures accurately and successfully. Thus, all engineering solutions provide opportunities for hand paralyzing patients to perform their physical therapy.

Last but not least, as the features recommendation the hardware of sensor glove will be made in smaller structure because to give the users feels lighter, comfortable and compatible while wearing the sensor glove. In this case, in future the sensor glove will be upgrade with exchange the linear slide potentiometer to flex sensor that will give more effectiveness and accuracy.

References

[1] T. Shiratori and J. K. Hodgins, "Accelerometer-based user interfaces for the control of a physically simulated character", ACM Transactions on Graphics, vol. 27, no. 5, (2008), pp. 1–9.

[2] P. K. Dick, S. Frank, "Minority Report", http://www.imdb.com/title/tt0181689/, (2009).

[3] T. Takahashi and F. Kishino, "hand Gesture Coding based on Experiments using a Hand Gesture Interface Device", SIGCHI Bull, vol. 23, no. 2, (1991), pp. 67-74.

[4] C. Lee and Y. Xu, "Online, Interactive Learning of Gestures for Human/Robot Interfaces", In IEEE International Conference on Robotics and Automation, (1996), pp. 2982-2987.

[5] The new media consortium, The Gesture-Based Computing, Horizon Report (2011) Edition.

[6] S. Sayeed, N. S. Kamel and R. Besar, "Virtual Reality Based Dynamic Signature Verification Using Data glove", in International Conference on Intelligent and Advanced Systems, IEEE, (2007), pp. 1260-1264.

[7] Tejashree Dhamapurkar, Neelakshi Ghag, Aarti Kamble, Krutika Bhiwapurkar, & Prof. Mrs. Shubhada Deshpande (2015). Wireless Animatronic Hand using Control Glove. International Journal of Innovative and Emerging Research in Engineering. Volume 2.

[8] Emi Tamaki, & Jun Rekimoto (2011). PossessedHand: Controlling Hand Movement with Computer Output.

[9] Shamsheer Verma (2013). Hand Gestures Remote Controlled Robotic Arm. Advance in Electronic and Electric Engineering. Volume 3.

[10] Elmar Rueckert, Rudolf Lioutikov, Roberto Calandra, Marius Schmidt, Philipp Beckerle, & Jan Peters. Low-cost Sensor Glove with Force Feedback for Learning from Demostrations using Probabilistic Trajectory Representations.

[11] Piyush Kumar, Jyoti Verma, & Shitala Prasad (2012). Hand Data Glove: A Wearable Real-Time Device for Human-Computer Interaction. International Journal of Advanced Science and Technology. Volume 43.

IOP Conf. Series: Materials Science and Engineering **341** (2018) 012008 doi:10.1088/1757-899X/341/1/012008

Design of 2.4Ghz CMOS Floating Active Inductor LNA using 130nm Technology

[1,2,a] **M. Muhamad, N. Soin[2], H. Ramiah[2]**

[1]Faculty of Electrical Engineering, Universiti Teknologi Mara (UiTM), [2]Faculty of Engineering, University Of Malaya, Kuala Lumpur.

a) Corresponding author: maizan@salam.uitm.edu.my

Abstract. This paper presents about design and optimization of CMOS active inductor integrated circuit. This active inductor implements using Silterra 0.13µm technology and simulated using Cadence Virtuoso and Spectre RF. The center frequency for this active inductor is at 2.4 GHz which follow IEEE 802.11 b/g/n standard. To reduce the chip size of silicon, active inductor is used instead of passive inductor at low noise amplifier LNA circuit. This inductor test and analyse by low noise amplifier circuit. Comparison between active with passive inductor based on LNA circuit has been performed. Result shown that the active inductor has significantly reduce the chip size with 73 % area without sacrificing the noise figure and gain of LNA which is the most important criteria in LNA. The best low noise amplifier provides a power gain (S21) of 20.7 dB with noise figure (NF) of 2.1dB.

1. Introduction

The LNA is widely used in wireless application such as phone. It is required LNA to operate in low voltage without compromising other factor such as high gain, low noise and linearity requirements at the operating frequency. In term of sizing of chip, it must design small as possible due to cost. The design of the LNA imposes a challenge in terms of meeting total functions. In order to reduce chip area or smaller size it is further challenging. The technology now days are racing towards nano-meter range. The conventional LNA is using passive inductor which are consume lot of area in chip [1,2]. For the reasons maintaining cost to market ratio is a bigger challenge. As the value of inductance climb it required lot space on a chip [3,4]. The consequences are that active inductor using CMOS is used. Several circuit topologies had been proposed in the past to implement active inductor based on gyrators design. Active inductor manages to constantly tune to protect temperature or process variation. The proses variation usually occur at fabrication foundry where naturally occurring variation in the attributes of transistors (length, widths, oxide thickness) when integrated circuits are fabricated. By implement the active inductor on LNA circuit it takes about 1% until 10% of passive active inductor which can reduce cost [5]. Other than that, active inductor has higher inductance value, wide frequency tuning range and higher quality factor which is important for circuit design. Indeed, active inductor may reduce the chip area, but it leads to high power dissipation [6,7].

IOP Conf. Series: Materials Science and Engineering **341** (2018) 012008 doi:10.1088/1757-899X/341/1/012008

Table 1 shows the design specification set for LNA following 802.11b/g/n IEEE Wireless Local Area Network standard. The LNA should be designed to run with the center frequency at 2.4 GHz.

Table 1. Design specification for LNA

Parameter	Specification
Process technology	RFCMOS 130nm
Center frequency (GHz)	2.4
Power supply (V)	≤ 1.2
S_{21} (dB)	> 15
S_{11} (dB)	< -12
S_{22} (dB)	< -12
S_{12} (dB)	< -20
Noise factor (dB)	< 4
Power dissipation (mW)	< 8

2. Active Inductor

Many applications required the inductance of active inductor can be tuneable from a small range up to larger range of inductance. The applications may include phase-locked loops, filters, voltage or current controlled oscillators and others. Thus, there are verity of active inductor design or technique to get the exeat inductance value. A floating active inductor using resistive feedback and single ended active inductor has been chosen.

2.1. Floating Active Inductor using Resistive Feedback

Figure 1 shown gyrator-C floating inductor differentially configured. Most of the transistor are in biased in the saturation region except transistor at M5 and M6. Both transistors were biased in triode region and behave as voltage-controlled resistors whose value are controlled by the gate voltage V_b [8]. M1, M2, M5, M6 are NMOS transistors and M3,M4 are PMOS transistors. Feedback resistor are added between M1 and M2.

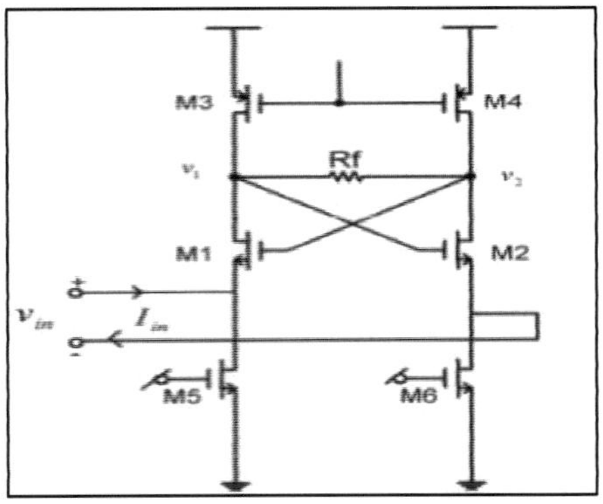

Figure 1. Proposed schematic floating active inductor using resistive feedback

In Figure 2, g_f is the conductance of feedback resistor R. C_L, gl C_2, g_2 are the effective parasitic capacitance and conductance at node VI and V2.

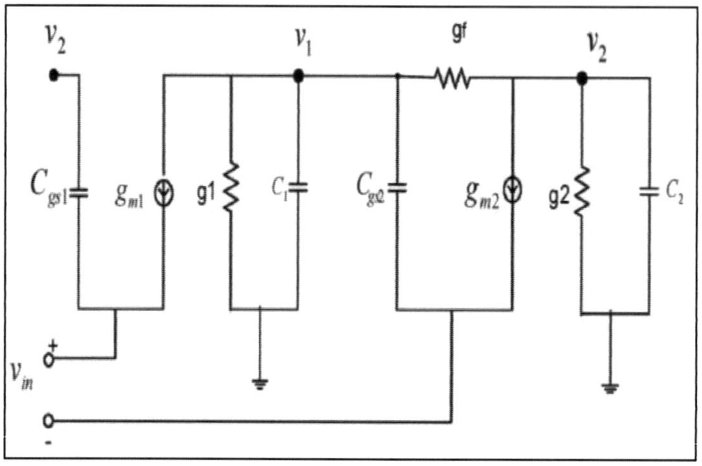

Figure 2. Small signal modal for floating active inductor

Input impedance equation:

$$Y_{in} = g_{m1} + \frac{g^2{}_{m1}(g_{m2} - g_f)}{(g_1+g_f+gC_1)+(g_1+g_f+gC_2)+(g_{m1} - g_f)(g_f - g_{m2})} \quad (1)$$

Quality Factor equation:

$$Q = \frac{\text{Im}[Z]}{\text{RE}[Z]} \quad (2)$$

Inductor equation:

$$L = \frac{\text{Im}(Z_{in})}{\omega} \quad (3)$$

Equation 1,2,3 indicate input impedance, quality factor and tuning inductor value respectively. These equations are required for tuning the value of inductor for floating active inductor using resistive feedback circuit.

2.2. Single Ended Active Inductor

Figure 3 indicate design of single ended active inductor. Transistor M_1 and M_2 represents positive transconductance G_{ml} between input and output. For transistor M_3 and M_4 it represents negative transconductance G_{ml}. These two transconductance will form gyrator –C which transform intrinsic capacitance into inductive behaviour.

IOP Conf. Series: Materials Science and Engineering **341** (2018) 012008 doi:10.1088/1757-899X/341/1/012008

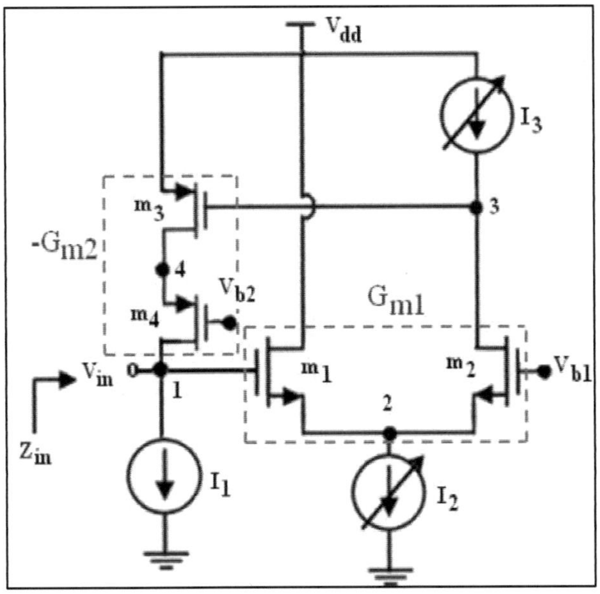

Figure 3. Schematic circuit of single ended active inductor

Input impedance equation:

$$Y_{in} = g_1 + sC_1 + \frac{g_{m4} + g_{m3} + g_{m2} + g_{m1}}{(G+sC2)(g_3+sC3)(g_{m4}+sC4+g_4)} \quad (4)$$

Quality Factor equation:

$$Q = \frac{\text{Im}[Z]}{\text{RE}[Z]} \quad (5)$$

Inductor equation:

$$L = \frac{Z_{in}}{j\omega} \quad (6)$$

Equation 4,5,6 indicate input impedance, quality factor and tuning inductor value respectively. These equations are required for tuning the value of inductor for single ended circuit. These works basically take the LNA inductively degenerated design technique and by using the same design active inductor is replaced with passive inductor.

The schematic as shown in Figure 4 is the combination technique between floating active inductor using resistive feedback and single ended active inductor. Both techniques are used in one circuit to optimize the result of the gain bandwidth and noise figure. This circuit basically is an inductive degenerated design and spiral inductor has been replacing with active inductor where it can give exact function as spiral inductor.

IOP Conf. Series: Materials Science and Engineering **341** (2018) 012008 doi:10.1088/1757-899X/341/1/012008

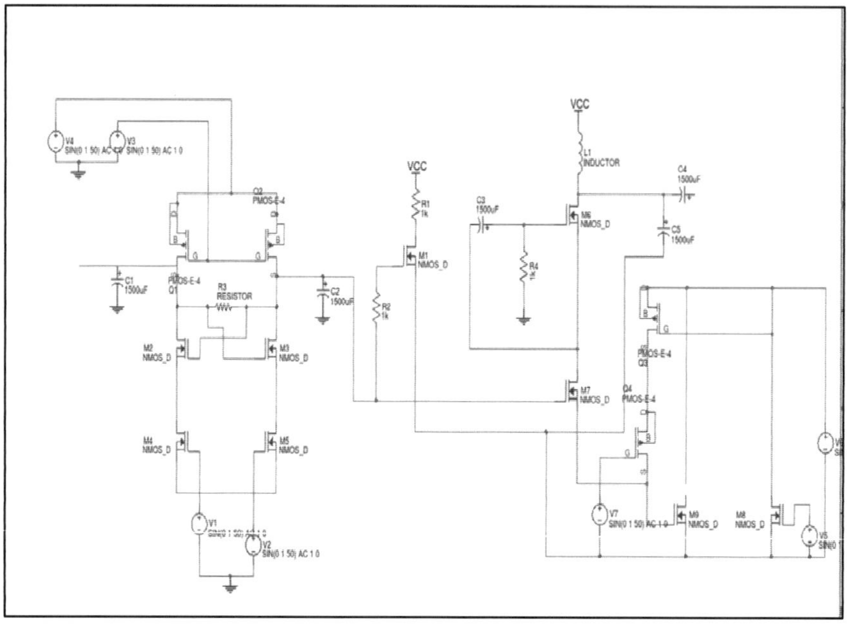

Figure 4. Inductively degenerated LNA circuit using active inductor

3. Results and Discussion

S-parameter measurements stated which is the characteristic impedance or system impedance. Figure 5 below indicate the s-parameters for LNA active inductor and passive inductor respectively. It includes all four s-parameters which are S_{11}, S_{12}, S_{21} and S_{22}. S_{21} is a power gain. Power gain should be higher than 15 dB to compensate for noise contribution of the next stage of the circuit. So with these LNA active inductor it can get value of 20.7dB and for the LNA passive inductor the value is 17.9dB. This value should have very small value in order to get high reverse isolation and it can give better stability to the LNA. By using LNA active inductor circuit the value at -42.39dB while passive inductor is -38.3dB. s_22 representing output matching. The LNA will produce the low output impedance. With this design it can give output matching of -19.8dB and -19.9dB for LNA passive inductor. The last s-parameters are s_11 which is input matching. With the LNA active inductor it archives value of -1dB and for LNA passive inductor the value is -19.3dB.

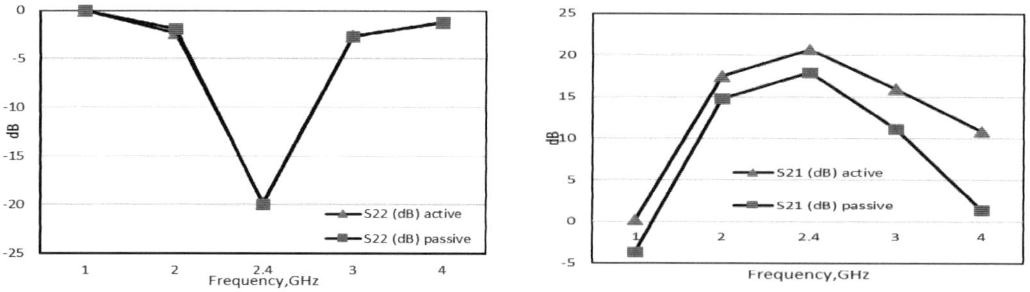

Figure 5. Comparison of S-Parameter LNA Performance using active and passive inductor

The value NF for active inductor are 2.1dB while the passive 1.66dB. There are 0.44dB increasing noise figure at active inductor as shown in Figure 6.

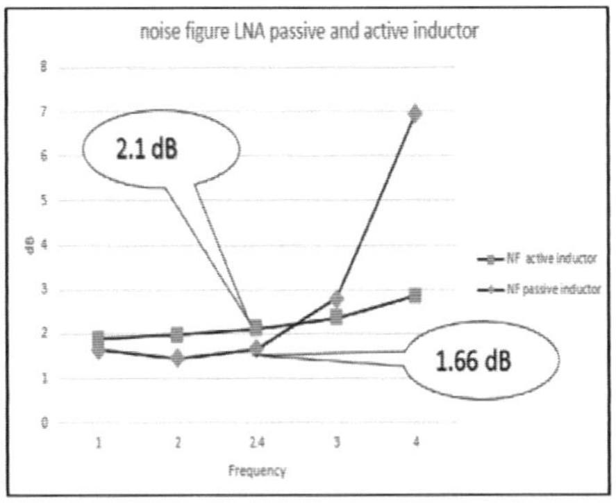

Figure 6. Noise Figure comparison between LNA active and passive inductor

The layout comparison between LNA active inductor and passive inductor respectively was shown in Figure 7. The sizing of layout can be reduced up to 73% when use active inductor to replace passive inductor. LNA active inductor also can reduce the cost of making chip since it consumed less area compare to LNA passive inductor.

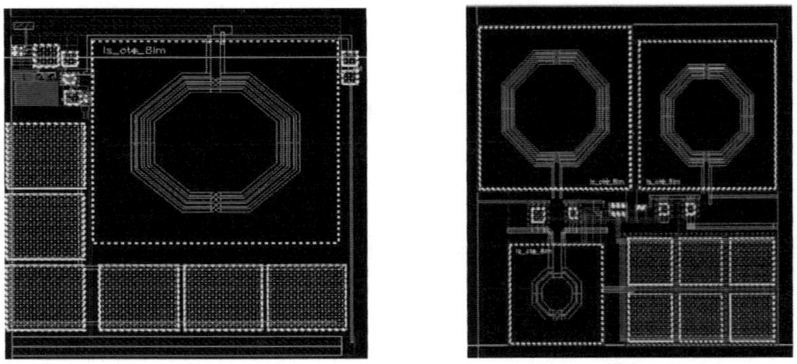

(a) (b)

Figure 7. LNA layout (a) using active inductor, (b) using passive inductor

Table 2. Comparison with others work

Parameter	Active Inductor [This work]	Passive Inductor	Other work [9]	Other work [10]	Other work [11]	Other work [12]
s_{11} (dB)	-1	-19.3	12.6	-10	-9	-10
s_{12} (dB)	-42.3	-38.3	-	-	-	-
s_{21} (dB)	20.7	17.9	14.6	13	16.4	14.18
s_{22} (dB)	-19.8	-19.9	-19.6	-10	-8	-16.21
NF (dB)	2.1	1.66	2.5	3	4	2.78
Power dissipation (mW)	24.9	4.19	8.14	62.5	14.5	10.36

4. Conclusion

A floating active inductor and single ended active inductor has been used in this project to replace the passive inductor at LNA circuit. LNA active inductor was design and tuning using Silterra 0.13µm CMOS technology. By designing active inductor, it can reduce the area of chip without compromising the two most important factors in LNA circuit which is noise figure, NF and power gain, S_{21} which is 2.1 dB and 20.7 dB respectively. The consequences of reduction area lead to reduction of cost in a chip making. The total power consumption for LNA active inductor is 24.9 mW. The future recommendation is to be able to replace all the active inductor since this work able to replace two of out three inductors without compromising the power gain and noise figure.

5. Acknowledgement

The authors wish to thank Research Management Center (RMC) UiTM and Ministry of Higher Education (MOHE) for the funding of this work under the Research Acculturation Collaborative Effort (RACE) grant No. 600-RMI/RACE 16/6/2(2/2015).

References

[1] C.Ling, L.Lin, X. Yang, W. Huang. Design of CMOS Inductor-Less LNA with Active Balun. *IEEE International Conference on Anti-Counterfeiting Security and Indentification*, Xiamen, China, pp,191-193,2011.

[2] C. Ke-Hou, L. Shen-Iuan, "Inductorless Wideband CMOS Low-NoiseAmplifiers Using Noise-Canceling Technique", *IEEE Transactionson Circuits and Systems I*: Regular Papers, vol. 59, no. 2, pp. 305–314, 2015.

[3] M. Park, S. Lee, C. S. Kim, H. K. Yu, K. S. Nam, "The Detailed Analysis of High Q CMOS-Compatible Microwave Spiral Inductors in Silicon Technology", *IEEE Trans. On Electron Device*, Vol. 45, No. 9, pp. 1953-1959, September 1998.

[4] J.Manjula, S. Malarvizhi, 2012. Design of low power low noise high quality factor single ended active inductor with better center frequency tuning capability, Proceedings of *IEEE ICECT* , 435 – 439, 2012.

[5] M. Aljarajreh, M.R Raffin, M.Mamun, S. Amin and D.W Low noise amplifier in RFID systems, *Journal of Engineering and Applied Science*, 8,pp. 21-28, 2013

[6] F. Belmas, F. Hameau and J.M. Fournier. A low power inductorless LNA with double Gm enhancementin 130nm CMOS. *IEE Journal of solid-state circuit*, 47(5),pp. 1094-1103, 2012.

[7] S.V Krishnamurthy, K. El-Sankary, E. El-Masry, Noise cancelling CMOS active inductor and its application in RF band-pass filter design, *International Journal Of Microwave Science and Technology*, 2010, pp. 1-8, 2010.

[8] Q. Lai and J. Mao, "A New floating active inductor using resistive feedback technique," *IEEE MTT-S International Microwave Symposium*, pp. 1-1, 2010.

[9] Chia-wei chang, Zhi-Ming Lin, "A 1-v 14.6-dB gain LNA for WiMAX 2~6GHz applications", *52nd IEEE International Midwest symposium on circuits and systems*, pp.1039-1041, 2009.

[10] M.Ben Amor, M.Loulou, S.Quintanel, D. Pasquet, "A wideband CMOS LNA design for WiMAX applications", *4th European conference on circuits and systems,* pp.93-96, July 2008.

[11] Man-Long Her, Hao-Chih Hsu, Hsin-Hung Lin, Tseng-Hsow Yuan, Shun-Ping Chan, "Design and implementation of LNA for 2 to 6GHz WiMax system", *15th Asia-Pacific Conference on Communications* APCC2009, pp.426-429, 2009.

[12] [5] K. S. Kumari and B. Bhuvan, "Compact WiMAX Low Noise Amplifier Using Active Inductor," *International Conference on Devices and Communications* (ICDeCom), 2011, pp. 1-5, 2011.

Restoration of Movement using FES: An Introductory Study I

M. Ahmed[1, 2,3], M. S. Huq[1,2] and B. S. K. K. Ibrahim[1,2]

[1]Department of Mechatronics and Robotic Engineering, Faculty of Electrical and Electronic Engineering, Universiti Tun Hussein Onn Malaysia, 86400 Parit Raja, Batu Pahat, Johor, Malaysia.
[2]Advance Mechatronic and Robotic Research Group (AdMiRe), Faculty of Electrical and Electronic Engineering, Universiti Tun Hussein Onn Malaysia, 86400 Parit Raja, Batu Pahat, Johor, Malaysia.
[3]Department of Electrical and Electronics Engineering, Faculty of Engineering and Engineering Technology, Abubakar Tafawa Balewa University, P. M. B. 0248 Bauchi, Bauchi State, Nigeria.

inunugoloma@yahoo.com

Abstract. FES has been applied for movement restoration, rehabilitation and therapy in spinal cord injury and nervous system failure subjects, whose number rise worldwide every year. Despite the increase, assist devices that are expected to aid healthcare delivery are not in abundance and could be as a result of standards imposed due to the sensitive condition of the subjects. Although closed loop control systems are expected to positively aid in that regards, the delicacy of the plant was a big constraint. An existing model from the works Ferrarin and Pedotti was elaborated from the knee swinging to the sit-to-stand movements and from the two mathematical models it can be inferred that even though similar higher level of excitation would be required for sit-to-stand manoeuvre. As part of research to improve on closed approach for the FES system, an appraisal was successfully done on the neuromuscular model together with fatigue models from the works on Lynch. The remodelled fatigue models to suit the research were presented.

1. Introduction

Functional electrical stimulation (FES) is a technique used for restoration of movements in humans lost due to health problems in the nervous system. Causes of the complications could be accidents or diseases [1-6]. The technique harnesses electrical signals of suitable form which are passed to appropriate muscles to trigger the desired movement [2, 4, 7, 8]. FES are also used for rehabilitation (for movement as well as gait restoration) and therapy [9-11].

Records showed global growth in the number of the patients annually [12, 13] without corresponding increase in assistive devices which could be due to laws imposed (www.cms.gov/CAG-00153R). Earlier devices in existence employ the open loop control systems. The predefined fixed nature, dependency on the subject for initiations of signals are among the properties of such systems that could be tackled with closed loop control systems. And were identified as promising tools (has the potential for solution for part of stringent requirements and old systems) towards reaching clinical approval for lower limb FES assisted movements [14-16].

The report basically is an initial stage of research which directed towards refining of models that could be used with relative ease for closed loop control designs in restoration of motion using FES in spinal cord injury subjects (paraplegics). It also examines the torque developed during stimulation and the effect of fatigue which is one of the major disturbances affecting accomplishment of the task. The remodeled fatigue model to be adopted for the research was also presented. The manuscript has five sections: introduction, movement models, fatigue effects, discussion and conclusion. The introduction provides information regarding background of the study and the second section discusses the transfer functions for both FES induced knee swinging and sit-to-standing movements from the experimental work conducted by Ferrarin and Pedotti. Next was the fatigue effects which briefly explains the Lynch models, the proposed approximation and the effect of fatigue in performing the task. The chapter on discussion was brief examination on the FES movement models, fatigue models and disturbances effects, while the conclusion rounds up major indices of the study.

2. Movement models

The Ferrarin's model describes the dynamics of the FES-induced movements of the knee joint, obtained from data obtained from both paraplegic and healthy subjects [17]. The knee angle θ during the FES-induced free swinging movement is equivalent to the thigh angle during FES-assisted sit-to-stand movement, so it is also denoted θ. The major difference during swinging opposite repetition of movement occur due to action of gravity which may even make the dynamics overshoot its initial resting position. Of course the action of gravity could result in reverse movement but not beyond the initial resting position which is sitting for the sit-to-stand, but is not part of interest for now. The FES sit-to-stand model is hypothesized based on the same assumptions in Ferrarin and Pedotti FES knee swinging model [17], that the ankle joint was fixed. Other assumptions were that the feet remain stationary and since same quadriceps muscles are stimulated, the damping, stiffness as well as the response of muscle to stimulator signals are same.

2.1. The FES-Indhced Knee Swinging Movement

Figure 1 describes arrangement used which also gave insight on the FES knee swinging dynamics. The extension of the knee is due to the stimulation and hence the torque developed by the shank ($\tau_{Shank}(t)$) is the summation of the torques developed due to the stimulation ($\tau_A(t)$), gravity ($\tau_G(t)$), damping ($\tau_D(t)$) and stiffness ($\tau_S(t)$) [18-21] . Equation (1) describes the statement.

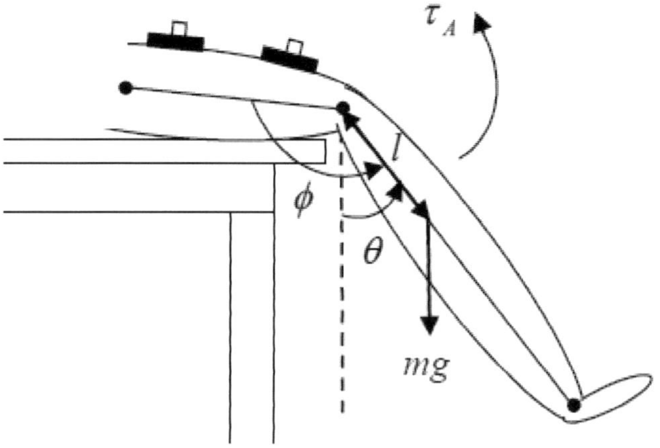

Figure 1: Description of the FES-induced knee swinging

$$\tau_{Shank}(t)=\tau_A(t)+\tau_G(t)+\tau_D(t)+\tau_S(t) \tag{1}$$

Flexion occur under the influence of gravity and therefore becomes the torque developed by the shank (τ_{Shank}) is the summation of the torques developed gravity (τ_G), damping (τ_D) and stiffness (τ_S), as illustrated by (2) [18-21].

$$\tau_{Shank}(t)=\tau_G(t)+\tau_D(t)+\tau_S(t) \tag{2}$$

$$\tau_{Shank}(t)=I\ddot{\theta} \tag{3}$$

$$\tau_A(s)=\frac{P_w(s)G}{1+s\tau} \tag{4}$$

$$\tau_G(t)=mgl\sin\theta \tag{5}$$

$$\tau_D(t)=\beta\dot{\theta} \tag{6}$$

$$\tau_S(t)=\lambda e^{-E\left(\theta+\frac{\pi}{2}\right)}\left(\theta+\frac{\pi}{2}-\omega\right) \tag{7}$$

The parameters: θ is the knee angle, I the moment inertia of the combination of the shank and foot, $\dot{\theta}$ the angular velocity of the combination of the shank and foot, $\ddot{\theta}$ the angular acceleration of the combination of the shank and foot, θ_{TS} the angle between the thigh and shank, β the coefficient of viscous friction, g acceleration due to gravity, m the mass of the combination of the shank and foot, P_w the stimulating signal (Pulse Width Modulated) and l the distance between the knee and the centre of mass the combination of the shank and foot.

Let $x_1=\theta$ (8)

$$x_2=\dot{\theta}=\dot{x}_1 \tag{9}$$

$$\dot{x}_2=\ddot{\theta} \tag{10}$$

$$x_3=\tau_A \tag{11}$$

$$\dot{x}_3=\dot{\tau}_A \tag{12}$$

From (4); $\dot{\tau}_A=\frac{GP_w}{\tau}+\frac{\tau_A}{\tau}$ (13)

Substituting (3)-(6) in (1) and making $\ddot{\theta}$ subject;

$$\ddot{\theta}=\frac{\tau_A}{I}-\frac{mgl\sin\theta}{I}-\frac{\beta\dot{\theta}}{I}-\frac{\lambda e^{-E\left(\theta+\frac{\pi}{2}\right)}\theta}{I}-\frac{\lambda e^{-E\left(\theta+\frac{\pi}{2}\right)}\left(\frac{\pi}{2}-\omega\right)}{I} \tag{14}$$

Let $g(x_1)=-\frac{mgl\sin\theta}{I}-\frac{\lambda e^{-E\left(\theta+\frac{\pi}{2}\right)}\theta}{I}-\frac{\lambda e^{-E\left(\theta+\frac{\pi}{2}\right)}\left(\frac{\pi}{2}-\omega\right)}{I}$ (15)

$$\ddot{\theta}=\frac{\tau_A}{I}+g(x_1)-\frac{\beta\dot{\theta}}{I} \tag{16}$$

$$\begin{pmatrix} \dot{x}_1 \\ \dot{x}_2 \\ \dot{x}_3 \end{pmatrix} = \begin{pmatrix} 0 & 1 & 0 \\ g(x_1) & -\frac{\beta}{I} & \frac{1}{I} \\ 0 & 0 & -\frac{1}{\tau} \end{pmatrix} \begin{pmatrix} x_1 \\ x_2 \\ x_3 \end{pmatrix} + \begin{pmatrix} 0 \\ 0 \\ \frac{G}{\tau} \end{pmatrix} P_w \tag{17}$$

$$\theta = \begin{bmatrix} 1 & 0 & 0 \end{bmatrix} \begin{pmatrix} x_1 \\ x_2 \\ x_3 \end{pmatrix} \tag{18}$$

Equations (17) and (18) forms the state space transfer function of the Ferrarin and Pedotti knee joint model. Note that; the model is nonlinear due the presence of the following terms: $\sin\theta$, $e^{-E\left(\theta+\frac{\pi}{2}\right)}$ terms, so also the τ_A (i.e. = $P_w G\ e^{-\eta t}$; from (4)) even though is somehow hidden in the final model, but in reality its dynamics is nonlinear.

Linearization is achieved using the well-known and popular linearization technique which also used in control systems, achieved representing the function in the Taylor's series form as shown in equation (19). At equilibrium, x_0, $A(x)$ equals zero. (That is $A(x_0)$= 0) and the higher order terms are neglected. Hence, everything is approximated as shown in equation (20) [22-24].

$$\dot{x} = \left| A(x) + \frac{f' A(x)}{1!}(x - x_0) + \frac{f'' A(x)}{2!}(x - x_0)^2 + \ldots \right|_{x=x_0} \tag{19}$$

$$\dot{x} = \left[f' A(x)(x - x_0) \right]_{x=x_0} \tag{20}$$

The linearized transfer function becomes as shown in (21) and (22), upon substituting the parameters of the paraplegic subject three (P3) from the work of Ferrarin and Pedotti [18] as shown in table 1.

$$\begin{pmatrix} \dot{x}_1 \\ \dot{x}_2 \\ \dot{x}_3 \end{pmatrix} = \begin{pmatrix} 0 & 1 & 0 \\ -41.82 & -0.73 & 2.54 \\ 0 & 0 & -4.65 \end{pmatrix} \begin{pmatrix} x_1 \\ x_2 \\ x_3 \end{pmatrix} + \begin{pmatrix} 0 \\ 0 \\ 0.07 \end{pmatrix} P_w \tag{21}$$

$$\theta = \begin{bmatrix} 1 & 0 & 0 \end{bmatrix} \begin{pmatrix} x_1 \\ x_2 \\ x_3 \end{pmatrix} \tag{22}$$

TABLE1. Parameters of the subject [18]

Parameter	Value
Moment inertia of shank and foot I (Kgm^2)	0.394
Mass of combined shank and foot m (Kg)	4.76
Distance between the knee and the center of mass combined shank and foot l (m)	0.233
Coefficient of exponential term λ (Nmrad^{-1})	15.352
Coefficient of exponential term E (rad^{-1})	1.644
Resting Elastic Knee angle ω (rad)	3.896

Parameter	Value
Time constant η (s)	0.215
Static gain G (Nms^{-1})	0.015
Stimulation frequency (Hz)	50
Mass of subject (Kg)	85
Age of subject (Yrs)	25
Height of subject (m)	1.78
Duration with injury (Yrs)	4
Level of leison	T3/T4
Moment of inertia for combined head, neck, trunk and thigh I_M (Kgm^2)	9.344
Mass of combined head, neck, trunk and thigh M (Kg)	49.13
Length of thigh L (m)	0.4361
Coefficient of viscous friction β ($Nmsrad^{-1}$)	0.289
Mass of thigh m_t (kg)	8.50
Center of mass of thigh l_t (m)	0.2472

Changing to s-domain, which is obtained using;

$$\frac{\theta(s)}{P_w(s)} = c(sI - A)^{-1} B + D \quad (23)$$

$$\frac{\theta(s)}{P_w(s)} = \begin{bmatrix} 1 & 0 & 0 \end{bmatrix} \left(s\begin{bmatrix} 1 & 0 & 0 \\ 0 & 1 & 0 \\ 0 & 0 & 1 \end{bmatrix} - \begin{bmatrix} 0 & 1 & 0 \\ -41.82 & -0.73 & 2.54 \\ 0 & 0 & 4.65 \end{bmatrix} \right)^{-1} \begin{pmatrix} 0 \\ 0 \\ 0.07 \end{pmatrix} + \begin{pmatrix} 0 \\ 0 \\ 0 \end{pmatrix} \quad (24)$$

$$\frac{\theta(s)}{P_w(s)} = \frac{0.18}{s^3 + 5.38s^2 + 45.21s + 194.46} \quad (25)$$

Equation (25) becomes the transfer function in the s-domain.

2.2. *The FES-Induced Sit-to-Stand Movement*

Figure 2 describe the arrangement for the FES induced sit-to-stand movement which also gave insight on its dynamics.

In the case of FES-induced sit-to-stand the extension is the most important movement which result upon stimulation of the quadriceps muscles. The extension of the knee is developed due to the stimulation but the torque is experience by the thigh (τ_{Thigh}), the ankle joint and the feet are assumed to be fixed.

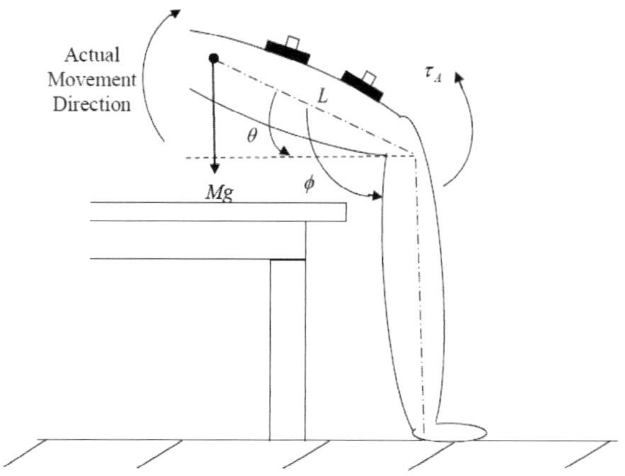

Figure 2: Description of the FES-induced sit-to-stand.

In the case of FES-induced sit-to-stand the extension is the most important movement which results upon stimulation of the quadriceps muscles. The extension of the knee is developed due to the stimulation but the torque resulted in the thigh ($\tau_{Thigh}(t)$), the ankle joint and feet are assumed to be fixed. The torque is the summation of the torques developed due to the stimulation ($\tau_A(t)$), gravity ($\tau_G(t)$), damping ($\tau_D(t)$) and stiffness ($\tau_S(t)$) [18-21] . Equation (26) is a description of the statement. The expressions for torques produced as a result of stimulation $\tau_A(s)$, damping $\tau_D(t)$ and stiffness $\tau_S(t)$ remain similar to that FES assisted knee swinging including the meanings of parameters involved and notations.

$$\tau_{Thigh}(t)=\tau_A(t)+\tau_G(t)+\tau_D(t)+\tau_S(t) \tag{26}$$

$$\tau_{Thigh}(t)=I_M\ddot{\theta} \tag{27}$$

$$\tau_A(s)=\frac{P_w(s)G}{1+s\eta} \tag{28}$$

$$\tau_G(t)=MgL\sin(90-\theta)+m_t g l_t \sin(90-\theta) \tag{29}$$

$$\tau_D(t)=\beta\dot{\theta} \tag{30}$$

$$\tau_S(t)=\lambda e^{-E\left(\theta+\frac{\pi}{2}\right)}\left(\theta+\frac{\pi}{2}-\omega\right) \tag{31}$$

The parameters: θ is the knee angle during swinging movement which is equivalent to the thigh angle during sit-to-stand movement and in reality the angular movements are with the case of swinging as shown in the figure but only that the movement seen is the opposite, I_M the moment inertia of the combination of the head, neck and thigh, $\dot{\theta}$ becomes the angular velocity of the combination of the head, neck and thigh, $\ddot{\theta}$ the angular acceleration of the combination of the shank and foot, ϕ the angle between the thigh and shank, β the coefficient of viscous friction, g acceleration due to gravity, m

the mass of the combination of the shank and foot, M the mass of the combination of the head, neck, trunk and thigh, m_t the mass of the thigh, l_t is the centre of mass of the thigh, P_w the stimulating signal (Pulse Width Modulated) L is the length of the thigh. Please note that; the gravity torque component in addition to its effect on M it also act on m_t though not mentioned on the figure.

Let;

$$x_1 = \theta \tag{32}$$

$$x_2 = \dot{\theta} = \dot{x}_1 \tag{33}$$

$$\dot{x}_2 = \ddot{\theta} \tag{34}$$

$$x_3 = \tau_A \tag{35}$$

$$\dot{x}_3 = \dot{\tau}_A \tag{36}$$

From(29)

$$\dot{\tau}_A = \frac{GP_w}{\eta} + \frac{\tau_A}{\eta} \tag{37}$$

Substituting (27)-(31) in (26) and making $\ddot{\theta}$ subject;

$$\ddot{\theta} = \frac{\tau_A}{I_M} - \frac{MgL\cos\theta}{I_M} - \frac{m_t g l_t \cos\theta}{I_M} - \frac{\beta\dot{\theta}}{I_M} - \frac{\lambda e^{-E\left(\theta+\frac{\pi}{2}\right)}\theta}{I_M} - \frac{\lambda e^{-E\left(\theta+\frac{\pi}{2}\right)}\left(\frac{\pi}{2}-\omega\right)}{I_M} \tag{38}$$

$$h(x_1) = -\frac{MgL\cos\theta}{I_M} - \frac{m_t g l_t \cos\theta}{I_M} - \frac{\lambda e^{-E\left(\theta+\frac{\pi}{2}\right)}\theta}{I_M} - \frac{\lambda e^{-E\left(\theta+\frac{\pi}{2}\right)}\left(\frac{\pi}{2}-\omega\right)}{I_M} \tag{39}$$

$$\ddot{\theta} = \frac{\tau_A}{I_M} + g(x_1) - \frac{\beta\dot{\theta}}{I_M} \tag{40}$$

$$\begin{pmatrix} \dot{x}_1 \\ \dot{x}_2 \\ \dot{x}_3 \end{pmatrix} = \begin{pmatrix} 0 & 1 & 0 \\ h(x_1) & -\frac{\beta}{I_M} & \frac{1}{I_M} \\ 0 & 0 & -\frac{1}{\eta} \end{pmatrix} \begin{pmatrix} x_1 \\ x_2 \\ x_3 \end{pmatrix} + \begin{pmatrix} 0 \\ 0 \\ \frac{G}{\eta} \end{pmatrix} P_w \tag{41}$$

$$\theta = \begin{bmatrix} 1 & 0 & 0 \end{bmatrix} \begin{pmatrix} x_1 \\ x_2 \\ x_3 \end{pmatrix} \tag{42}$$

The linearized transfer function becomes as shown in (43) and (44), after substituting the parameters of the paraplegic subject three (P3) from the work of Ferrarin and Pedotti [18] as shown in table 1.

$$\begin{pmatrix} \dot{x}_1 \\ \dot{x}_2 \\ \dot{x}_3 \end{pmatrix} = \begin{pmatrix} 0 & 1 & 0 \\ -24.71 & -0.03 & 0.11 \\ 0 & 0 & -4.65 \end{pmatrix} \begin{pmatrix} x_1 \\ x_2 \\ x_3 \end{pmatrix} + \begin{pmatrix} 0 \\ 0 \\ 0.07 \end{pmatrix} P_w \tag{43}$$

$$\theta = \begin{bmatrix} 1 & 0 & 0 \end{bmatrix} \begin{pmatrix} x_1 \\ x_2 \\ x_3 \end{pmatrix} \tag{44}$$

Changing to s-domain, which is obtained using;

$$\frac{\theta(s)}{P_w(s)} = c(sI - A)^{-1} B + D \tag{45}$$

$$\frac{\theta(s)}{P_w(s)} = \begin{bmatrix} 1 & 0 & 0 \end{bmatrix} \left(s \begin{bmatrix} 1 & 0 & 0 \\ 0 & 1 & 0 \\ 0 & 0 & 1 \end{bmatrix} - \begin{bmatrix} 0 & 1 & 0 \\ -24.71 & -0.03 & 0.11 \\ 0 & 0 & 4.65 \end{bmatrix} \right)^{-1} \begin{pmatrix} 0 \\ 0 \\ 0.07 \end{pmatrix} + \begin{pmatrix} 0 \\ 0 \\ 0 \end{pmatrix} \tag{46}$$

$$\frac{\theta(s)}{P_w(s)} = \frac{0.0077}{s^3 - 4.62s^2 + 24.57s - 114.90} \tag{47}$$

Equation (47) becomes the transfer function in the s-domain.
It can be seen that more amount of torque would be rquired for moving the knee joint in order to achieve the goal which is standing. The effort from the upper limbs is used to provide the additional necessary support.

3. Effect of fatigue

Fatigue contributes to the nonlinear nature of the plant which is time variant in nature [20, 25, 26]. It is one of the disturbances intended to be suppressed as well as evaluated in the study. The model proposed by Lynch et al. [20, 25, 26] would be adopted which were obtained from experimental by the authors. The models were examples of the nature of the fatigue as shown in figure 3, figure 4 and figure 5, due to unavailability of the mathematical models for analysis remodeling was done.

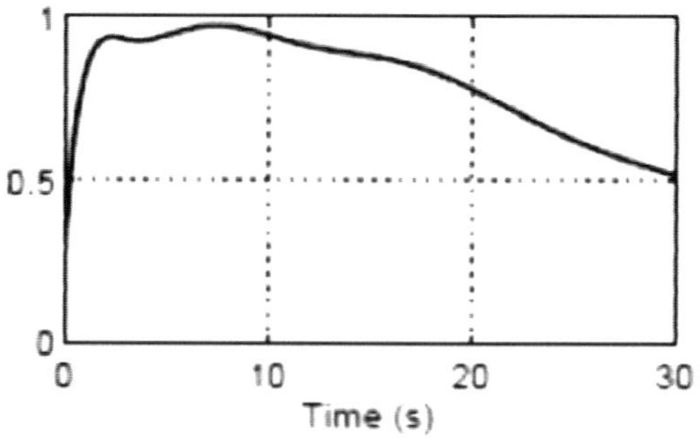

Figure 3: Mild fatigue [20].

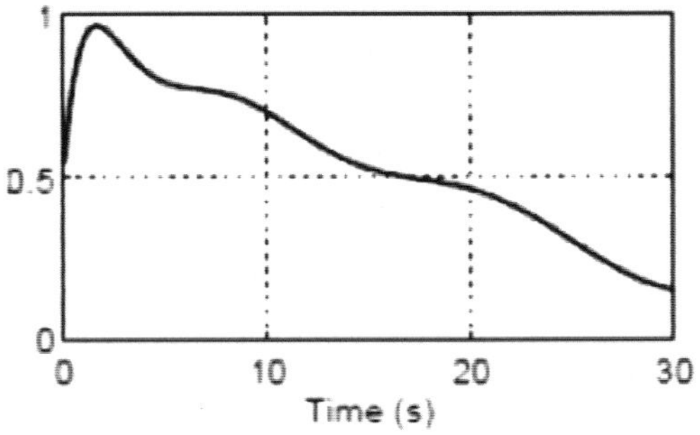

Figure 4: Moderate fatigue [20].

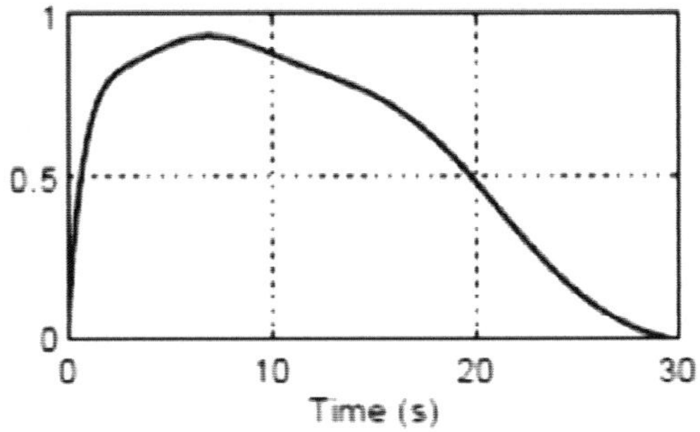

Figure 5: Severe fatigue [20].

The mild ($F_i(t)$), moderate ($F_o(t)$) and severe fatigue ($F_s(t)$) models were approximated using 6 order polynomials as given by (48), (49) and (50) with regression errors of 0.9655, 0.9951 and 0.9618 respectively.

$$F_i(t) = -0.0000001x^6 + 0.00001x^5 - 0.0004x^4 + 0.008x^3 - 0.0775x^2 + 0.3446x + 0.4125 \quad (48)$$

$$F_o(t) = -0.00000008x^6 + 0.000006x^5 - 0.0003x^4 + 0.006x^3 - 0.0597x^2 + 0.2834x + 0.4107 \quad (49)$$

$$F_s(t) = -0.00000006x^6 + 0.000007x^5 - 0.0003x^4 + 0.0058x^3 - 0.0579x^2 + 0.2219x + 0.6405 \quad (50)$$

4. discussion

As earlier mentioned the study is part of work towards developing a novel control scheme using sliding mode wavelet networks approach for FES induced movements in subjects with impaired nervous systems. A good representation (model) of the plant is required for such application. Considering the delicacy of the situation, it would be hectic to carry out measurements on each of such individuals which may even be impossible in some cases. The work tries to obtain suitable FES induced sit-to-stand model (transfer function) for control design as well as analysis. It was based on experimental works of Ferrarin and Pedotti, where the developed the swinging model of the knee joint. An elaboration of the model was done and also extended for the sit-to-stand movement. In further studies, the model would be combine with fatigue, tremor and spasm models from the works of Lynch et al. for analysis.

Figures 6, 7, 8 and 9 portray the torques developed at frequencies of 20, 25, 33 and 50Hz respectively due to stimulation pulse widths of 50, 100, 150 and 200 μ s. It can be seen that the level of torques produced increased as the stimulation currents increase.

Figure 10, Figure 11, Figure 12 and Figure 13 show the torques developed with stimulation pulse widths of 50, 100, 150 and 200 μ s respectively at frequencies of 20, 25, 33 and 50Hz. It can be seen that the level of the torques produced increase as the stimulation frequencies increase.

Figure 14, Figure 16, Figure 18 and Figure 20 indicate the effect of attenuating disturbance at frequencies of 20, 25, 33 and 50Hz respectively on the system in open loop configuration. Figure 15, Figure 17, Figure 19 and Figure 21 express what happens when the closed loop system (step response) was employed. It can be seen that the level of error was more severe for the open loop.

Figure 22 was an indication of the response when the fatigue is the disturbance to the system. It can be seen that apart from reduction of the response, nonlinearity was introduced due to the nature of the perturbation.

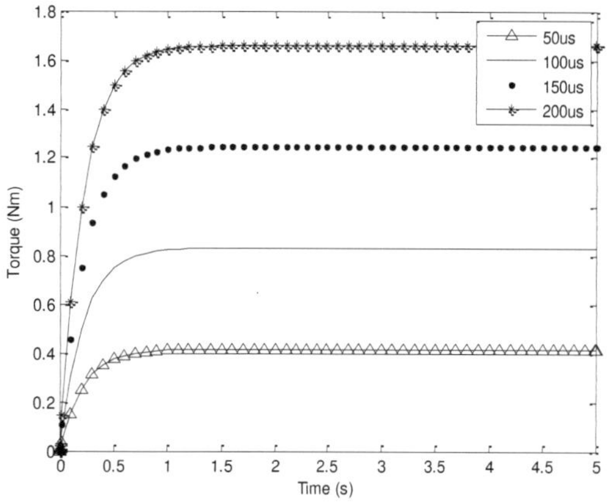

Figure 6: Torque, 20Hz and different currents.

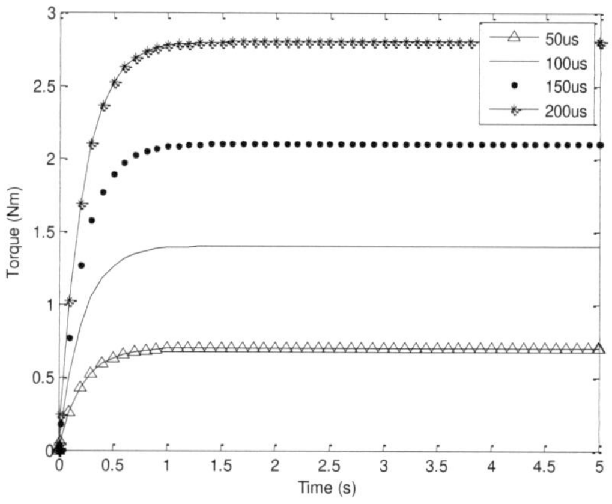

Figure 8: Torque, 33Hz and different currents.

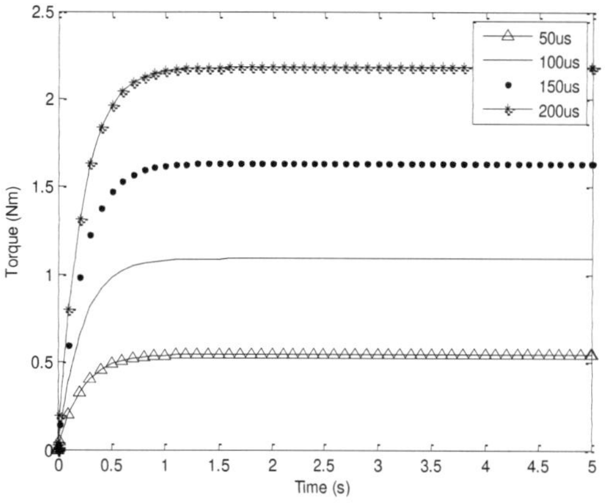

Figure 7: Torque, 25Hz and different currents.

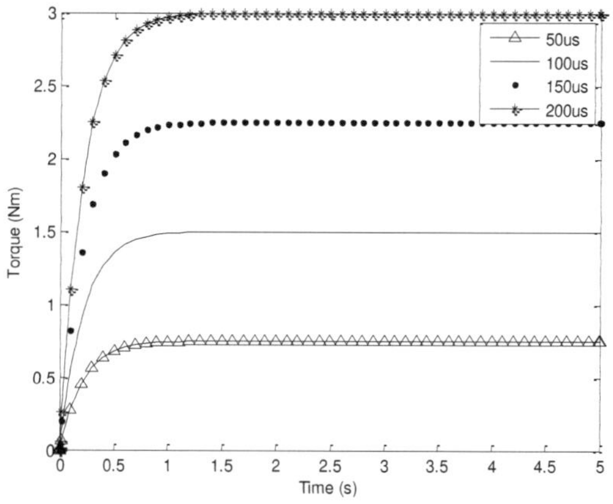

Figure 9: Torque, 50Hz and different currents.

IOP Conf. Series: Materials Science and Engineering **341** (2018) 012009 doi:10.1088/1757-899X/341/1/012009

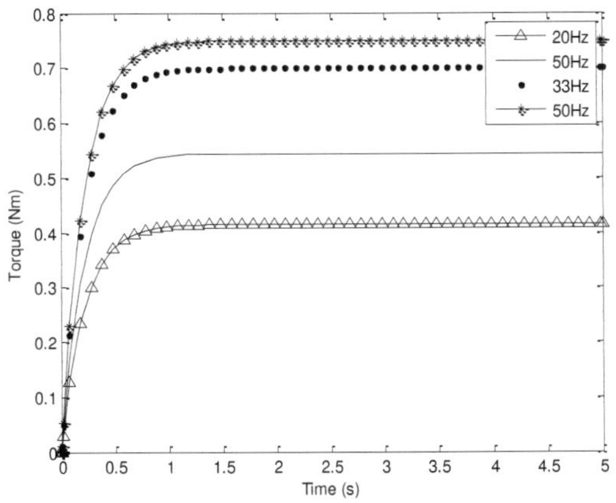

Figure 10: Torque, 50us PW and different frequencies.

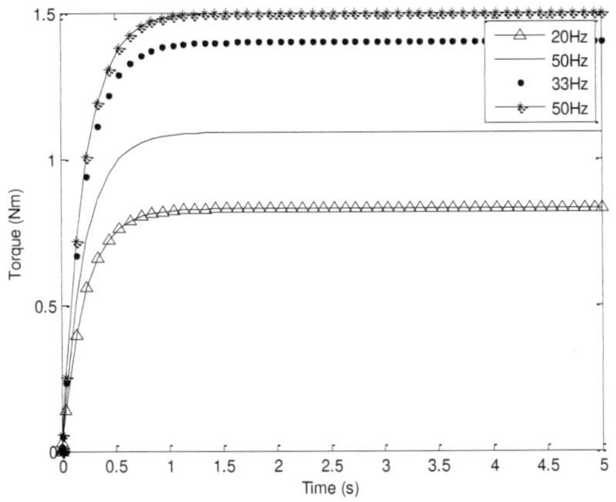

Figure 11: Torque, 100us PW and different frequencies.

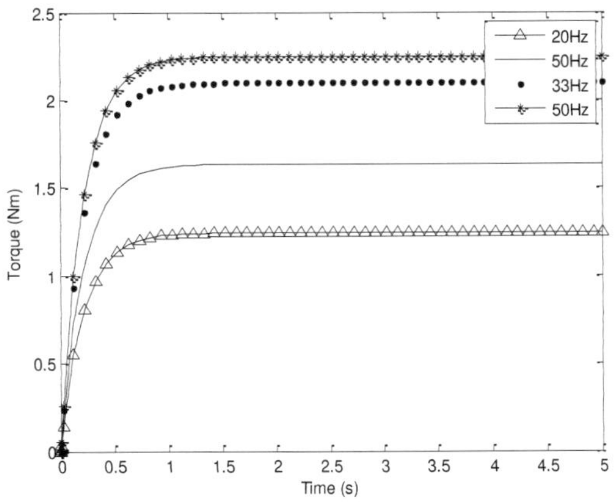

Figure 12: Torque, 150us PW and different frequencies.

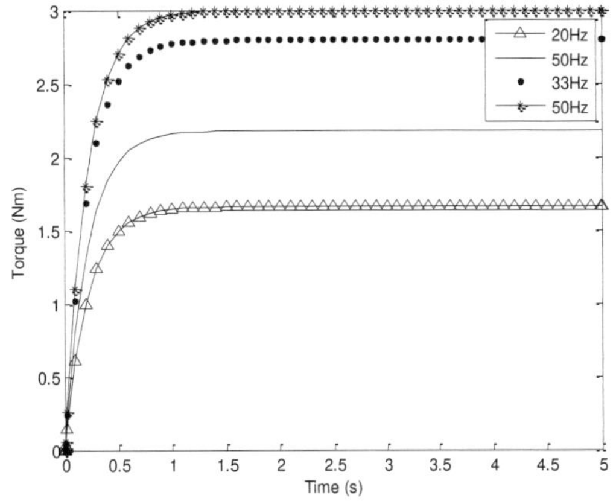

Figure 13: Torque, 200us PW and different frequencies.

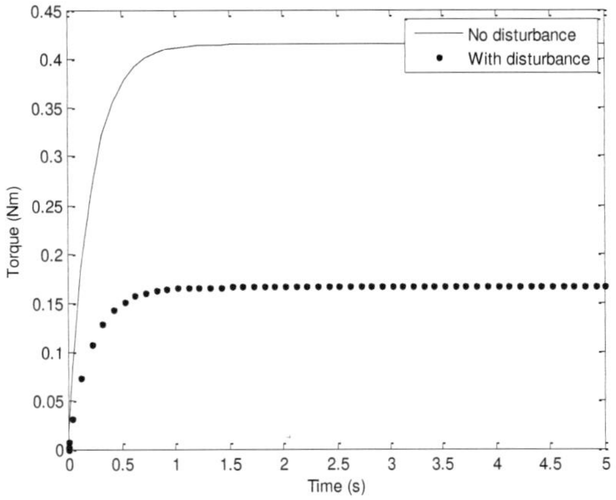

Figure 14: Effect of disturbance on torque 20Hz open loop.

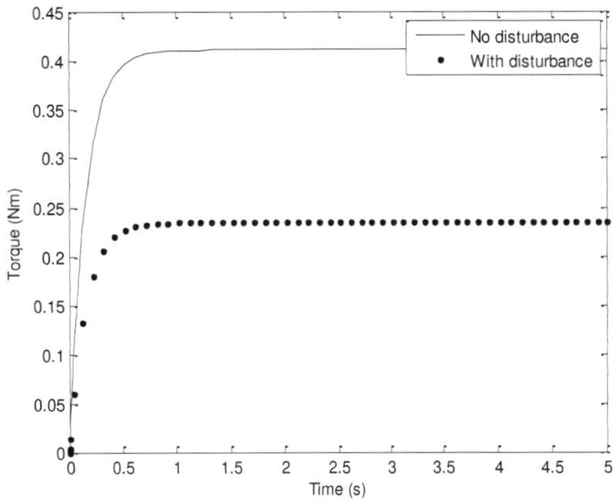

Figure 15: Effect of disturbance on torque 20Hz closed loop.

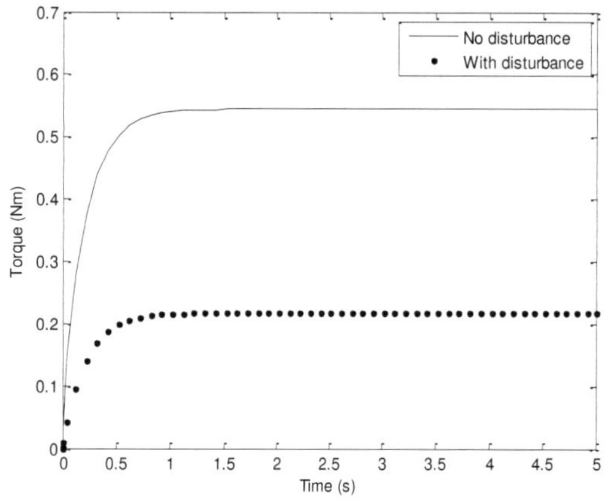

Figure 16: Effect of disturbance on torque 25Hz open loop.

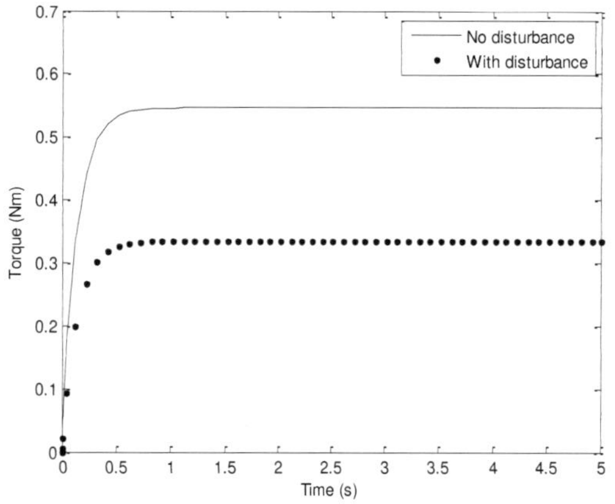

Figure 17: Effect of disturbance on torque 25Hz closed loop.

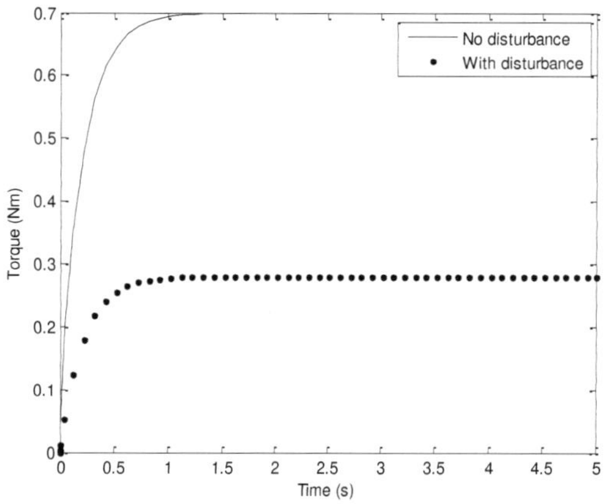

Figure 18: Effect of disturbance on torque 33Hz open loop.

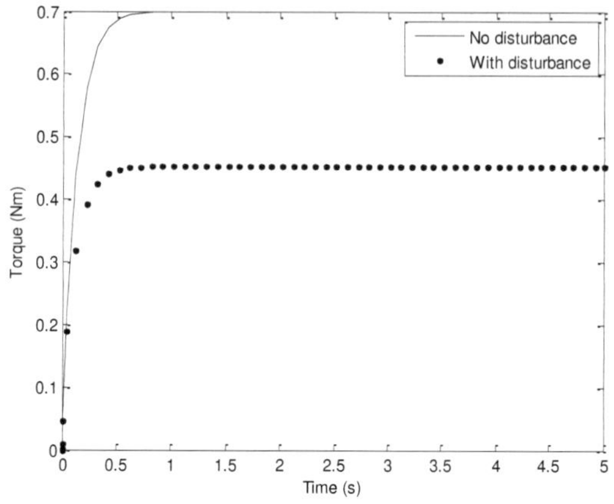

Figure 19: Effect of disturbance on torque 33Hz closed loop.

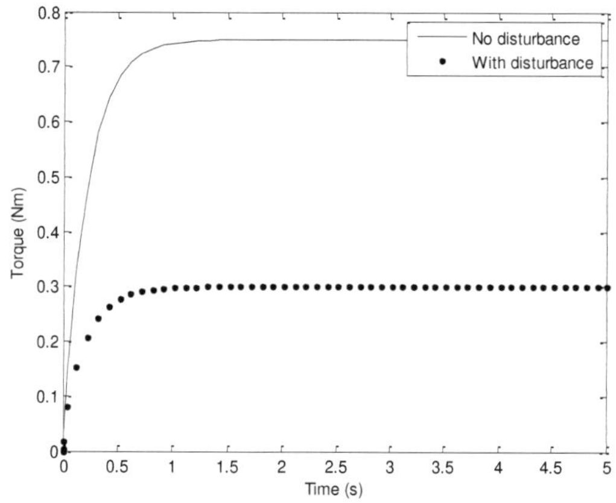

Figure 20: Effect of disturbance on torque 33Hz open loop.

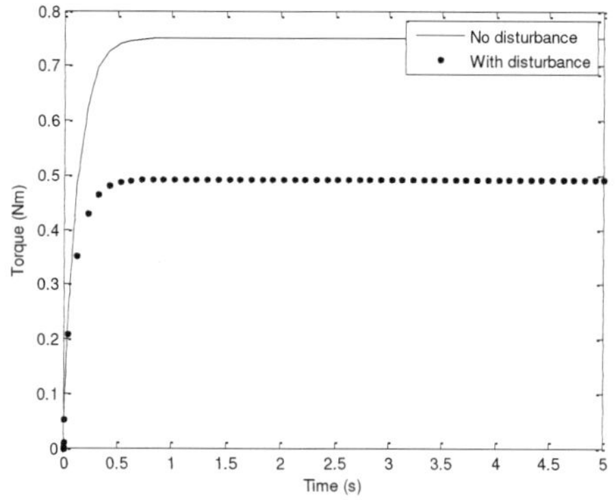

Figure 21: Effect of disturbance on torque 50Hz closed loop.

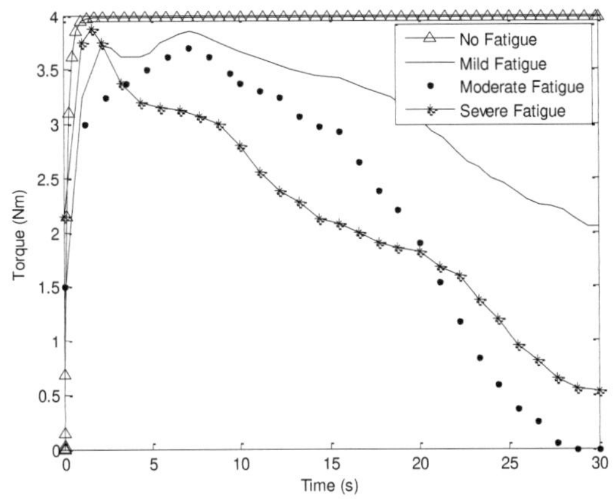

Figure 22: Effect of fatigue on torque developed.

5. Conclusion

FES is used for movement restoration, rehabilitation and therapy. Availability of abundant assist devices could improve healthcare delivery in the area which is currently lacking due to mostlikely standards compliance issues. Closed loop control systems could be harnessed to achieve that, but the plant is delicate and has to be carefully handled. Suitable modification of the Ferrarin and Pedotti knee swinging model was made in order to achieve the sit-to-stand movement. And from the models it can be seen that higher effort (stimulation) would be required for the sit-to-stand scenario.

The work is new in the sense that there is no known work to the best knowledge of the authors that: applied or extend the Ferrarin and Pedotti model for sit-to-stand, or portrays intention towards harnessing the model together with fatigue, tremor and spasm models proposed by Lynch, as well as the sensor error for robustness and stability analysis.

FES is a technique used for revival of movement in different forms but existing devices whose operation utilize that together with other gadgets are scarce. The effects of the stimulation current and frequency, closed loop arrangement and fatigue on the models were appraised and found to be in conformity with literature. The torque produced due to electrical stimulation was the main component in achieving the induced motion and is associated with fatigue. Fatigue causes decline in the torque produced as well as introduce nonlinearity in the behavior of the overall system dynamics.

Acknowledgement

"This paper was partly sponsored by the Centre for Graduate Studies UTHM".

References

[1] S. Jezernik, P. Inderbitzin, T. Keller, and R. Riener, "A novel sliding mode controller for functional electrical stimulation," in *Proceedings of the 15th IFAC World Congress on Automatic Control, Barcelona Session slot T-Mo-M20: Modelling and Control in Biomedical Systems*, 2002.

[2] W. K. Durfee, "Gait restoration by functional electrical stimulation," in *Climbing and Walking Robots*, ed: Springer, 2006, pp. 19-26.

[3] Y. Chen, J. Hu, L. Peng, and Z.-g. Hou, "The FES-assisted control for a lower limb rehabilitation robot: simulation and experiment," *Robotics and Biomimetics,* vol. 1, pp. 1-20, 2014.

[4] N. Arantes, D. Vaz, M. Mancini, M. Pereira, F. Pinto, and T. Pinto, "Effects of functional electrical stimulation applied to the wrist and finger muscles of hemiparetic subjects: a

systematic review of the literature," *Brazilian Journal of Physical Therapy,* vol. 11, pp. 419-427, 2007.

[5] G. P. Braz, M. Russold, and G. M. Davis, "Functional electrical stimulation control of standing and stepping after spinal cord injury: A review of technical characteristics," *Neuromodulation: Technology at the Neural Interface,* vol. 12, pp. 180-190, 2009.

[6] H. Kern, U. Carraro, N. Adami, D. Biral, C. Hofer, C. Forstner, *et al.*, "Home-based functional electrical stimulation rescues permanently denervated muscles in paraplegic patients with complete lower motor neuron lesion," *Neurorehabilitation and neural repair,* 2010.

[7] A. Papachristos, "Functional Electrical Stimulation in Paraplegia," in *Topics in Paraplegia*, Y. Dionyssiotis, Ed., ed, 2014, pp. 109-126.

[8] J. F. Tan, K. Masani, A. H. Vette, J. Zariffa, M. Robinson, C. Lynch, *et al.*, "Inverted pendulum standing apparatus for investigating closed-loop control of ankle joint muscle contractions during functional electrical stimulation," *International Scholarly Research Notices,* vol. 2014, 2014.

[9] P. J. Grahn, G. W. Mallory, B. M. Berry, J. T. Hachmann, D. A. Lobel, and J. L. Lujan, "Restoration of motor function following spinal cord injury via optimal control of intraspinal microstimulation: toward a next generation closed-loop neural prosthesis," *Frontiers in neuroscience,* vol. 8, 2014.

[10] A. K. Vafadar, J. N. Côté, and P. S. Archambault, "Effectiveness of functional electrical stimulation in improving clinical outcomes in the upper arm following stroke: a systematic review and meta-analysis," *BioMed research international,* vol. 2015, 2015.

[11] K. Masani and M. R. Popovic, "Functional electrical stimulation in rehabilitation and neurorehabilitation," in *Springer handbook of medical technology*, ed: Springer, 2011, pp. 877-896.

[12] A. Officer, T. Shakespeare, P. von Groote, J. Bickenbach, and W. H. Organization, *International perspectives on spinal cord injury*: World Health Organization, 2013.

[13] W. H. Organization, *Neurological disorders: public health challenges*: World Health Organization, 2006.

[14] M. O. Ibitoye, N. A. Hamzaid, N. Hasnan, A. K. A. Wahab, and G. M. Davis, "Strategies for rapid muscle fatigue reduction during FES exercise in individuals with spinal cord injury: a systematic review," *PloS one,* vol. 11, p. e0149024, 2016.

[15] M. O. Ibitoye, N. A. Hamzaid, J. M. Zuniga, N. Hasnan, and A. K. A. Wahab, "Mechanomyographic parameter extraction methods: an appraisal for clinical applications," *Sensors,* vol. 14, pp. 22940-22970, 2014.

[16] M. O. Ibitoye, N. A. Hamzaid, J. M. Zuniga, and A. K. A. Wahab, "Mechanomyography and muscle function assessment: A review of current state and prospects," *Clinical Biomechanics,* vol. 29, pp. 691-704, 2014.

[17] M. Ferrarin and A. Pedotti, "The relationship between electrical stimulus and joint torque: A dynamic model," *IEEE transactions on rehabilitation engineering,* vol. 8, pp. 342-352, 2000.

[18] M. Ferrarin and A. Pedotti, "The relationship between electrical stimulus and joint torque: a dynamic model," *Rehabilitation Engineering, IEEE Transactions on,* vol. 8, pp. 342-352, 2000.

[19] N. M. d. Santos, R. Gaino, M. R. Covacic, M. C. M. Teixeira, A. A. d. Carvalho, E. Assunção, *et al.*, "Robust control of the knee joint angle of paraplegic patients considering norm-bounded uncertainties," *Mathematical Problems in Engineering,* vol. 2015, 2015.

[20] C. L. Lynch, "Closed-Loop Control of Electrically Stimulated Skeletal Muscle Contractions," PhD, University of Toronto, 2011.

[21] C. L. Lynch and M. R. Popovic, "A comparison of closed-loop control algorithms for regulating electrically stimulated knee movements in individuals with spinal cord injury," *Neural Systems and Rehabilitation Engineering, IEEE Transactions on,* vol. 20, pp. 539-548, 2012.

[22] J.-J. E. Slotine and W. Li, *Applied nonlinear control* vol. 199: prentice-Hall Englewood Cliffs, NJ, 1991.

[23] K. Ogata, *Modern control engineering*: Prentice Hall PTR, 2010.
[24] R. C. Dorf and R. H. Bishop, *Modern control systems*: Pearson, 2011.
[25] C. L. Lynch, G. M. Graham, and M. R. Popovic, "Including nonideal behavior in simulations of functional electrical stimulation applications," *Artificial organs,* vol. 35, pp. 267-269, 2011.
[26] C. L. Lynch, G. M. Graham, and M. R. Popovic, "A generic model of real-world non-ideal behaviour of FES-induced muscle contractions: simulation tool," *Journal of neural engineering,* vol. 8, p. 046034, 2011.

IOP Conf. Series: Materials Science and Engineering **341** (2018) 012010 doi:10.1088/1757-899X/341/1/012010

A Low-Cost Hand Trainer Device Based On Microcontroller Platform

Muhammad Akmal Mohammad Sabor, Norashikin M. Thamrin[1]

[1]Faculty of Electrical Engineering, Universiti Teknologi Mara, 40450 Shah Alam, Selangor, Malaysia

[1]norashikin.mthamrin@gmail.com

Abstract. Conventionally, the rehabilitation equipment used in the hospital or recovery center to treat and train the muscle of the stroke patient is implementing the pneumatic or compressed air machine. The main problem caused by this equipment is that the arrangement of the machine is quite complex and the position of it has been locked and fixed, which can cause uncomfortable feeling to the patients throughout the recovery session. Furthermore, the harsh movement from the machine could harm the patient as it does not allow flexibility movement and the use of pneumatic actuator has increased the gripping force towards the patient which could hurt them. Therefore, the main aim of this paper is to propose the development of the Bionic Hand Trainer based on Arduino platform, for a low-cost solution for rehabilitation machine as well as allows flexibility and smooth hand movement for the patients during the healing process. The scope of this work is to replicate the structure of the hand only at the fingers structure that is the phalanges part, which inclusive the proximal, intermediate and distal of the fingers. In order to do this, a hand glove is designed by equipping with flex sensors at every finger and connected them to the Arduino platform. The movement of the hand will motorize the movement of the dummy hand that has been controlled by the servo motors, which have been equipped along the phalanges part. As a result, the bending flex sensors due to the movement of the fingers has doubled up the rotation of the servo motors to mimic this movement at the dummy hand. The voltage output from the bending sensors are ranging from 0 volt to 5 volts, which are suitable for low-cost hand trainer device implementation. Through this system, the patient will have the power to control their gripping operation slowly without having a painful force from the external actuators throughout the rehabilitation process.

1. Introduction

The National Stroke Association of Malaysia (NASAM) recently has reported that stroke is now the third largest cause of death in Malaysia. It can abruptly change the life of a healthy and vibrant individual in an instant. The patients need to learn how to do basic daily tasks, such as self-feeding, and this activity can be an overwhelming physical hurdle to them. When they have lost the dexterity of their hands due to the stroke, the hand movements such as grasping and releasing objects for daily activities seem like insurmountable challenges. However, there are some helpful exercises provided by the rehabilitation center with the help of the therapist to help the stroke patients to reclaim their dexterity with these hand exercises during stroke recovery.

With the help of the positive and rapid growing of the technology, the machine-guided hand trainer or hand therapy device has been invented to help the stroke patients in exercising the essential muscles to bring back their dexterity and the use of their fingers. These low-impact exercise tools can be performed either in the rehabilitation center or at home in helping them to restore the motion of the hands after a stroke. In normal practice, the existing hand exercising equipment in the most rehabilitation is using pneumatic pressure during the valve or joint movement operation of this assisted hand machine. One of the disadvantages of the pneumatic-operated hand trainer is that the handling of the system is difficult as the trainer need to be there to assist the movement of the patient all the time as the harsh movement of the hand trainer could harm the patient due to the pneumatic actuator. Furthermore, this machine is also heavy and difficult to be lifted and moved from one space to another. The used of a soft pneumatic actuator for hand trainer has caught a lot of interest recently. However, the device is not yet widely commercial across the globe and the source to get it from is quite limited. Furthermore, by equipping the hand trainer device with game and music can make the therapy session become more interesting and the rehab less boring. In fact, the clinically testing has proven that by linking the hand movement with the music or game motivates more repetitive hand movement during the rehab session, which has improved the finger dexterity faster [20]. Figure 1 shows the main parts and joints of a hand that has been researched for the hand trainer device invention.

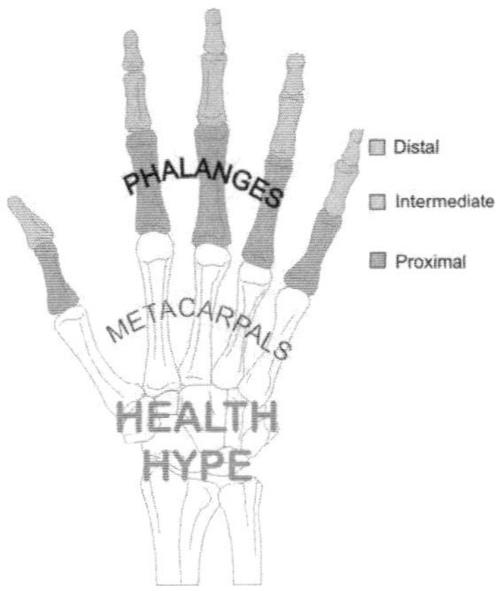

Figure 1. The skeleton segmentation of a hand.

Referring to Figure 1, the skeleton of a human hand consists of 27 bones that are organized into proximal row and distal row, which articulates the bases of the metacarpal bones. This bone will each turn articulates with the phalanges of a thumb and the fingers. The thumb only consists of three phalanx bones, which are the distal and proximal phalanx. Whereas, the other four fingers consist of three phalanx bones, namely, proximal, intermediate and distal. To move them, the muscles of the forearm are used for this purpose. Due to this unique structure of a hand, researchers have invented a hand glove that consists of bending sensors such as soft actuators and flex sensors to detect the bending angle of the fingers for the rehabilitation purpose.

Therefore, the objective of this work is to propose a prototype of hand glove with the used of bend sensors to detect the movement of the fingers for rehab purpose. Secondly, the movement of this hand glove will synergize the movement of the artificial hand which copies the motion of the hand.).

2. Previous work

In this section, the available research works in hand trainer device for rehabilitation purpose are discussed. There are a number of machines that available commercially to aid in the rehabilitation of the hand. For an instance, the MRC-Glove, which is a Function Magnetic Resonance Imaging (fMRI) compatible soft robotic glove for hand rehabilitation [2] as shown in Figure 2. It has been designed and fabricated by using soft robotic glove which can be used with fMRI during the hand continuous passive motion (CPM) in the rehabilitation process. This device used two major components, namely, a) soft pneumatic actuator and b) a glove. The soft pneumatic actuators are made from silicon elastomer. It generates a bending motion and actuates the finger joints upon pressurization. The fMRI has been widely used to investigate the brain activity and reorganization in response to dynamic environments. Recent advances in neuroscience research by using fMRI have increased the understanding of brain neural plasticity and its relationship to motor recovery [5]. It gives important insights into brain mechanisms controlling voluntary movement as well as provides supporting evidence for optimal rehabilitation interventions and the development of rehabilitation robotics which can assist in the execution of motor task and study of brain responses using the fMRI machine. Unfortunately, the fMRI is a huge machine and will need high maintained continuously and costly. This is clearly become the main disadvantage for portable and low-cost hand trainer system.

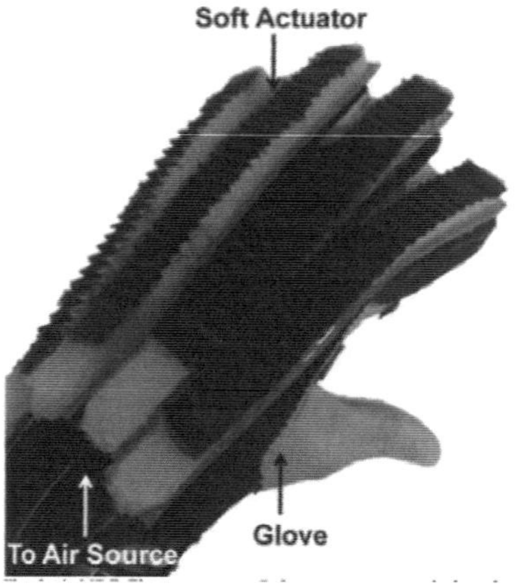

Figure 2. An MRC-Glove prototype [2].

Therefore, the MRC-Glove is used together with the fMRI machine to investigate the brain responses and understand the motor performances of the brain during the hand CPM exercises during the rehabilitation period. The soft actuators have drawn increasing research interest due to their low inherent stiffness as well as high customizability and compliance [6]. Several research groups have adopted a soft robotic approach to design soft rehabilitation devices that are more lightweight and wearable [7-8]. Another interesting characteristic of soft actuators is that they are made from soft elastomeric materials which are non-ferromagnetic, thus they remain functional and do not affect the MR environment as shown in Figure 3.

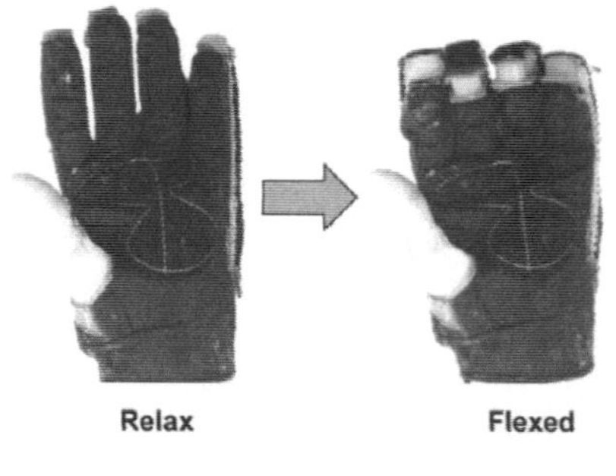

Figure 3. Actuator bent and actuated the finger joint upon pressurization [2].

Based on Figure 3, the glove is attached with a Velcro hook straps with four soft pneumatic actuators on it. These actuators are corresponded to the index, middle, ring and small fingers, respectively. Silicone connecting tubes of the actuators are then connected to an external pneumatic source. The entire prototype weighs less than 200g. Upon pressurization, the prototype actuates and conforms to a hand grasping profile. This design is really useful in realizing the portable hand trainer, unfortunately, the used of external pneumatic source might jeopardize its portability as the weight of it could hinder it to be carried away freely.

An Actuated Finger Exoskeleton (AFX) which has been invented by the researchers from Northwestern University [3], has built the hand trainer to facilitate patient with the movement of a pinch, by making an exoskeleton to permit independent actuation of each of the three joints of the index finger as depicted in Figure 4. Separate actuators are used for flexion and extension, with the closed-loop control of either force or position. As the hand is exceedingly complex, which is more than 20 degrees-of-freedom, it focuses on the important function of the hand such as pinching movement. This specific hand trainer consists of three parts, namely, the mechanical exoskeleton, actuation with motors, and a local control using feedback loop provided by the sensors.

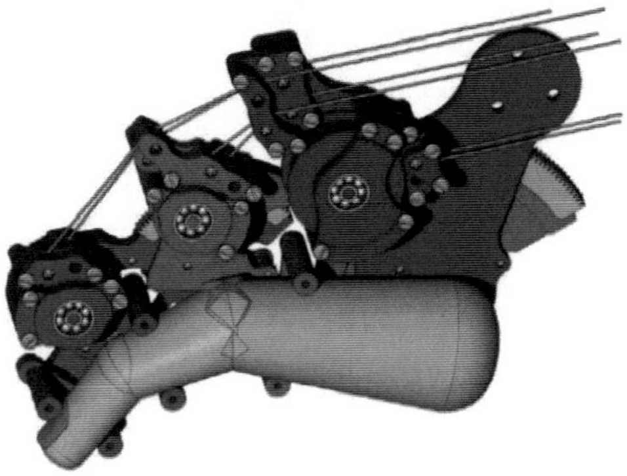

Figure 4. Pairs of metal rods connect to the AFX to each of the finger segment [3].

According to Figure 4, AFX is a metal exoskeleton, which is built from aluminum and D2 steel and is connected to the proximal, middle, and distal segments of the finger through pairs of rods. Specifically, pairs of rods contact the ventral and dorsal sides of each segment of the digit to control the movement of that segment about the proximal joint. The AFX also has three pin joints which match the metacarpophalangeal (MCP) proximal interphalangeal (PIP), and distal interphalangeal (DIP) joints of the finger. To avoid the potential difficulties associated with remote centers of rotation, each of the rotational axes of the exoskeleton is aligned with the anatomical joint of the digit as shown in Figure 5.

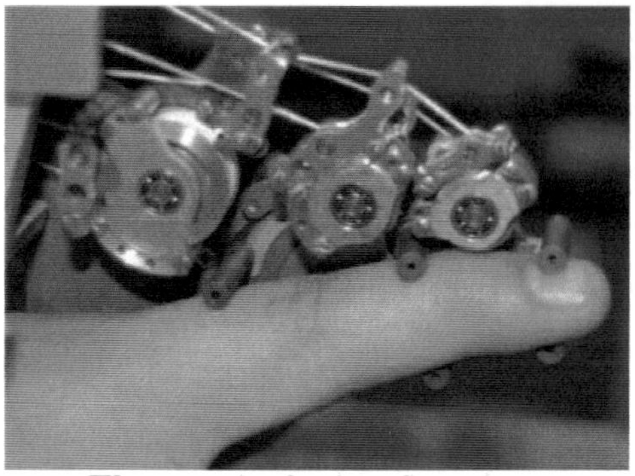

Figure 5. AFX located radial of the index finger with parallel post- interfacing with each finger segment [3].

All components are fabricated from aluminum or steel to withstand the relatively high torques required of the device. Thrust bearings at the MCP and PIP joints serve to accommodate potentially significant off-axis moments. The exoskeleton was designed to allow large ranges of motion: -15 to 75°, 0 to 90° and 0 to 90° for the MCP, PIP and DIP joints, respectively. The angle of the joint of the fingers is manipulated by the used of optical encoders as shown in Figure 6. This angle is computed by using the data obtained from these optical encoders that are attached to the DC motors. As the pulley sizes and gear ratios at each joint are fixed, motor shaft rotation can be translated into the joint angle. Sensors will also be used to measure the tension of the cable. As the cable is spooled from the motor, it runs over a small pulley mounted on a cantilever beam.

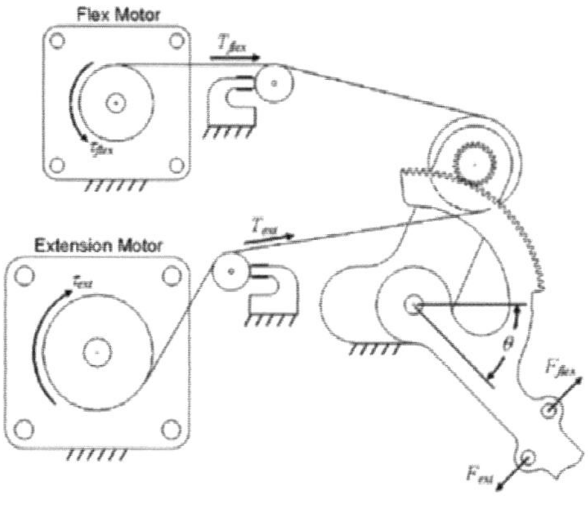

Figure 6. Schematic of one joint of the manipulator [3].

This particular beam is instrumented with strain gages to determine the force applied by the cable. The joint torque is calculated from the height information of the above the joint and the gear ratio. Unfortunately, there could be a backlash due to imperfect mating between the gears and therefore, it could put a patient in danger situation.

3. Methodology

In this work, flex sensors have been used to detect the bending angle of the fingers. These sensors are stitched and attached to the glove for this purpose as depicted in Figure 7. Only four fingers are used in this experiments to move an artificial hand as shown in Figure 8. The artificial hand is used as the hand movement emulator which will imitate the movement made by the hand trainer device.

IOP Conf. Series: Materials Science and Engineering **341** (2018) 012010 doi:10.1088/1757-899X/341/1/012010

The flex sensor and 22kΩ resistor are connected in series to the Arduino Mega 2560 development board as illustrated in Figure 9. The analog data, which is the bending force voltage value from these sensors are fed to the Arduino. This signal is converted into digital values through the Analog-To-Digital (ADC) converter for further processing operation in the microcontroller-based platform.

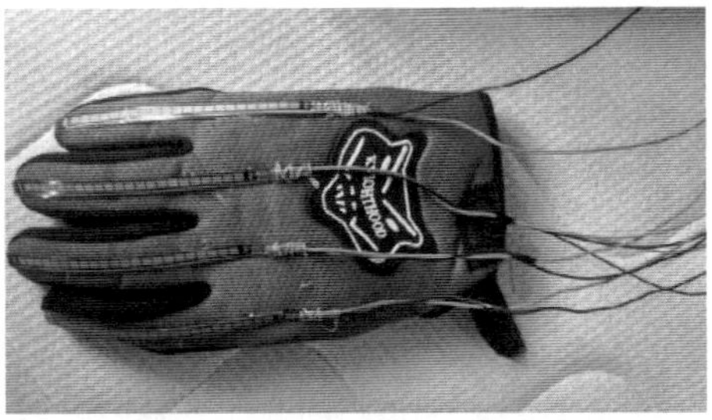

Figure 7. The flex sensors are stitched and attached to the hand trainer glove.

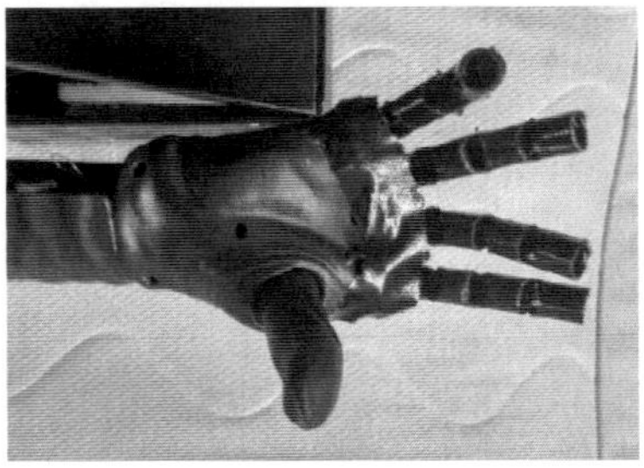

Figure 8. An artificial hand is used to emulate the movement detected by the flex sensors on the hand trainer device.

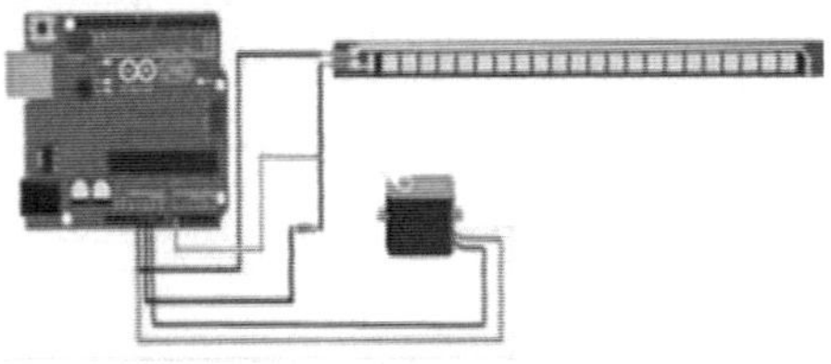

Figure 9. The flex sensor is connected to the Arduino development board.

Finally, the fingers on the artificial hand will be stretched and moved according to the rotation of the servo motor. Figure 10 shows the architecture of the artificial fingers. It uses cylindrical plastic pipes which have been segmented into phalanges part of the fingers. A nylon thread is filling through the drilled hole of each segmented phalange to act as the joint of the fingers. This nylon thread is connected to the servo motor and will be extended and contracted during this movement.

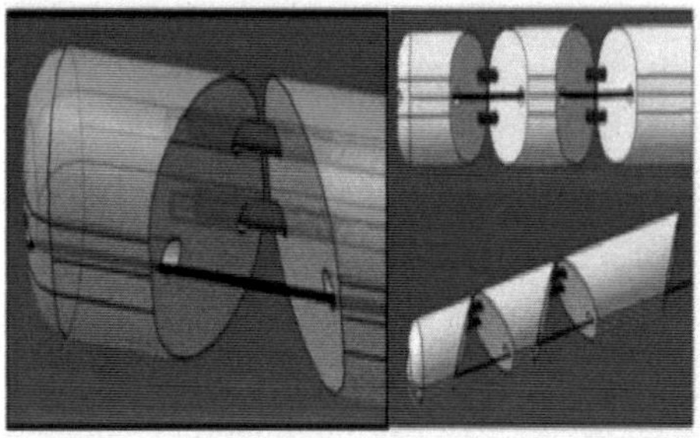

Figure 10. The connection of the fingers on the artificial hand.

4. Result and discussion

The voltage value received at this point depends on the position of the flex sensors as shown in Table 1. When the flex sensor is straight or unbend, the obtained digital value is 720 V. The value is decreasing whenever the flex sensors are bent. The digital value for 45, 90, 135 and 180 degrees are 702, 695, 682 and 670, respectively. These values are really useful in energizing the servo motor that is attached on the artificial hand to be rotated at the same angle, simultaneously. To do this, the pulse-width signals are generated by the Arduino platform. Table II shows the pulse width modulation required by the servo motor to be rotating in various angles.

Table 1. The digital value of ADC output and Motor angle.

Digital Value	Motor Angle(°)
720	0
702	45
695	90
682	135
670	180

Table 2. Relation between pulse width and angle of rotation motor.

Pulse Width(ms)	Rotation of motor(°)
1	0
1.5	90
2	180

Table 3. Bending of flex sensor and rotation servo motor.

Bending of flex sensor (°)	Rotation of servo motor (°)
0	0
30	60
45	90
60	120
75	150
90	180

The relationship between the degrees of the bent flex sensor on the hand trainer glove with the rotation of the servo motor at the artificial hand is stated in Table III, respectively. The initial angle for both of flex sensor and the servo motor is 0 degree. From this table, it can be analyzed that, the total degree turn of the servo motor is doubled than the bending angle of the flex sensor. This discrepancy could be attributed to the nylon string attached to each servo motor at the artificial hand. The tension produced by the strings is quite firm and therefore, more effort is needed by the servo motor to tight or to release it during hand releasing and grabbing movements. The linear relationship between the bending angle of the flex sensors and rotating angle of the servo motors is graphically plotted in Figure 11 for a clearer view.

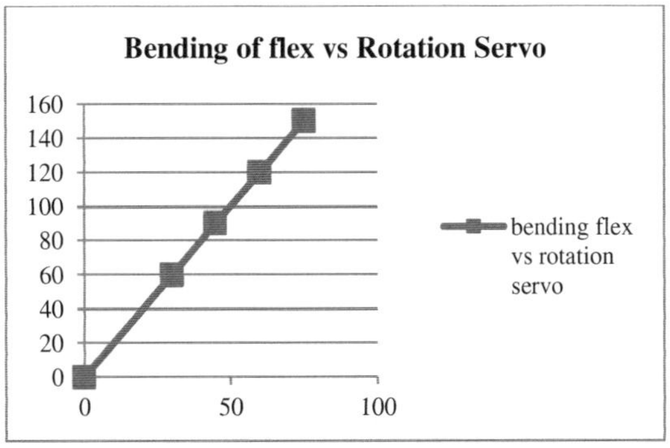

Figure 11. Plot bending sensor with respect to rotation.

When the sensor is bending, the values of the resistor and output voltage change immediately. Table IV shows the values of the flex sensor resistance and the obtained output voltage with the bending angle of the flex sensors. The value of the resistor is comparatively increased whenever the flex sensor is relatively bending. Similar to the output voltage. When the flex sensors are bending, the upper body of it is expanded. More voltage is needed to across this resistance. And therefore, the output voltage is incremented directly with the increment of the resistance as shown Figure 12 and Figure 13.

Table 4. Bending of flex sensor and flex sensor resistor.

Flex sensor resistor (kΩ)	Flex sensor bending angle.	Vout
2.8	0	0.564 V
3.283	45	0.6492 V
4.022	60	0.7728 V
4.769	90	0.8907 V
5.013	135	0.9279 V

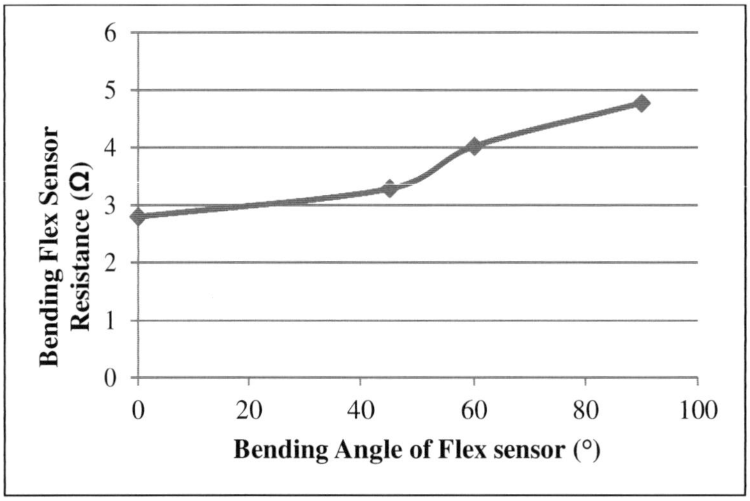

Figure 12. The relationship between the bending of the flex sensors with respect to the resistance.

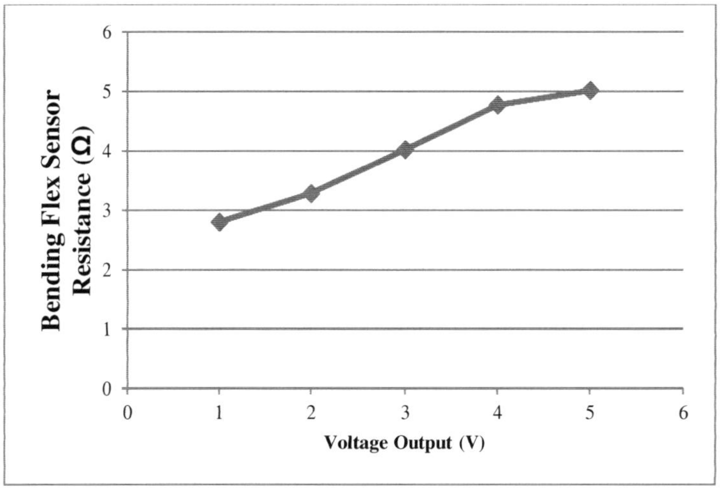

Figure 13. The relationship between the bending of the flex sensors with respect to the voltage output.

5. Conclusion

This project has come out with the working glove for hand trainer device by using flex sensors as the main input. The movement of the hand, either releasing or grabbing movements are mimicking by the artificial hand through the manipulation of the servo motor rotations and the nylon string that attached to each single artificial finger. It is one way to indicate the patient whether their hand movement is strong and sufficient by looking at the movement of the artificial hand during the rehabilitation process. This prototype could be enhanced by replacing the servo motor with the higher torque motor so that the artificial hand could grab and lift a heavier material such as ball during the hand exercise process and making the rehabilitation more interactive and interesting.

References

[1] Nor Azlin MN, Syed Aljunid SJ, Noor Azahz A, Amrizal MN, Saperi S. *Direct medical costs of stroke: findings from a tertiary hospital in malaysia*. Med J Malaysia. 2012;67:473–7.

[2] Hong Kai Yap,Jeong Hoon Lim,Fatima Nasrallah,Fan-Zhe Low, James C. H. Goh, Raye C. H. Yeow *MRC-glove: A fMRI compatible soft robotic glove for hand rehabilitation application*, 2015 IEEE International Conference on Rehabilitation Robotics (ICORR), pp 735 – 740

[3] Worsnopp TT, Peshkin MA, Colgate JE, Kamper DG, *An Actuated Finger Exoskeleton for Hand Rehabilitation Following Stroke*, InRehabilitation Robotics, 2007. ICORR 2007. IEEE 10th International Conference on 2007 Jun 13, pp. 896-901.

[4] *Monitoring functional arm movement for home-based therapy after stroke.*, vol. 7, Jan. 2004. [Online]. Available: http://www.ncbi.nlm.nih.gov/pubmed/17271381

[5] N. S. Ward, M. M. Brown, A. J. Thompson, and R. S. J. Frackowiak, *Neural correlates of motor recovery after stroke: a longitudinal fMRI study*, Brain, vol. 126, pp. 2476-2496, 2003.

[6] R. V. Martinez, C. R. Fish, X. Chen, and G. M. Whitesides, *Elastomeric Origami: Programmable Paper-Elastomer Composites as Pneumatic Actuators*, Advanced Functional Materials, vol. 22, pp. 1376-1384, 2012.

[7] Y. Kadowaki, T. Noritsugu, M. Takaiwa, D. Sasaki, and M. Kato, *Development of soft power-assist glove and control based on human intent*, Journal of Robotics and Mechatronics, vol. 23, pp. 281-291, 2011.

[8] P. Polygerinos, S. Lyne, W. Zheng, L. F. Nicolini, B. Mosadegh, G. M. Whitesides, et al., *Towards a soft pneumatic glove for hand rehabilitation*, in 2013 IEEE/RSJ International Conference on Intelligent Robots and Systems (IROS), 2013, pp. 1512-1517.

[9] M. Gibb, New media art, design, and the Arduino microcontroller: A malleable tool. PhD thesis, Pratt Institute, 2010.

[10] M. Margolis, *Arduino cookbook*. O'Reilly Media, Inc.", 2011.

[11] Arduino, *Arduino uno* Last visited on 21/05/2016.

[12] J. Boxall, *Arduino Workshop: A Hands-on Introduction with 65 Projects*. No Starch Press, 2013.

[13] *Intro to arduino*, January 2014.

[14] Srinivasan C.R .,Sen S. ,Kumar A. & Saibabu C.(2014). *Measurement of Flow Using Bend Sensor*. Advance in Energy Conversion Technologies (ICAECT),2014 Conference on Manipal.

[15] D. G. Kamper, R. L. Harvey, S. Suresh, and W. Z. Rymer, *Relative contributions of neural mechanisms versus muscle mechanics in promoting finger extension deficits following stroke*, Muscle Nerve, vol. 28, pp. 309-18, Sep 2003.

[16] D. G. Kamper and W. Z. Rymer, *Impairment of voluntary control of finger motion following stroke: role of inappropriate muscle coactivation,* Muscle Nerve, vol. 24, pp. 673-81, May 2001.

[17] C. G. Canning, L. Ada, R. Adams, and N. J. O'Dwyer, *Loss of strength contributes more to*

physical disability after stroke than loss of dexterity, Clin Rehabil, vol. 18, pp. 300-8, May 2004.

[18] J. G. Colebatch and S. C. Gandevia, *The distribution of muscular weakness in upper motor neuron lesions affecting the arm*, Brain, vol. 112 (Pt 3), pp. 749-63, Jun 1989.

[19] J. Roh, W. Z. Rymer, E. J. Perreault, S. B. Yoo, and R. F. Beer, *Alterations in upper limb muscle synergy structure in chronic stroke survivors*, J Neurophysiol, vol. 109, pp. 768-81, Feb 2013.

[20] K. M. Triandafilou, H. C. Fischer, J. D. Towles, D. G. Kamper, and W. Z. Rymer, *Diminished capacity to modulate motor activation patterns according to task contributes to thumb deficits following stroke*, J Neurophysiol, vol. 106, pp. 1644-51, Oct 2011.One, vol. 8, p. e68745, 20

[21] MusicGlove available at https://www.flintrehab.com/2017/best-hand-exercise-equipment-stroke-patients/

Acknowledgment

This research is sponsored by Faculty of Electrical Engineering, Universiti Teknologi MARA. We thank UiTM, FKE for their support in terms of financial, facilities and knowledge.

Transimpedance Amplifier for MEMS SAW Oscillator in 1.4GHz

N Kamarudin[1], J Karim[1], and H Hussin[1]

1 Faculty of Electrical Engineering, University Teknologi MARA, 40450, Shah Alam, Selangor, Malaysia

e-mail: lizawati9242@gmail.com

Abstract. This work is to design a transimpedance amplifier for MEMS SAW resonator to achieve low power consumption at desired frequency. A transimpedance amplifier is designed and characterized for MEMS SAW resonator in 0.18μm CMOS process. The transimpedance amplifier achieves gain is 31 dBΩ at 176°.The power consume by oscillator is 0.6mW at V_{DD} 1.8V while phase noise at -133.97dB_c/Hz at 10kHz.

1. Introduction

Oscillator is a key component in electronic circuit. They provide "heartbeat" to circuit in processor, digital-to-analog converter, analog-to-digital converter, memory function, communication ports and many others function. Many types of oscillator are available in electronic industrial, such as quartz oscillator, LC Oscillator, RC oscillator and MEMS oscillator. Commonly used is crystal oscillator since it provides higher stability. Although crystal oscillator can provide clean and stable signal, unfortunately it cannot be fabricate in system on chip since quartz crystal resonator has to be fabricate on piezoelectric substrates instead of silicon substrate.

The demand of technology today required miniaturized sizes, low power consume, reliability, system on chip and low cost during fabrication and CMOS-MEMS oscillator offer entire requirement. Previous work indicates that dealing with passive components, extra sources at bias circuit and sizing the transistor are challenging. External sources give a circuit not fully integrated since it has more than one voltage sources. Transimpedance amplifier proposed in this work consists of minimum passive component, single source at higher frequency and low power dissipation. Passives component are reduce and circuit design used only single source, V_{DD} to power up the CMOS MEMS oscillator.

2. Literature Review

2.1. SAW Resonator

Vibrating mechanical tank components, such as crystal and surface acoustic wave (SAW) resonators, are widely used for frequency selection in communication sub-systems because of their high quality factor (Q's in the tens of thousands) and exceptional stability against thermal variations and aging[1]. MEMs SAW resonator was chosen in this work due to following fact that it can provide system integration capabilities, high precision fabrication, low cost and suitable for high volumes[2]. Besides, it also can provide fully integrated oscillator in electronic application.

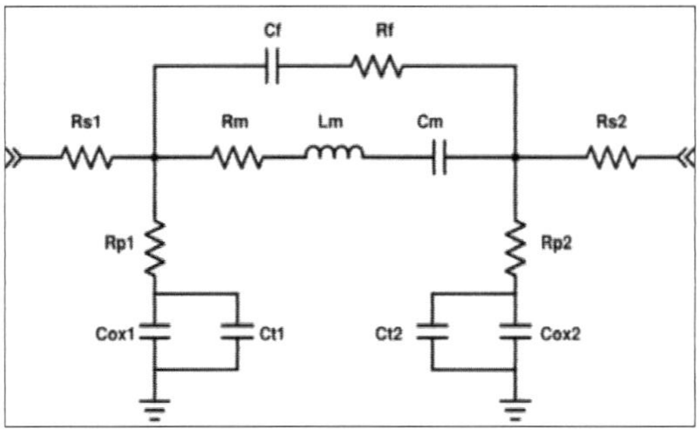

Figure 1. SAW resonator circuit.

Figure 1 shows equivalent circuit model of two port of SAW resonator. It develops by RLC circuit. It consists of resistor, capacitance and inductor in circuit.

Table 1. Parameter in SAW Resonator.

Parameter	Value	Unit
Rm	512	Ω
Lm	2.13	μH
Cm	6.0	fF
Rf	91.8	Ω
Cf	295.8	fF
Rs1	493	Ω
Rs2	412	Ω
Rp1	22.1	Ω
Rp2	19.4	Ω
Cox1+Ct1	0.2365	pF
Cox2+Ct2	0.2725	pF

Table I shows summarize parameter for SAW Resonator. Figure 2 show the S_{21} simulation for SAW resonator with insertion loss is -26.56dB of frequency response 1.396GHz.

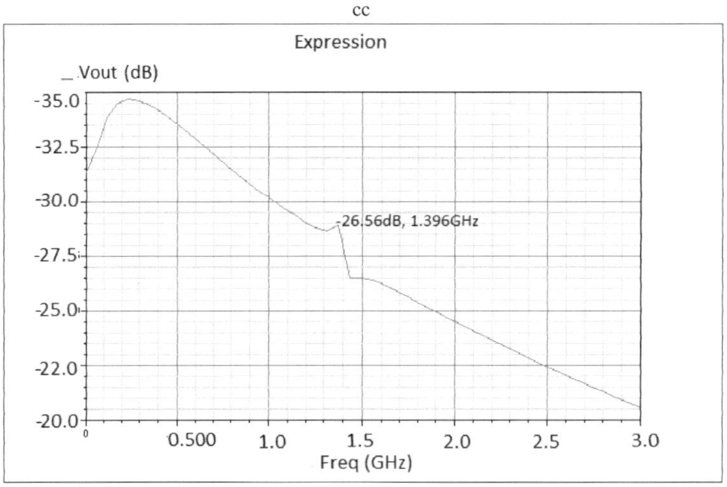

Figure 2. Transmission characteristic SAW Resonator.

3. TIA Amplifier

This work indicates on designing transimpedance amplifier for an oscillator. Transimpedance amplifier was selected to keep low current with sufficient gain-bandwidth in feedback compensation. This simulation performed in Cadence Tools 0.18µm CMOS technology.

There are two major step required in design process. Firstly, conception of design ,and secondly optimization of design[3]. To have accurate parameter during sizing the transistor, g_m/I_D methodology was chosen. g_m/I_D methodology strongly related to performance of analog circuit, give an indication point of device operating region and provide tools for calculating transistor dimension.

$$\frac{gm}{I_D} = \frac{1}{I_D}\frac{\partial I_D}{\partial V_{GS}} = \frac{\partial I_D}{\partial V_{GS}}\log(I_D) \quad (1)$$

Equation 1 shows derivation at maximum weak inversion region. The transconductance efficiency, g_m/I_D decreases when the operating points move toward strong inversion region. To have low power consume in oscillator, the transconductance efficiency in oscillator should be higher.

$$\text{Tranconductance efficiency} = \frac{g_m}{I_D} \quad (2)$$

Design specification

- Bandwidth >3GHz
- Amplifier gain <32dB
- Power consume by oscillator <1mW
- L=0.18µm

Since the design circuit is to achieve high frequency amplifier, the minimum length of transistor is selected and all transistor working in minimum current in saturation region

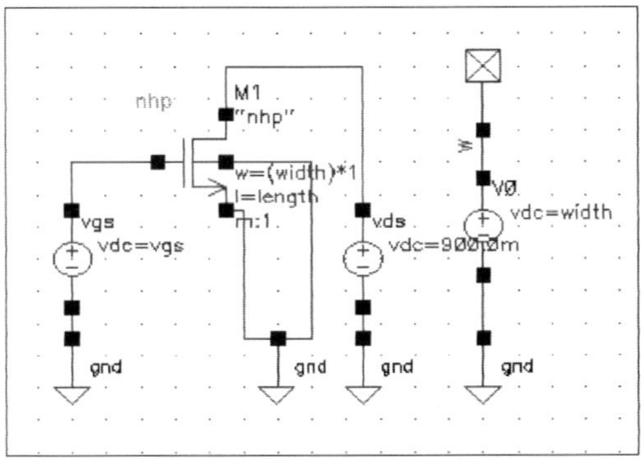

Figure 3. Simulation of NMOS transistor.

Figure 3 shows NMOS transistor circuit. The purpose of this simulation is to obtain characteristic of the transistor. Characterization can be made by plotting curves for NMOS and PMOS transistor based on our requirement.

- g_m/I_D vs V_{OV}
- $I_D(W/L)$ versus g_m/I_D
- F_t vs g_m/I_D

Low current mean higher transconductance efficiency required. In order to have higher transconductance efficiency, transistor must be biased in lower overdrive voltages, V_{OV}. g_m/I_D is

IOP Conf. Series: Materials Science and Engineering **341** (2018) 012011 doi:10.1088/1757-899X/341/1/012011

obtained by using V_{OV}. Then, the width (W) of the mosfet is determined by using the following formula:

$$\frac{I_D}{\left(\frac{W}{L}\right)} = \frac{I_D design\ spec}{\left(\frac{Wactual}{L}\right)} \tag{3}$$

Table II show parameter of transimpedance amplifier and the value was obtained from simulation in cadence.

Table II. Parameter Transimpedance Amplifier.

Parameter		**0.18µm Cadence**	
	g_m/I_D	**Value**	**Unit**
R_1	-	500	Ω
R_2	-	10	kΩ
R_3	-	2	kΩ
R_4	-	2.9	kΩ
R_5	-	63	kΩ
W_1/L_1	6.304	1.0/ 0.18	µm/µm
W_2/L_2	1.064	0.5/ 0.18	µm/µm
W_3/L_3	8.704	0.5/ 0.18	µm/µm
W_4/L_4	6.034	1.0/ 0.18	µm/µm
W_5/L_5	12.91	0.22/ 0.18	µm/µm
W_6/L_6	13.37	70.0 /0.18	µm/µm
W_7/L_7	0.965	0.22/0.18	µm/µm
W_8/L_8	6.137	0.22 /0.18	µm/µm
C_1	-	70	fF
C_2	-	1.5	pF
C_3	-	135	fF

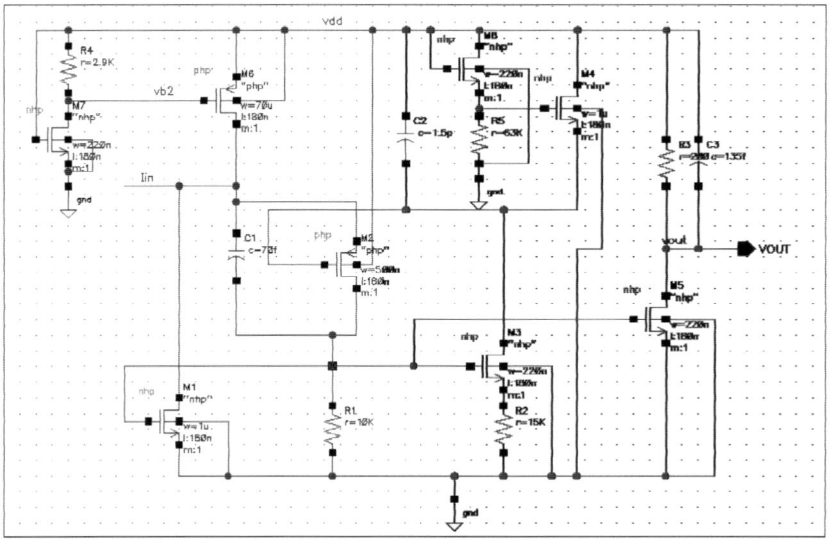

Figure 4. Transimpedance amplifier schematic.

Figure 4 shows detail schematic at the designed TIA. Three circuit has been divided into three main block: an input stage, a positive feedback loop, and the final output[4]. The input stages consist of resistor R_1 and transistor M_1 and M_2[4]. Transistor M7 and R4 provide sufficient current to M6 to 'turn on' input stage while M8 and R5 is used to give sufficient current to positive feedback loop. C_1 has be added to the circuit and implements at feedforward compensation to enlarges bandwidth to more than 2GHz. At high frequencies, above singularities introduced by C_1 the feedback loop is closed by capacitance;M1 works in triode configuration and TIA functionality is preserved[5].

This circuit biasing current at very low to keep power consume less than 1mW and all the MOSFET in strong inversion region. Strong inversion region give maximize cut off frequency and give extra advantages for TIA working at high frequency application. The transconductance of transistor in circuit design cannot be increase to keep minimize power dissipated in oscillator. Only one source was used in the design to obtain fully integrated circuit. Besides, second stage is added to keep low input impedance. Second stage which is positive feedback amplifier is chosen to ensure stability in circuit [4]. Without this part, the gate of M_2 were fixed, the input voltage will rise when the input current rises[4].

Fig.5 show transimpedance gain of transimpedance amplifier in 31.3dB with phase shift 176°. The bandwidth of this amplifier extends to 16.531GHz.

c

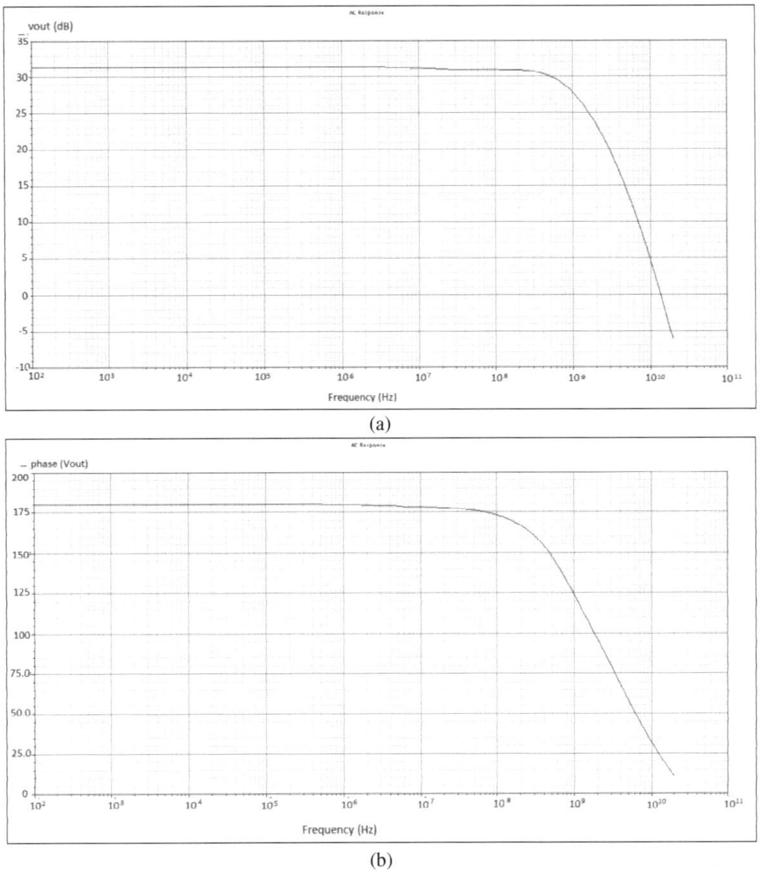

Fig 5. Open loop response of transimpedance amplifier. a) Gain plot b) Phase plot.

4. Simulation Results and Discussion

The simulation of TIA circuit connected with MEMS SAW resonator has been conducted. The bandwidth of transimpedance amplifier extends to 16.531GHz with 31dB gain and phase shift 176°. The gain obtained is able to overcome losses of the resonator. Meanwhile in Figure 6 shows transient analysis of CMOS MEMS oscillator. The oscillation produced after close loop gain at 1.396GHz. The oscillator achieved phase noise at -133.97dB$_c$/Hz of 10kHz in Figure 7. The power consume by oscillator calculate by using equation 3. The power consume by oscillator is 0.6mW.

$$\begin{aligned} P_{consume} &= V_{DD} \times I_D \\ P_{consume} &= 1.8 \times 328.1021\mu \\ &= 0.6mW \end{aligned} \quad (4)$$

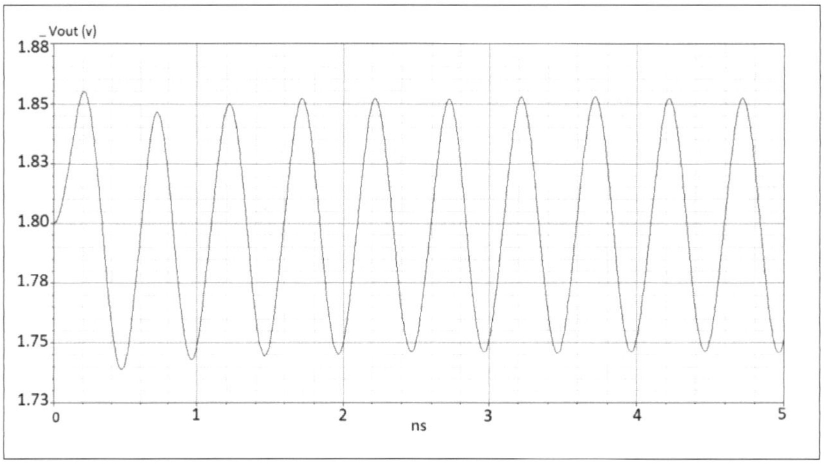

Figure 6. Close loop feedback simulation of transimpedance amplifier with SAW resonator.

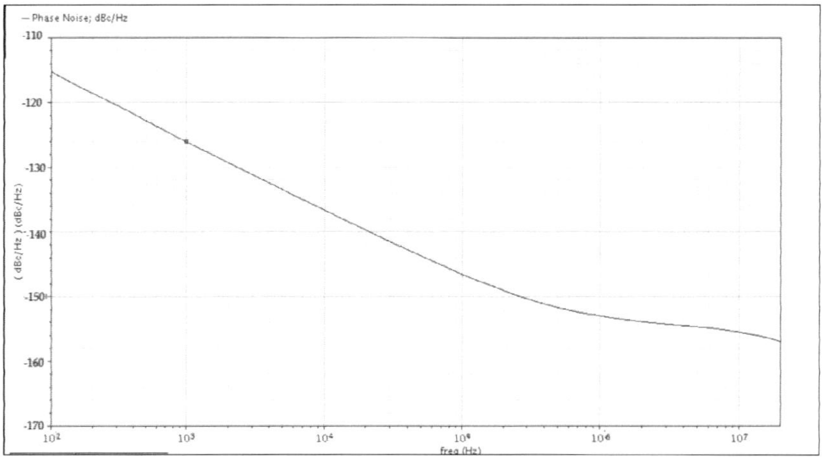

Figure 7. Simulated phase noise of transimpedance amplifier with SAW resonator.

Table 3. Design performance comparison.

Ref.	Year	Tech. (μm)	Power Consume, (W)	Gain (dB)	Phase Noise	F_t (Hz)
[6]	2011	0.18	7.2m	76	-122dB$_c$/Hz at 10kHz	1.006G
[7]	2012	0.18	1.57m	99	-128dB$_c$/Hz at 1KHz	48M
[5]	2015	0.18	6.4m	120	-102 dB/Hz at 10kHz	17.22M
[8]	2016	0.18	6.24m	122	-97dB/Hz at 10kHz	17.22M
This work		0.18	0.6m	31	-133.97dB$_c$/Hz at 10kHz -124.3dB$_c$/Hz at 1kHz	1.396G

Table 3 demonstrate comparison between previous works and propose design. CMOS MEMS oscillator in this work show lowest power consumes resultant low gain in amplifier. Although it has lowest gain compare to previous work, the frequency response provide by TIA at highest frequency. Therefore, it give an advantages for proposed design to operate at higher frequency with low power consume.

5. Conclusion

The objective of this work is to obtain low power consume for sustaining circuit in oscillator. The transimpedance amplifier was selected as a sustaining circuit for CMOS-MEMS oscillator since it can provide low power consume with desired frequency and gain. The oscillator consumes 0.6mW of frequency response 1.39GHz at phase noise -133.97dB$_c$/Hz at 10kHz. Further improvement is needed for the next work by reducing passive component and transistor in circuit design. Besides, obtain more sustaining amplifier also required to variety application among oscillator.

References

[1] C. T and C.Nguyen, "Microelectromechanical Devices for Wireless Communications," in *Proceedings MEMS 98. IEEE. Eleventh Annual International Workshop on Micro Electro Mechanical Systems. An Investigation of Micro Structures, Sensors, Actuators, Machines and Systems*, 1998, pp. 1–7.

[2] A. A. Md Ralib, A. N. Nordin, and U. Hashim, "Equivalent circuit modeling of two-port Al-doped zinc oxide CMOS SAW resonator using MATLABTM," in *ICCAS 2013 - 2013 IEEE International Conference on Circuits and Systems: "Advanced Circuits and Systems for Sustainability,"* 2013, pp. 140–145.

[3] R. C. Sayan Bandyopadhyay, Deep Mukherjee, "Design Of Two Stage CMOS Operational Amplifier in 180nm Technology With Low Design Of Two Stage CMOS Operational Amplifier in 180nm Technology With Low Power and High CMRR," vol. 11, no. November 2016, pp. 239–247, 2014.

[4] M. Crotti, I. Rech, G. Acconcia, A. Gulinatti, M. Ghioni, and S. Member, "A 2-GHz Bandwidth , Integrated Transimpedance Amplifier for Single-Photon Timing Applications," vol. 23, no. 12, pp. 2819–2828, 2015.

[5] S. Roy, H. Ramiah, and A. Wasif, "Designing a new high gain CMOS ampli fi er towards a 17 . 22 MHz MEMS based Si oscillator for a cost effective clock generator IC," *IEICE Electron. Express*, vol. 12, no. 10, pp. 1–8, 2015.

[6] H. M. Lavasani, W. Pan, B. Harrington, R. Abdolvand, and F. Ayazi, "A 76 dBΩ 1.7 GHz 0.18 μm CMOS tunable TIA using broadband current pre-amplifier for high frequency lateral MEMS oscillators," *IEEE J. Solid-State Circuits*, vol. 46, no. 1, pp. 224–235, 2011.

[7] M. Li, C. Li, L. Hou, Y. Liu, and S. Li, "A 1.57mW 99dBΩ CMOS Transimpedance Amplifier for VHF Micromechanical Reference Oscillators," in *2012 IEEE International Symposium on Circuits and Systems*, 2012, pp. 209–212.

[8] S. Roy, H. Ramiah, A. W. Reza, and C. C. Lim, "Design of a MEMS-Based Oscillator Using 180nm CMOS Technology," *PLOS ONE 11(7)*, pp. 1–19, 2016.

Acknowledgment

This work was supported by the Fundamental Research Grant: FRGS/1/2015/TK04/UITM/02/19 provided by the Ministry of Higher Education of Malaysia through Universiti Teknologi MARA.

IOP Conf. Series: Materials Science and Engineering **341** (2018) 012012 doi:10.1088/1757-899X/341/1/012012

Aging analysis of high performance FinFET flip-flop under Dynamic NBTI simulation configuration

M. F. Zainudin[1,2], H. Hussin[1,2,a], A. K. Halim[1] and J. Karim[1]

[1]Faculty of Electrical Engineering, Universiti Teknologi Mara (UiTM), 40450 Shah Alam, Selangor, Malaysia, [2]Integrated Microelectronics Systems and Applications Research Group, Universiti Teknologi Mara (UiTM), 40450 Shah Alam, Selangor, Malaysia
[a)]Corresponding author: hanimh@salam.uitm.edu.my

Abstract. A mechanism known as Negative-bias Temperature Instability (NBTI) degrades a main electrical parameters of a circuit especially in terms of performance. So far, the circuit design available at present are only focussed on high performance circuit without considering the circuit reliability and robustness. In this paper, the main circuit performances of high performance FinFET flip-flop such as delay time, and power were studied with the presence of the NBTI degradation. The aging analysis was verified using a 16nm High Performance Predictive Technology Model (PTM) based on different commands available at Synopsys HSPICE. The results shown that the circuit under the longer dynamic NBTI simulation produces the highest impact in the increasing of gate delay and decrease in the average power reduction from a fresh simulation until the aged stress time under a nominal condition. In addition, the circuit performance under a varied stress condition such as temperature and negative stress gate bias were also studied.

1. Introduction

Negative-bias Temperature Instability (NBTI) is one of the major physical mechanisms which are responsible for the undesired effects in circuit reliability and robustness. This mechanism alters the main electrical parameters of the stressed device especially in long-term device lifetime. According to [1], [2], the increasing in the threshold voltage shift is the main reason that contributes to NBTI degradation. Other parameters such as device transconductance (g_m), channel mobility (μ_{eff}), linear and saturation current (I_{DLIN} and I_{DSAT}) are decrease due to NBTI effect. Overall, NBTI degrades the main electrical parameters in circuit reliability analysis.

Not only that, NBTI results in poor circuit performance which results in shorter device lifetime [3]–[8]. In circuit design, there is always a trade-off between the delay and the power to achieve the current needs in the industry. Observation in [9] and [7] shown that NBTI results in increasing in the device timing delay at gate-level device which later disrupt the timing sequences of the memory devices such as flip-flop. Meanwhile, observation in [6] shown that the NBTI reduce the average power consumption due to the reduction of the leakage current.

The objective of this work was to explore the circuit performance of FinFET flip-flop under a present of the NBTI degradation based on the different HSPICE simulation commands. Not only that, the NBTI characterization technique based on variation of the supply voltage and temperature were used to identify which stress condition contributes to the NBTI degradation.

This paper is divided into five sections. The first section begins with the introduction to the NBTI mechanism. This is followed by the second section which describes the details on NBTI model provided by MOSFET Reliability Analysis (MOSRA) and how this model is related to the circuit performance. The third section explained the simulation setup for the tested device. The evaluation of the obtained results will be analyzed in the fourth section followed by the overall conclusion in the last section.

2. Literature Review

In MOSRA, there are two physical mechanisms for the NBTI degradation which are related to the contribution of the interface traps and the traps deep inside the dielectric layer [10]. Equation 1 is related to the interface traps generation while Equation 2 is related on the oxide traps of the stressed device. By using both equation, the threshold voltage shift (ΔV_{TH}) is calculated which indicates the NBTI degradation. The NBTI impact on the threshold voltage can be expressed as follow:

$$\Delta V_{TH,IT} \sim exp\left(-\frac{E_a}{K.T}\right).\left[\frac{\varepsilon}{t_{ox}}\left(V_{gs}-V_{TH}\right)\right]^{TITCE}.exp\left[TITFD.E\left(V_{gs},V_{ds}\right)\right]t^{NIT} \quad (1)$$

$$\Delta V_{TH,OT} \sim \exp\left[-\frac{TOTFD+\frac{TOTTD}{T}}{E\left(V_{gs},V_{ds}\right)}\right]t^{NOT} \quad (2)$$

From the equation, there are two parameters which can contribute to the change of the threshold voltage shift. According to [1], NBTI increases in negative stress gate bias (V_{GS}) at elevated temperature. Thus, the threshold voltage shift is mainly affected by supply voltage and temperature.

3. Experimental setup

In order to predict a circuit performance lifetime, an accurate modeling of NBTI-induced degradation depends on intensive SPICE simulation. Synopsys HSPICE simulator [11] with NBTI model provided in MOSFET Reliability Analysis (MOSRA) was used to simulate the circuit performance under a different simulation measurement under a different NBTI parameter conditions. A High Performance of Predictive Technology Model (PTM) of 16nm was used in the SPICE simulation. The simulation results were verified with a Transmission Gate D flip-flop design as shown in Figure 1.

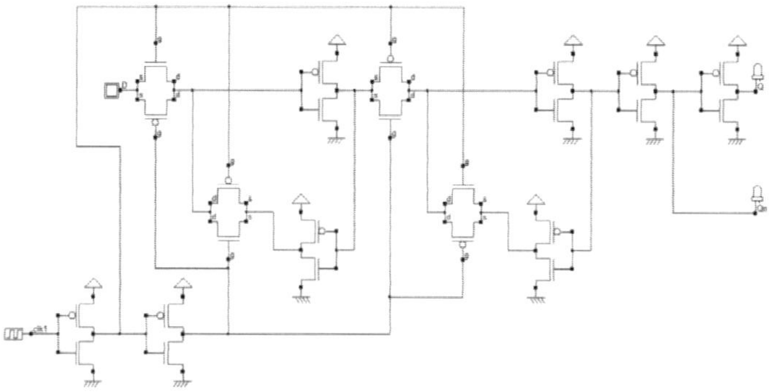

Figure 1. Transmission Gate D flip-flop.

In this study, the performance of D flip-flop is simulated under three different dynamic NBTI simulation. Next, a variation of operation conditions such as supply voltage and temperature are simulated on each of the dynamic NBTI simulation. The V_{DD} were varied from 0.6V to 0.8V while the temperature was varied from 25°C to 125°C. The simulation measurement were varied based on three different measurement setup as shown in Table 1. The circuit was simulated under 10 years of the device lifetime with 2fF of load capacitance at the output, Q. The circuit delay time and average power are recorded for every 1 year from fresh simulation (0 year) until aged simulation (10 years).

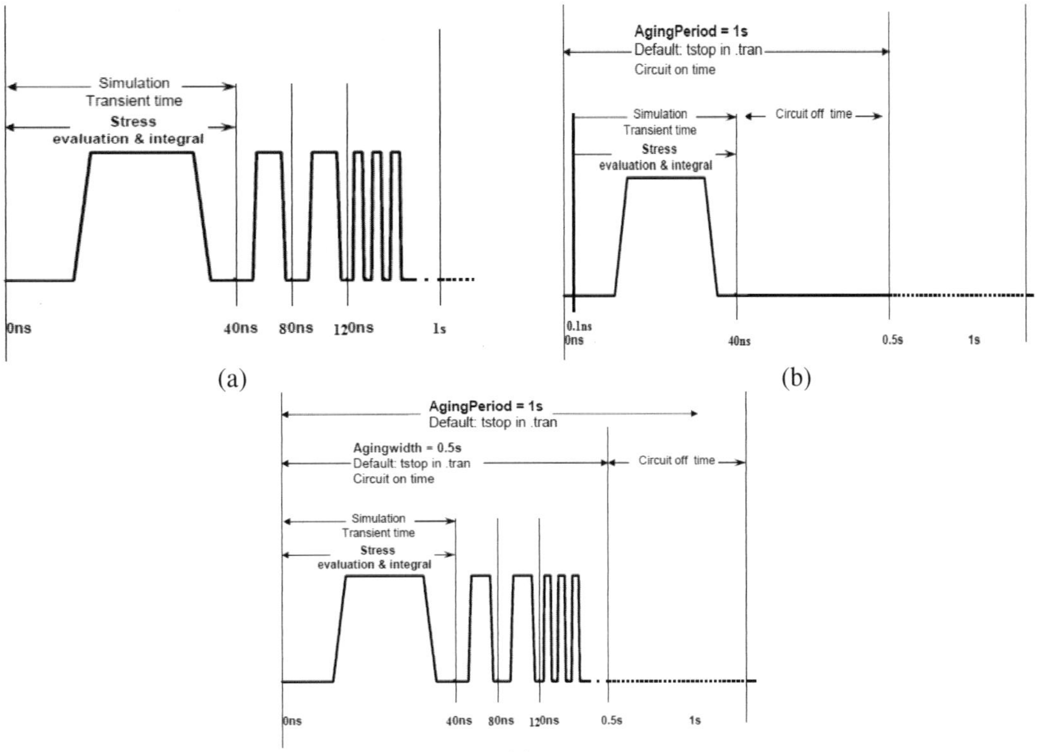

(c)

Figure 2. Graphic Illustration for each commands.

Table 1. Different measurement setup for HSPICE reliability simulation

Measurement Setup	Commands
(a) Command 1	.MOSRA Reltotaltime = 3.154e+8 RelStep = 3.154e+7 .Tran 0.01n 4ns
(b) Command 2	.MOSRA Reltotaltime = 3.154e+8 RelStep = 3.154e+7 +AgingStart=10ns AgingStop=100nS +AgingPeriod=1s .Tran 0.01n 4ns
(c) Command 3	.MOSRA Reltotaltime = 3.154e+8 RelStep = 3.154e+7 + AgingPeriod=1s Agingwidth= 0.5s .Tran 0.01n 4ns

Based on Figure 2 and Table 1, the dynamic NBTI simulation were divided into three types of commands. The first command illustrates the normal dynamic NBTI simulation as shown in Figure 2 (a). Next, the second command in Figure 2 (b) shows that the cut-off time (recovery time) for the dynamic NBTI simulation is longer than the stress time. Finally, the third command shows that the circuit undergoes a normal dynamic NBTI simulation in the first half cycle and followed by the cut-off time on the other half cycle under one second time period.

4. Simulation Results and Discussion

This section discusses the aging effect on the performance of D flip-flop circuit and analysed the changes in circuit performance with the presence of the NBTI degradation.

4.1. The aging effect on the circuit performance

A different measurement setup based on HSPICE reliability commands as shown in Table 1 are simulated under a nominal voltage supply Vdd = 0.7V with temperature of 25°C. Figure 3 shows the gate delay C-to-Q under a different HSPICE reliability commands. Based on the results, the gate delay for command 1 and 3 increases every year while the gate delay for command 2 remains constant. This is due to shorter stress evaluation and integral in command 2 compared to the other commands which causes the threshold voltage shift, ΔV_{TH} to increase slowly over time. Since the ΔV_{TH} for command 2 is too small, the gate delay C-to-Q is not affected by NBTI degradation. The result is in agreement with observation in [12] and [13]. However, the ΔV_{TH} for command 1 shows the highest NBTI degradation which results in the increase of the gate delay C-to-Q. Thus, the circuit with longer stress evaluation and integral results in increasing of ΔV_{TH} which later effect the gate delay.

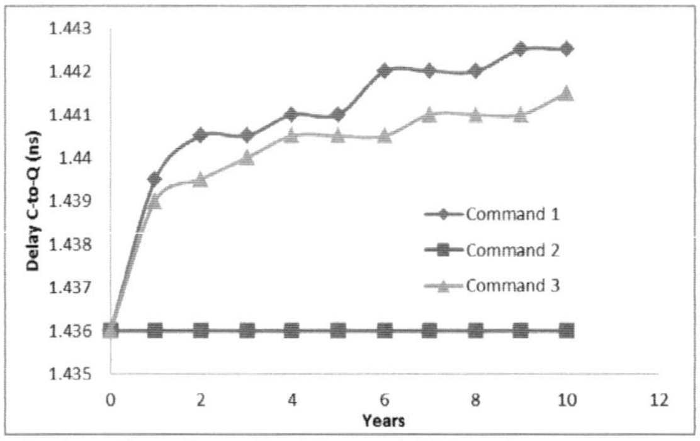

Figure 3. Delay C-to-Q under V_{DD} = 0.7V and temperature of 25°C.

Next, the average power of the D flip-flop for each commands are observed. The results obtained from Figure 4 display the average power under V_{DD} = 0.7V and temperature of 25°C. As shown in Figure 4, the average power for each command reduces from fresh simulation until the aged simulation. However, the average power in command 2 only decreases in a small value compared to command 1 and 3. NBTI not only increases the ΔV_{TH}, other parameters like sub threshold current (I_{SUB}) decreases since I_{SUB} is proportional to exp ($-V_{TH}/mkT$) [6]. Since the ΔV_{TH} for command 2 is too small, the I_{SUB} for command 2 only decreases in a small value. Compared to the ΔV_{TH} produce by command 1, the I_{SUB} for command 1 decrease gradually over time. Thus, this leads to decrease the leakage power (L_{eakage}) which later results in the average power.

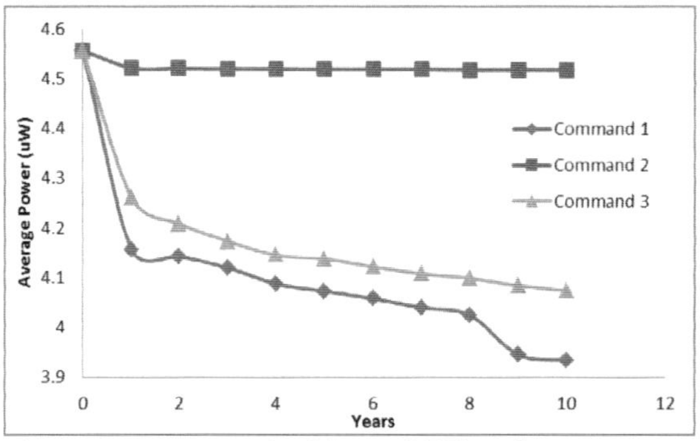

Figure 4. Average power under V_{DD} = 0.7V and temperature of 25°C.

Overall, the gate delay C-to-Q increases with a longer NBTI stress while the average power decreases with a longer NBTI stress as shown in Figure 3 and Figure 4. Shorter stress evaluation such as command 2 does not affect the circuit performance. Thus, the circuit with a longer NBTI stress degrades the circuit performance.

IOP Conf. Series: Materials Science and Engineering **341** (2018) 012012 doi:10.1088/1757-899X/341/1/012012

4.2. The aging effect based on supply voltage and temperature variation

In this section, we consider the variation of operating conditions such as V_{DD} and temperature in order to test the result accuracy. For this reason, we examine the gate delay C-to-Q and the average power for each HSPICE commands from fresh simulation until the aged simulation to predict the circuit performance.

Figure 5 shows the gate delay C-to-Q for each HSPICE command under a variation of V_{DD} and temperature. Based on Figure 5, the gate delay increases with an increase of temperature and a decrease in V_{DD}. NBTI increases with temperature and negative stress gate bias that causes the increasing of ΔV_{TH}. Not only that, the charging current of the switching capacitances in the circuit is decreased which lead to the increasing of the gate delay through the logic [14]. Thus, the overall NBTI effect based on the operating variation also degrade the circuit performance.

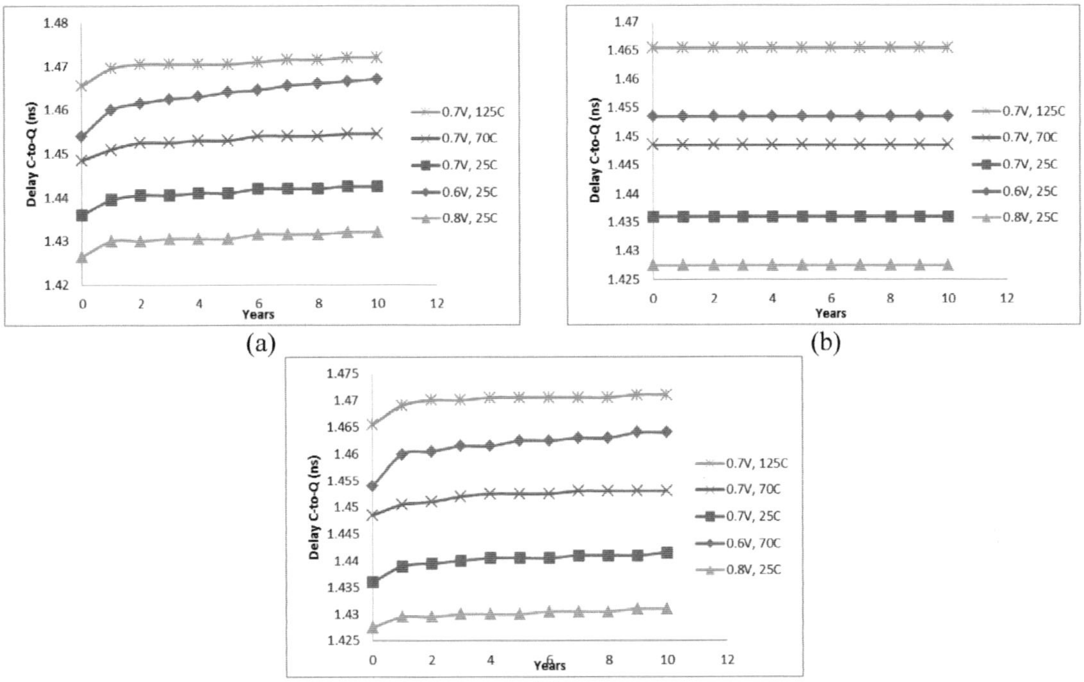

Figure 5. Delay C-to-Q under the variation of operating condition. (a) Command 1, (b) Command 2, (c) Command 3

Figure 6 highlights the average power for each HSPICE command under a variation of V_{DD} and temperature. The result shows that the average power decrease as the V_{DD} decrease and the temperature increase. The average power can be reduce by lowering the V_{DD}. Meanwhile, the increasing in temperature will decrease the switching current which later result in the average power reduction.

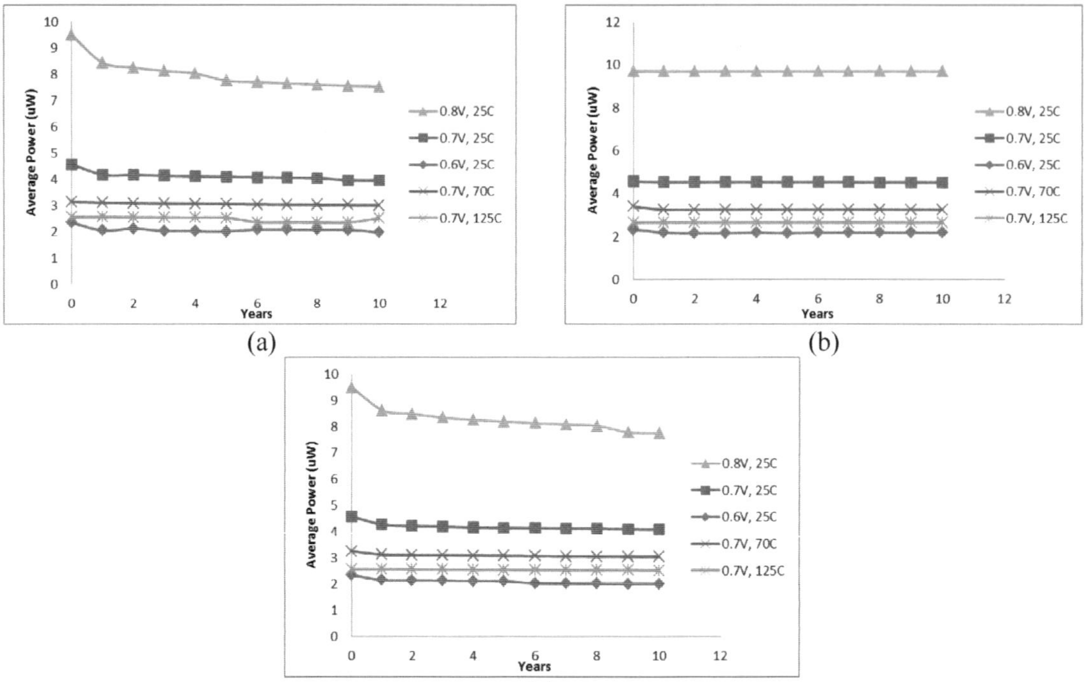

Figure 6. The average power under the variation of operating condition. (a) Command 1, (b) Command 2, (c) Command 3

5. Conclusion

This paper presents comprehensive analyses of the performance of PTM 16nm High Performance flip-flop under a present of the NBTI degradation based on the different HSPICE simulation commands. The result shown that the longer NBTI stress produce the highest increase in the gate delay and a reduction in the average power. Not only that, the effect of the circuit performance under a different operation conditions such as V_{DD} and temperature were also included to provide the SPICE result accuracy.

Acknowledgment

This work was supported by Fundamental Research Grant (FRGS: 600-RMI/FRGS 5/3 (31/2015)) and 600-RMI/DANA 5/3/ARAS (18/2015).

References

[1] S. Mahapatra, N. Goel, S. Desai, S. Gupta, B. Jose, S. Mukhopadhyay, K. Joshi, A. Jain, A. E. Islam, and M. A. Alam, "A comparative study of different physics-based NBTI models," *IEEE Transactions on Electron Devices*, vol. 60, no. 3. pp. 901–916, 2013.

[2] J. Franco, B. Kaczer, and G. Groeseneken, *Degradation Mechanisms, Reliability of High Mobility SiGe Channel MOSFETs for Future CMOS Application*. Springer, 2014.

[3] S. Taghipour and R. N. Asli, "Aging comparative analysis of high-performance FinFET and CMOS flip-flops," *Microelectron. Reliab.*, vol. 69, pp. 52–59, 2016.

[4] C. Nunes, P. F. Butzen, A. I. Reis, and R. P. Ribas, "BTI, HCI and TDDB aging impact in flip-flops," *Microelectron. Reliab.*, vol. 53, no. 9–11, pp. 1355–1359, 2013.

[5] E. L??te, A. A. Vatanjou, T. Ytterdal, and S. Aunet, "Extended Comparative Analysis of Flip-Flop Architectures for Subthreshold Applications in 28 nm FD-SOI," *Microprocess. Microsyst.*, vol. 48, pp. 11–20, 2016.

[6] M. Houshmand Kaffashian, R. Lotfi, K. Mafinezhad, and H. Mahmoodi, "Impacts of

NBTI/PBTI on performance of domino logic circuits with high-k metal-gate devices in nanoscale CMOS," *Microelectron. Reliab.*, vol. 52, no. 8, pp. 1655–1659, 2012.

[7] M. Bahadori, M. Kamal, A. Afzali-Kusha, and M. Pedram, "A comparative study on performance and reliability of 32-bit binary adders," *Integr. VLSI J.*, vol. 53, pp. 54–67, 2016.

[8] H. K. Alidash, A. Calimera, A. Macii, E. Macii, and M. Poncino, "On-chip NBTI and PBTI tracking through an all-digital aging monitor architecture," *Lect. Notes Comput. Sci. (including Subser. Lect. Notes Artif. Intell. Lect. Notes Bioinformatics)*, vol. 7606 LNCS, pp. 155–165, 2013.

[9] P. Alegre, T. Copetti, G. Medeiros, L. B. Poehls, F. Vargas, S. Kostin, M. Jenihhin, J. Raik, and R. Ubar, "Gate-Level Modelling of NBTI-Induced Delays Under Process Variations," pp. 75–80, 2016.

[10] W. Paper, "MOS Device Aging Analysis with HSPICE and CustomSim," no. August, 2011.

[11] Synopsys, "HSPICE ® User Guide : Simulation and Analysis," no. September, 2009.

[12] S. Desai, S. Mukhopadhyay, N. Goel, N. Nanaware, B. Jose, K. Joshi, and S. Mahapatra, "A comprehensive AC / DC NBTI model: Stress, recovery, frequency, duty cycle and process dependence," in *IEEE International Reliability Physics Symposium Proceedings*, 2013.

[13] C. Yilmaz, L. Heiss, C. Werner, and D. Schmitt-Landsiedel, "Modeling of NBTI-recovery effects in analog CMOS circuits," in *IEEE International Reliability Physics Symposium Proceedings*, 2013.

[14] F. Keivanian, N. Mehrshad, and A. Bijari, "Multi - objective optimization of MOSFETs channel widths and supply voltage in the proposed dual edge - triggered static D flip - flop with minimum average power and delay by using fuzzy non - dominated sorting genetic algorithm - II," *Springerplus*, 2016.

IOP Conf. Series: Materials Science and Engineering **341** (2018) 012013 doi:10.1088/1757-899X/341/1/012013

Induction motor speed control using varied duty cycle terminal voltage via PI controller

A Azwin[1] and S. Ahmed[2]

[1]Electronics and Communication Engineering Department,University Tenega Nasional, Bangi, Selangor, 43000
[2]Electronics and Communication Engineering Department,University Tenega Nasional, Bangi, Selangor, 43000
Email: Aidilazwin@uniten.edu.my

Abstract. This paper deals with the PI speed controller for the three-phase induction motor using PWM technique. The PWM generated signal is utilized for voltage source inverter with an optimal duty cycle on a simplified induction motor model. A control algorithm for generating PWM control signal is developed. Obtained results shows that the steady state error and overshoot of the developed system is in the limit under different speed and load condition. The robustness of the control performance would be potential for induction motor performance improvement.

1. Introduction

Three phase induction motor (TIM), considering its features of ruggedness, high efficiency, simple architecture, and low inertia is the most appropriate solution for industries, which is also long lasting and economic[1]. However, speed control of TIM especially in high performance demand in industrial applications are challenged due its non-linearity and dynamic behaviour. However, to overcome the constraints for the speed control application for TIM, numerous control techniques has been designed and developed depending on its sector of applications among others scaler control is the simplest one [2]–[4]. In scalar control variables are controlled only in magnitude and not in direction. Traditionally, PI control system for the TIM speed response is the most extensively used control technique because it is less complicated in design and economic. Moreover, PI control system provides the system with higher overshoot, maintains an acceptable level of steady stare error and keeps the oscillation of speed response and torque if any changes occurs in mechanical load[5][6]. Pulse width modulation (PWM) technique is being extensively used for induction motor speed control for a long time [7][8] and its popularity has been dramatically increased in the recent times because of the raising demands in the low cost energy conversion system. In general, main goal of PWM technique is optimized control of switching power semiconductor switches i.e. IGBT, FET etc. in the inverter circuit to provide the balanced three phase power to the terminal of TIM at a required magnitude, phase and frequency [9][10]. Numerous PWM techniques have already being developed for speed control of induction motor implemented through the Voltage Source Inverter(VSI) [11][12]. Apart from these conventional methods, Duty cycle or duty ratio based control technique is another, approach of delivering converted power in the terminal of TIM [9] and torque control as well [13][14]. Duty Cycle based control algorithm for PWM control is mostly implemented for inverters application for various operations regarding control of power [15]–[17]. In this paper, a duty cycle based PWM technique is been proposed to introduce a scaler control for TIM by using PI control technique. The PI control system have been applied to control the speed of TIM by adjusting the duty cycle of the terminal voltage from the inverter output during the load changing conditions. To obtain a duty cycle based speed control a relationship between the duty cycle and the magnitude of each harmonics is established. Proposed system's validity is established via simulation in MATLAB/Simulink environment. Obtained results shows the strength and weakness of robustness and performance response of the speed control system.

2. Methodology and Procedure

2.1. Model of Induction Motor

The mathematical model of TIM is considered based on traditional and simplified per phase equivalent circuit as shown in Figure 1, where voltage applied to the stator winding would directly affect the induced torque and hence slip and speed.

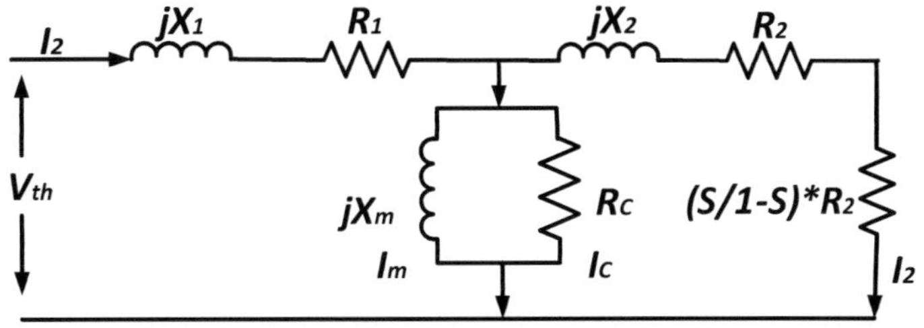

Figure 1. Simplified model of TIM

The relation between the torque and the voltage magnitude, slip and synchronous speed as shown in (1) and (2)

$$\tau_{ind} = \frac{3V_{th}^2 R_2 / S}{\omega_{sync}[(R_{th} + R_2 / s)^2 + (X_{th} + X_2)^2]} \tag{1}$$

Where, S is the mechanical slip and S can be defined as-

$$S = \frac{\omega_{sync} - \omega_{rotor}}{\omega_{sync}} \tag{2}$$

So, the synchronous speed of the stator frequency is,

$$Ns = \frac{120 * f}{p} \tag{3}$$

And the rotor speed can be defined from the above as –

$$N = N_s(1 - s) \tag{4}$$

In (1), the variable V_{th} represents the voltage applied to the stator windings. The values of R_2 and R_{th} depend on the resistance values of the motor windings and are not frequency dependent. The variable s is the slip of the motor and would depend on the rotor speed and also the speed of the rotating magnetic field of the respective harmonic. This relation can be seen in (2). The variables X_2 and X_{th} are locked rotor reactance values related to the motor windings which depend on the inductance of the motor windings and the synchronous frequency of each harmonics.

2.2. Proposed Controller Design

The main purpose of the proposed control system is to operate the TIM at a desired speed during changeable load condition or to run the TIM at different level of reference speed. Proposed speed control system is designed based on PI control algorithm as shown in Figure 2. So, the control system takes input from a reference speed to operate it at that speed. In other words, the control system supplies necessary amount of supply voltage at the motor terminal through the voltage source inverter. The terminal voltage sets the synchronous speed of the stator's magnetic field according to the reference speed. The rotor follows the stator's magnetic field and runs closer to the speed of that synchronous magnetic field. During the disturbances such as changes occurring in the reference speed or change occurring in the mechanical load the rotor speed goes down from the desired frequency. So, the difference between the reference speed and actual speed increase. Here, PI controller contributes to the system by, using its proportional and integral properties to detect the error quickly and resolves the error to obtain the desired speed by using a PI control algorithm. The difference between the desired speed and the actual speed gives the error. PI controller processes the error and generates an optimal duty cycle for the PWM control signal.

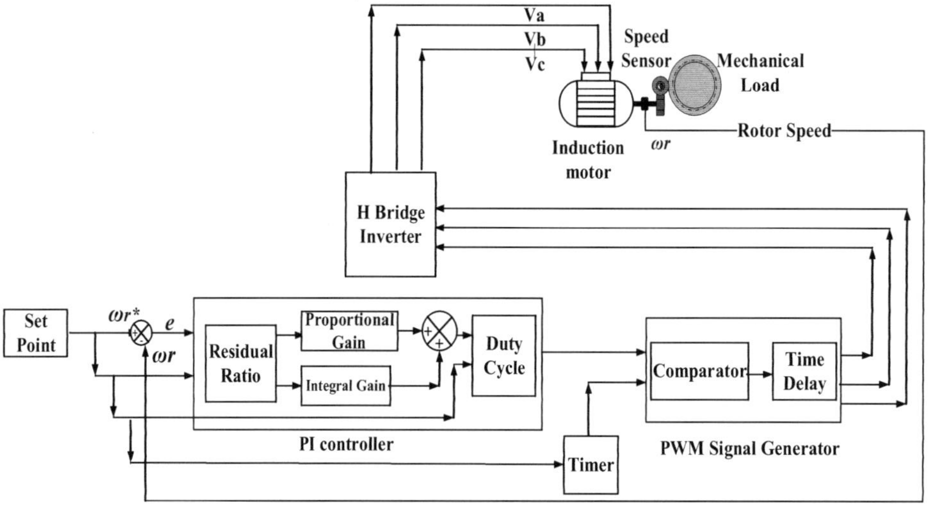

Figure 2. Diagram of the Proposed Speed Control System

Here, PI control system is designed to take the error value by mathematical operation to obtain the difference between reference speed and the actual speed. Then a residual ratio is taken from the error value to the set point value for the input to the proportional and integral action of the controller. Main purpose of the residual ratio is to track the deviation of the actual speed from the desired speed so that, when actual speed i.e. slip is deviated from the desired then controller takes necessary actions to reduce the gap by using PI controller properties. Proportional gain and integral gain of the PI controller is estimated via iterative method. Also, a discrete time integration is performing the integral action of the

PI controller by summing up the error values of instantaneous time until the error gets to zero. Whereas proportional part of the controller ensures that control system's response is proportional to the error signal. Proportional part of the controller doesn't let the deviation between the reference speed and the actual speed into a large quantity. However, output of both proportional and integral part is added and kept between a limit of 0 to 0.5. Hence, PI controller is providing the right duty cycle the at which voltage source inverter is operated to provide the terminal voltage.

2.3. PWM and Voltage Source Inverter Model

PWM technology is usually implemented for controlling the average power to the load through an induction motor control circuit. The simplest way to get a PWM is to compare a DC signal with a saw tooth signal. In this proposed speed control system PWM signal is generated from comparing the PI controller output and the time period of set point frequency in the PWM signal generator block as shown in the Figure 2. Here, the necessary carrier waveform for the comparator block is being generated from the set point frequency, which is basically converted into a saw tooth signal. The required DC signal is set the PI controller, where the output of the PI controller limit is set between 0 to 0.5. With the variation from speed due to change in reference or load disturbances the output of the controller is varied which alters the pulse width of the PWM wave as shown in the Figure 3(a). In figure 3(b) actual PWM control signal output from the simulation model is shown.

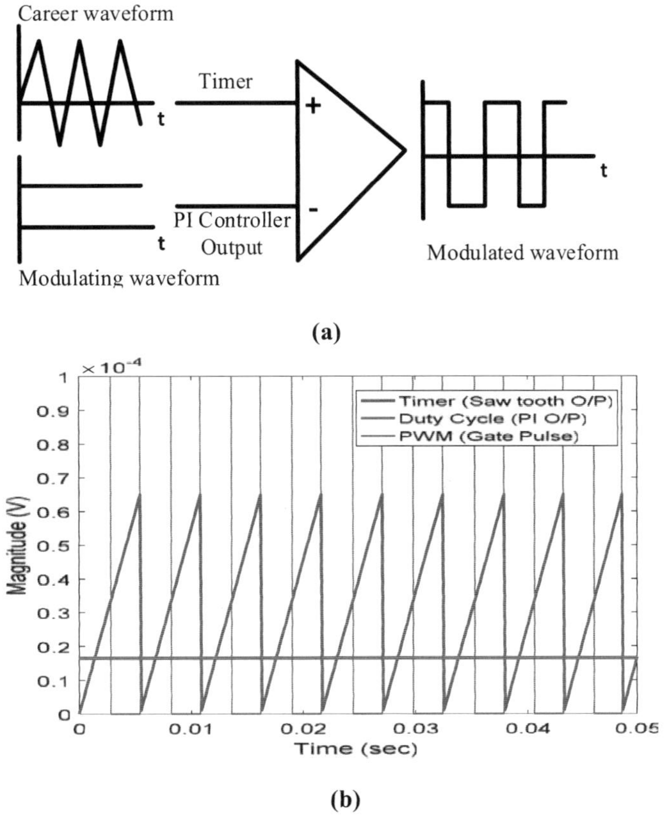

Figure 3. Modulated wave generation by the controller comparator action.

The generated PWM control signal from the controller determines the switching signal for the power electronic switches and the necessary duty cycle for this speed control system. In this proposed speed control system, PWM is implemented via a single stage H bridge inverter. Since, target of the PWM is to generate an output voltage from the voltage source inverter according to the reference speed. Moreover, a balanced set of three phase voltage of equal magnitude with 120-degree phase shift from each other is needed for AC machine to operate. So, PWM control signal is divided into three parts with a time delay of 1, 2/3 and 1/3 respectively to get a balanced three phase output from the inverter with 120-degree phase shift. The power inverter topology used in this controller is a single stage H bridge for all the phases of the induction motor is shown in the figure below. In this voltage source inverter, to produce a balanced three phase PWM voltage to the motor terminal a balanced three phases set of PWM wave input is considered and output of inverter's duty cycle is provided by the PWM generator and voltage is supplied by a DC source.

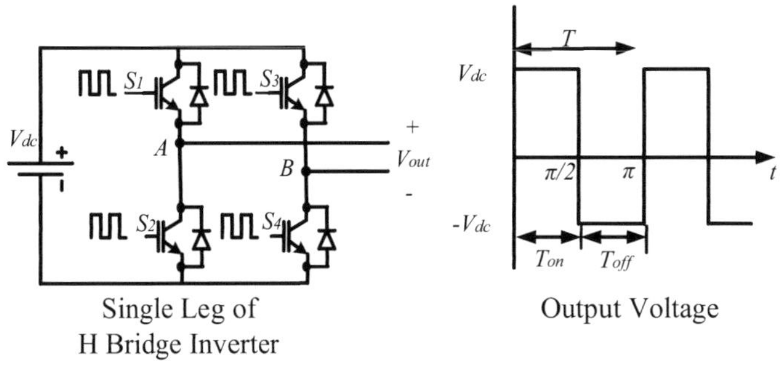

Figure 4. Single phase voltage output from H Bridge circuit

This would mean that the inverter can be able change the voltage polarity on the load given to it according to the individual switching time of the IGBTs. Hence, PWM voltage source inverter system is designed in Simulink as a subsystem based on an operating principle of a single stage H- bridge inverter. An optimal three phase balanced set of PWM wave is generating the optimal switching pulses for the inverter.

2.4. Duty Cycle and Speed Control Relationship

Duty cycle can be expressed as the ratio of duration of pulse i.e. pulse width and the time period of a rectangular waveform, which is usually expressed in percentage (%). Since, the DC voltage is the input voltage of the inverter circuit so, the input voltage has a constant value of V_{dc}. Duty cycle is the measure of how long the inverter output is set for on function i.e. T_{on} and a total time period of operation T where T_{off} is the total time period when the inverter is off in as shown in (6) and can be referred to the Figure 4 above.

$$\text{So, Total time period, } T = T_{on} + T_{off} \qquad (5)$$

$$\text{And Duty Cycle} = \frac{T_{on}}{T_{on} + T_{off}} = \frac{T_{on}}{T} \qquad (6)$$

From the equation 1 above, Variable V_{th} represents the supply voltage applied to the stator windings of the TIM. Moreover, TIM slip is proportional to rotor resistance, stator voltage and frequency. Since, supply voltage would depend on the duty cycle applied to the system so, duty cycle is directly related to the motor speed. The PWM control algorithm which is generating duty cycle percentage is provided below in **Figure 5**.

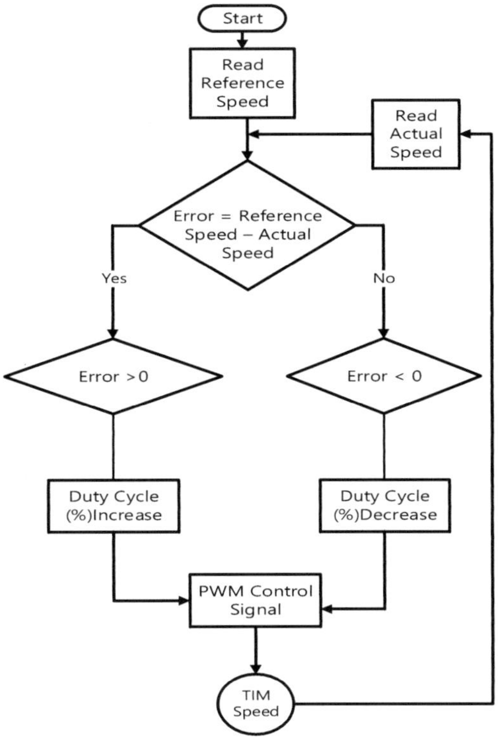

Figure 5. Control Algorithm for PWM control signal generation

3. Result and Discussion:

The effect of the duty cycle based PI control on TIM has been presented based on evaluating the performance of speed response of the system by varying the reference speed in different interval and also by varying the load in different intervals. A duty cycle percentage measurement is also calculated while evaluating the speed response for both two cases respectively reference speed variation and load variation. Robustness of the control system performance is included to evaluation system by testing the system with abrupt changes in reference speed at a constant load and sudden changes mechanical load at a constant speed. The results of variation in speed with at a constant load and variation is load with a constant speed is presented below.

3.1. Speed variation with constant load:

The speed response is evaluated in a range of set point speed from a lower value to a higher value with a different disturbance level in a time interval. To evaluate the performance of the duty cycle based speed controller, changes in different level of speed is introduced at a no-load condition. At the start of the operation, TIM is operated at its rated speed, 1430rpm for 5 seconds when with sudden deviation the speed goes down to 75% of the rated speed until 5 seconds, 1073rpm, then again speed changed to 50% of rated speed, 715rpm for next 5 seconds and finally it goes down to 25% of the rated value for another 5 seconds. Then in similar way speed is gradually increased back to the rated speed value at 5

IOP Conf. Series: Materials Science and Engineering **341** (2018) 012013 doi:10.1088/1757-899X/341/1/012013

second interval as shown in Figure 6. The good results of speed response performance are summarized in the Table 1.

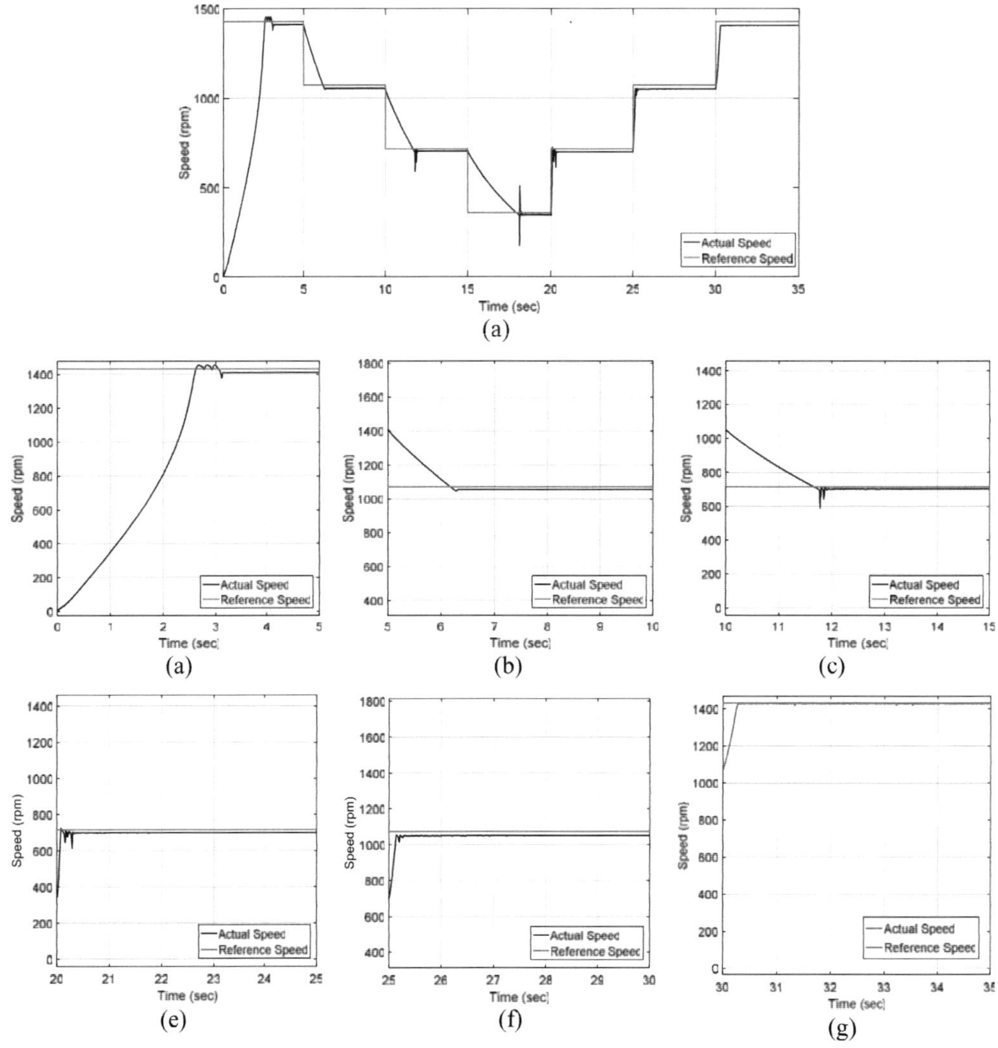

Figure 6. Speed Response with Reference Speed Variation

Table 1. Speed response performance with variation in reference speed and constant load.

Time Interval (sec)		Reference Speed Variation (rpm)		Steady State Error (%)	Overshoot (%)	Duty cycle variation Percentage
From	To	From	To			
0	5	0	1430	1.2	0.016	37.60
5	10	1430	1073	1.68	0.026	20.69
10	15	1073	715	1.60	0.021	17.45
15	20	715	357.5	2.51	0.464	9.83
20	25	357.5	715	2.35	0.036	17.783
25	30	715	1072	1.96	0.003	21.59
30	35	1072	1430	1.68	0.0003	24.11

The variation in duty cycle of PWM at these variable speed is also found as such - each time the changes in reference speed occurs the duty cycle varies with the changes. When the speed gradually goes down with the decrease in reference speed duty cycle goes down to zero i.e. no PWM control signal is generated from the PI controller. Where, when the speed is being increased gradually then of duty cycle increases. The variation of duty cycle with the speed variation is shown in Figure 7 below.

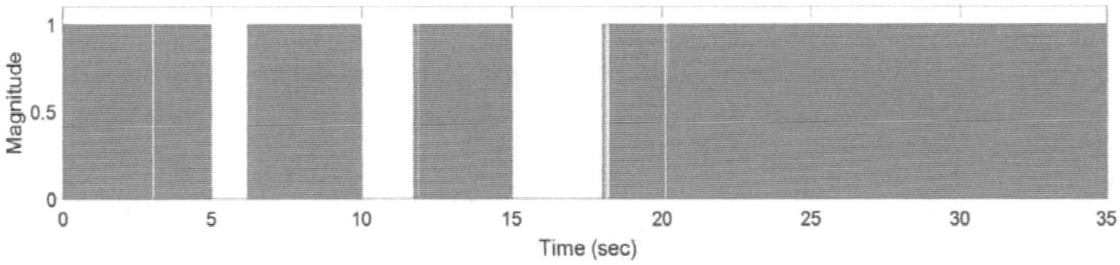

Figure 7. Duty cycle vs Time during Speed Variation with a Constant Load

3.2. Load variation with a constant speed:

Now, the TIM is operated based on a constant speed with a variable mechanical load condition. Here, as constant speed 75% of rated speed is taken as a reference where per unit mechanical load is being gradually increased and decreased at 5 second interval as shown in Figure 8(a). In Figure 8(b) the speed response in being plotted against the reference speed where the variation of actual speed from the reference speed is visible with the attachment of the higher per unit torque value. Since the controller vary the duty cycle of the PWM control signal to reach the desired speed the variation in duty cycle (%) of the modulated pulses are also being observed during the load variation.

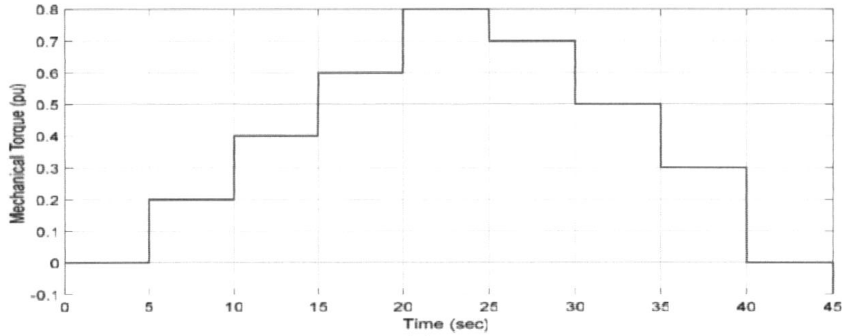

(a)

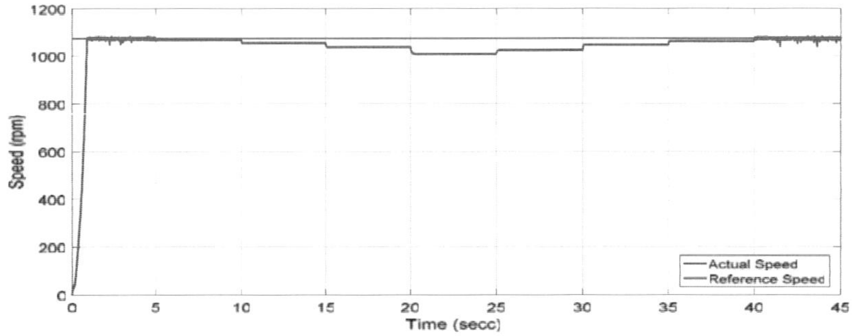

(b)

Figure 8. Speed response Performance with load variation

At the starting condition, motor is operated at no load speed for a 5 second interval at 75% of its rated speed value. Then the mechanical load is being increased 0.2pu in every 5 second interval to a maximum value of 0. 8pu.With the engagement of higher value of load the deviation in actual speed from the desired speed increases and the height value of deviation is found to be around 6% from at 0.8pu mechanical load. The mechanical load is then gradually decreased back to no load again while the speed of the motor gradually increases and eventually catches up to the reference value at no load condition. The speed response of the controller with the disturbance in load and the changes in the duty cycle (%) is summarized in Table 2, where the good results are stated.

Table 2. Speed response performance with variation of mechanical load with a constant speed.

Time Interval (sec)		Mechanical Load Variation (pu)	Steady State Error (%)	Overshoot (%)	Duty cycle (%)
From	To				
0	5	0	0.005	0.84	33.25
5	10	0.2	0.37	0.84	49.9
10	15	0.4	1.72	1.32	50
15	20	0.6	3.31	1.73	50
20	25	0.8	5.87	2.77	50
25	30	0.7	4.42	0.68	50
30	35	0.5	2.47	0.38	50
35	40	0.3	1.08	0.09	50
40	45	0	0.05	0.93	27.4

4. Conclusion

In this paper, a PWM based speed controller is utilized for TIM drive, where the duty cycle of PWM is determined by PI control algorithm to control the terminal voltage of the motor. The proposed controller is executed under different reference speed and load conditions. The obtained results show that the developed controller perform well on the TIM speed tracking under the conditions above. The simulation results also demonstrate that the speed response is also achieved with a reduced overshoot and state error as explained in Table1 and Table2. Thus, the robustness of the proposed controller would be very potential to improve the performance of TIM.

5. References

[1] R. G. M. Scholar, "Performance Analysis of Induction Motor For Sudden Load Disturbance Using PI And FUZZY Controller," no. June, pp. 6–9, 2015.

[2] J. Fonseca, J. L. Afonso, J. S. Martins, and C. Couto, "Fuzzy logic speed control of an induction motor," *Microprocess. Microsyst.*, vol. 22, no. 9, pp. 523–534, 1999.

[3] K. S. Rao and R. Mishra, "Comparative study of P, PI and PID controller for speed control of VSI-fed induction motor," *Int. J. Eng. Dev. Res.*, vol. 2, no. 2, pp. 2321–9939, 2014.

[4] X. Lu, S. Chen, C. Wu, and M. Li, "The Pulse Width Modulation and its Use in Induction Motor Speed Control," *2011 Fourth Int. Symp. Comput. Intell. Des.*, pp. 195–198, 2011.

[5] H. M. Kojabadi, "A comparative analysis of different pulse width modulation methods for low cost induction motor drives," *Energy Convers. Manag.*, vol. 52, no. 1, pp. 136–146, 2011.

[6] M. R. Khan and I. Town, "Duty cycle based volt/hertz inverter control," pp. 1717–1721, 1994.

[7] M. Fakhizan, M. F. Pesol, and K. N. Hasan, "PWM Technique to Control Speed of Induction Motor using Matlab / xPC Target Box," no. PECon 08, pp. 2–5, 2008.

[8] K. V. Kumar, P. A. Michael, J. P. John, and S. S. Kumar, "Simulation and Comparison of Spwm and Svpwm Control for Three Phase Inverter," *ARPN J. Eng. Appl. Sci.*, vol. 5, no. 7, pp. 61–74, 2010.

[9] K. Marouani, F. Khoucha, L. Baghli, D. Hadiouche, and A. Kheloui, "Study and harmonic analysis of SVPWM techniques for VSI-fed double-star induction motor drive," in *2007 Mediterranean Conference on Control and Automation, MED*, 2007.

[10] Y. Zhang and H. Yang, "Model predictive torque control of induction motor drives with optimal duty cycle control," *IEEE Trans. Power Electron.*, vol. 29, no. 12, pp. 6593–6603, 2014.

[11] A. M. Dakhil, "Minimization of Torque Ripple in DTC of Induction Motor Using Fuzzy Mode Duty Cycle Controller," pp. 237–244, 2010.

[12] Y. Zhang, W. Xie, Z. Li, and Y. Zhang, "Model predictive direct power control of a PWM rectifier with duty cycle optimization," *IEEE Trans. Power Electron.*, vol. 28, no. 11, pp. 5343–5351, 2013.

[13] Z. Nouman, B. Klima, and J. Knobloch, "Generating PWM Signals With Variable Duty From 0% to 100% Based FPGA SPARTAN3AN," *Elektrorevue*, vol. 4, no. 4, pp. 75–79, 2013.

[14] Y. Gangui, L. Wenhua, C. Yuanhua, W. Yucheng, and L. Junhui, "Duty-cycle-regulation based PWM control of five-level flying-capacitor inverter," in *Power Electronics and Motion Control Conference, 2004. IPEMC 2004. The 4th International*, 2004.

[15] K. H. Nam, *AC Motor Control and Electrical Vehicle Applications*. 2010.

K-Band Substrate Integrated Waveguide (SIW) Coupler

N Khalid[1], S Z Ibrahim[1] and W F Hoon[1]
[1]Advanced Communication Engineering Centre (ACE)
School of Computer and Communication Engineering,
Universiti Malaysia Perlis,
Arau, Perlis, Malaysia

nurehansafwanah@studentmail.unimap.edu.my, sitizuraidah@unimap.edu.my,
fhwee@unimap.edu.my

Abstract. This paper presents a designed coupler by using substrate Roger RO4003. The four port network coupler operates at (18-26 GHz) and designed by using substrate integrated waveguide (SIW) method. Substrate Integrated Waveguide (SIW) are high performance broadband interconnects with excellent immunity to electromagnetic interference and suitable in microwave and millimetre-wave electronics applications, as well as wideband systems. The designs of the coupler are investigated using CST Microwave Studio simulation tool. These proposed couplers are capable of covering the frequency range and provide better performance of scattering parameter (S-parameter). This technology is successfully approached for millimetre-wave and microwave applications. Designs and results are presented and discussed in this paper. The overall simulated percentage bandwidth of the proposed coupler is covered from 18 to 26 GHz with percentage bandwidth of 36.36%.

Index terms: *coupler, substrate integrated waveguide (SIW), using CST Microwave Studio simulation tool, millimetre-wave*

1. Introduction

Coupler is a passive device which couples for generating the transmission power with properties requirements such as operational bandwidth, frequencies, and size [1]. Communication services and wireless systems boost up with the presence of wideband coupler which is used to isolate the receiver port from the transmitters. The isolation in coupler is significant since poor isolation will result in reduced receiver sensitivity and higher DC offset in the system [1]. There are four available types of coupler which are directional, hybrid, rat-race and coupled line coupler. Each type of them exhibits different advantage [2]. In its basic configuration, a conventional coupler is formed by implementing multi-section approach. However, the conventional type has a narrow bandwidth which is lower than 25%, beside size of the coupler is large as well.

Recently, Substrate Integrated Waveguide (SIW) is a technology of an emerging method for the implementation of millimeter-wave and microwave field [4] such as the wireless base-stations satellites and the earth stations. These technologies consists of two layers of metal surfaces that covered both layers of substrate and the metallic via that connect two ground planes of substrate. Substrate Integrated Waveguide (SIW) is the highest frequency that can come up with very good frequency performance on low dielectric permittivity substrate materials [5]. This technology had already proven that structure of Substrate Integrated Waveguide relatively obtained less loss, high Q-factor waveguide in compact size and highly integrated structural [6-7]. The compact structure could reduce space and metal loss. There are many components that applied Substrate Integrated Substrate technique such as power divider [8], filters [7–10], antennas [7–9], and couplers [16]. In this paper, there are one (single-layer) coupler that proposed by using Printed Circuit Board (PCB) substrate materials which to evaluate the performance of coupler in terms of S-parameter in new square metallic via.

2. Design Descriptions

2.1 Configuration of coupler

The basic Structure for the coupler is shown in Figure 1. A directional coupler is a four-port network that is used to divide and distribute power. Moreover, coupler may have 3-port or 4-port components which denoted as input, through (transmitted), coupled and isolated [2]. The mechanisms of coupling are divided into two which depends on the interspacing between them, and the coupled line [3]. Directional coupler defines to operate an input so that the other output signals are functioning [1]. The basic concept of directional coupler is that the input power to port 1 (input) is coupling to port 3 (coupled) , meanwhile the remaining input power is to port 2 (through) and port 4 (isolated) which is where no power is delivered [1]. All port are matched there which means that the percentage of power entering Input port (port 1) is 100%. The 100% of power is equally divided into two, which is then transmitted to port 2 and port 3.

.

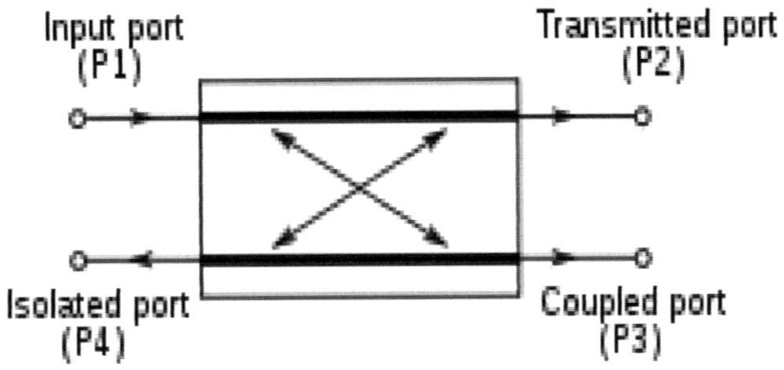

Figure 1. Basic Structure of Coupler

From Figure 1, The coupling value for S1,1 and S3,1=-3dB as shown in equation (1) below.

$$S_{21} = S_{31} = 10Log\frac{50}{100} = -3dB \tag{1}$$

The coupler is designed to have a good matching and isolation such that reflection and isolation is set to be less than 10% by using expression (2) below, so that all input power will be transmitted to the desired port only.

$$S_{11} = S_{41} = 10Log\frac{10}{100} = -10dB \tag{2}$$

2.2 Configuration of SIW Coupler

The basic configuration of K-Band Substrate Integrated Waveguide (SIW) coupler is shown in Figure 2 with parameters.

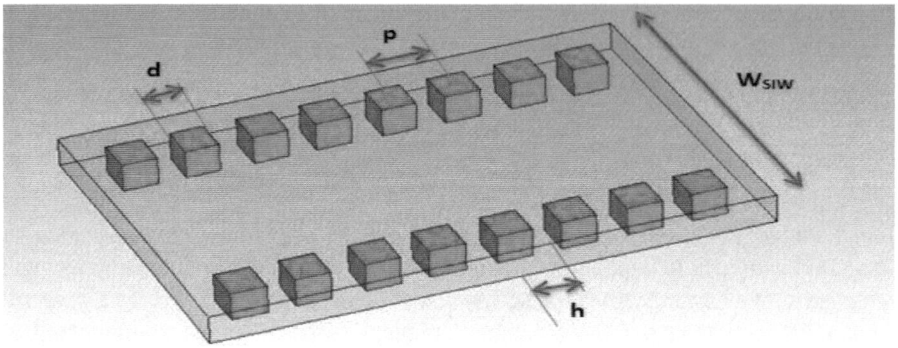

Figure 2. Basic Structure of Substrate Integrated Waveguide (SIW) [4-16]

Figure 2 shows the physical parameters that are important for designing the diameter of metal square holes (d) and the distance between adjacent via holes (p). Based on the Figure 1, distance between two rows of square holes (W_{siw}) and the thickness of substrate (h) are calculated by using equation [19].

$$Weq = Wsiw - \frac{d^2}{0.95p} \tag{3}$$

$$p < \frac{\lambda_0}{2}\sqrt{\varepsilon_r} \tag{4}$$

$$p < 4d \tag{5}$$

$$\lambda_0 = \frac{c}{f} \tag{6}$$

Where, λ_0 is the space wavelength.

The spacing between adjacent via holes (p) must be small to reduce the loss between metallic holes.

2.3 Microstrip Tapered Design

The microstrip is interconnected with the transmission lines by using tapered microstrip. By using tapered Transition the Substrate Integrated Waveguide (SIW) is to match the 50 Ω microstrip lines. The parameters involved are length of taper (L) and width of taper (W) [17–19]. Figure 3 shows the microstrip taper and the line model.

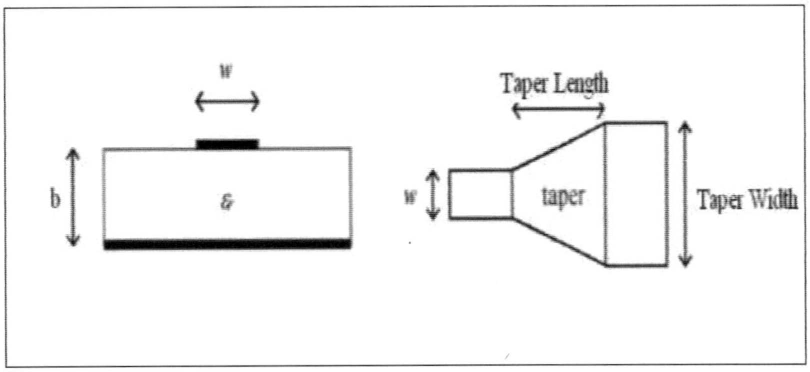

Figure 3. Microstrip taper and line model

Microstrip to waveguide transition is a tapered microstrip line connecting microstrip of width **'w'** to waveguide of width **'Ag'** as shown in Figure 3. Taper width **'Wt'** and taper length **'Lt'** are other two essential parameters for microstrip taper design. **'Wt'** and **'Lt'** can be determined by the relation given below from [20-21]:

$$\frac{W_t}{A_g} \approx 0.4 \tag{7}$$

$$(\frac{\lambda_g}{2}) < L_t < \lambda_g \tag{8}$$

Where,

$$\lambda_g = \lambda_c / \sqrt{\varepsilon_r} \tag{9}$$

$$L_t = \frac{n\lambda_g}{2}, n = 1,2,3,4... \tag{10}$$

3. Parameter Design of Substrate Integrated Waveguide (SIW) Coupler

The couplers are known as passive devices in microwave system which is applied for routing, dividing and combining the signals. By using equation from (4) and (5) the value of metallic via diameter and the distance between adjacent via were calculated. This coupler used a type of substrate Roger RO4003 with dielectric permittivity of substrate (ε_r) of 3.55, and thickness of 0.508 mm. The copper thickness of 0.035 mm with loss tangent of 0.0027 which is small means small signal loss making the substrate suitable for coupler design. Table 1 shows the values of parameters involved for the designed coupler. The microstrip transition is represented using tapered transition as shown in Figure 4.

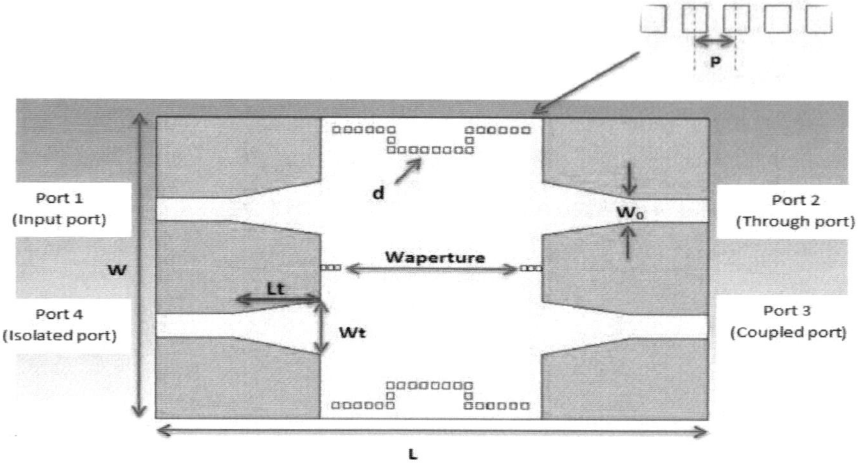

Figure 4. The configuration of coupler

Table I
The parameters for Substrate Integrated Waveguide (SIW) coupler

Parameters	Descriptions	Dimensions (mm)
L	Length of R04003 Substrate	25.00
W	Width of R04003 Substrate	15.00
d 1	Diameter of side wall square holes	0.40
d 2	Diameter of centre square holes	0.25
p	Distance between square holes	0.60
Wt	Width transition taper	2.63
Lt	Length transition taper	4.00
$W_{aperture}$	Width of aperture	8.52
W_{siw}	Width of siw	6.58
W_o	Width of microstrip TL	1.15

4. Results And Discussion

This section paper involves the analysis of coupler performance in terms of S-parameters such as reflected coefficient, isolation coefficient, bandwidth, E-field distribution and phase difference between S_{21} and S_{31} of the coupler. The K-Band Substrate Integrated Waveguide coupler is simulated by using CST software. The frequency obtained is from 18 GHz up to 26 GHz.

4.1 Simulation Result of S-Parameters vs Frequency

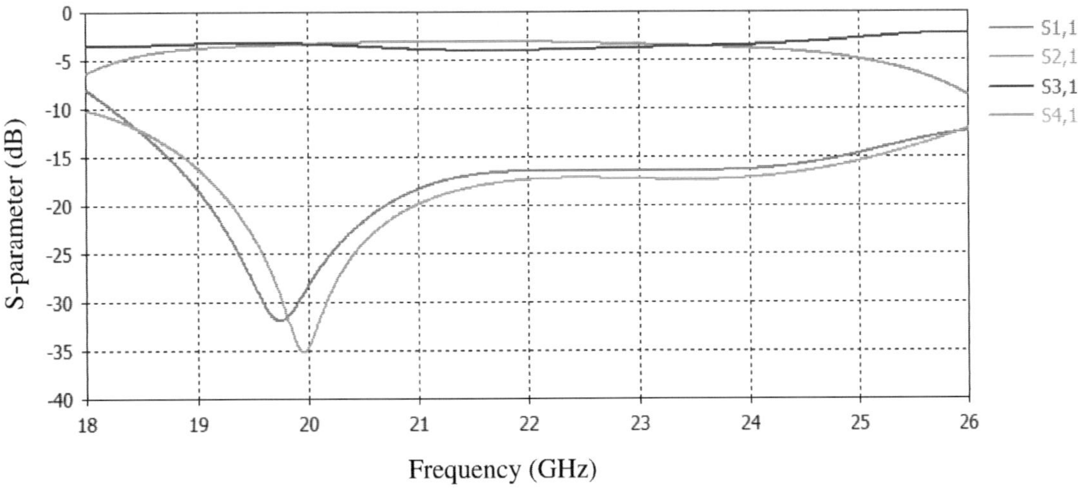

Figure 5. Simulated Result of SIW coupler

From Figure 5 the result analysis it shows the input port with reflection coefficient (S_{11}) of -28.13 dB. The isolation coefficient (S_{41}) value of simulation is -34.51 dB at frequency of 20 GHz. The output coupling port 2 and port 3 for insertion loss (S_{21}) and Coupling Coefficient (S_{31}) is are approximately -3.36 dB and -3.35 dB with the operating frequency of 18 GHz to 26 GHz. Therefore, S_{21} and S_{31} are in the range of best requirement results which are -3 ± 1.5 dB meanwhile S_{11} and S_{41} are less than -10 dB. The reflection and isolation are covering more than 36.36% of the bandwidth. By referring to Figure 5 based on this investigation of designed coupler, it is concluded that the coupling between port 1 and port 3 is affected by via and $W_{aperture}$.

4.2 E-Field Distribution of the Substrate Integrated Waveguide (SIW) coupler

Figure 6 shows the power that travel through the whole part of the coupler. The power flows from port 1 (input port) and coupled to port 3 (coupled port). The other input power goes to port 2 (through port). The design is in ideal condition because there are small amount of power that passed through ~~of~~ port 4 (isolated port).

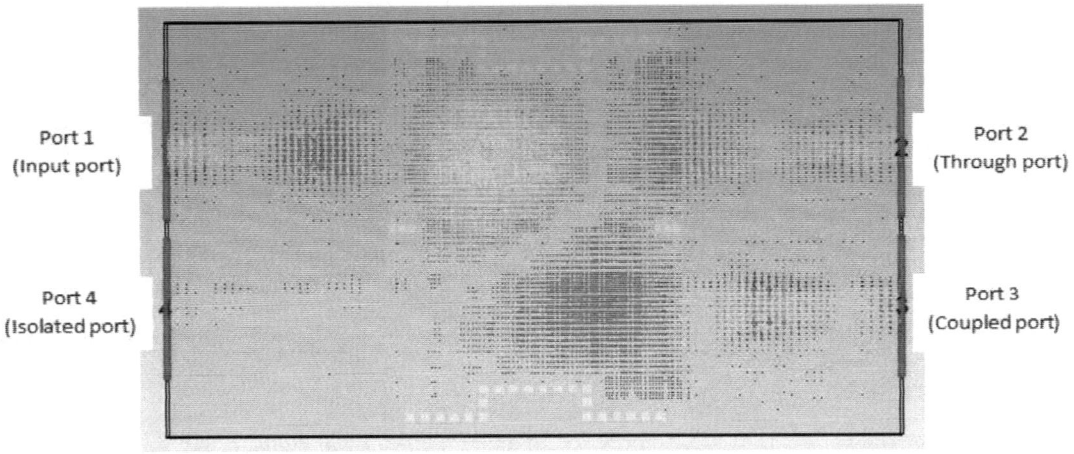

Figure 6. Electric field of SIW coupler

4.3 Simulated Phase of coupled between S_{21} and S_{31}

Figure 7 shows the graph of phase difference between port 2 and port 3 are calculated. As the simulation results, the phase difference is nearly 90 degree within frequency (19-25 GHz).

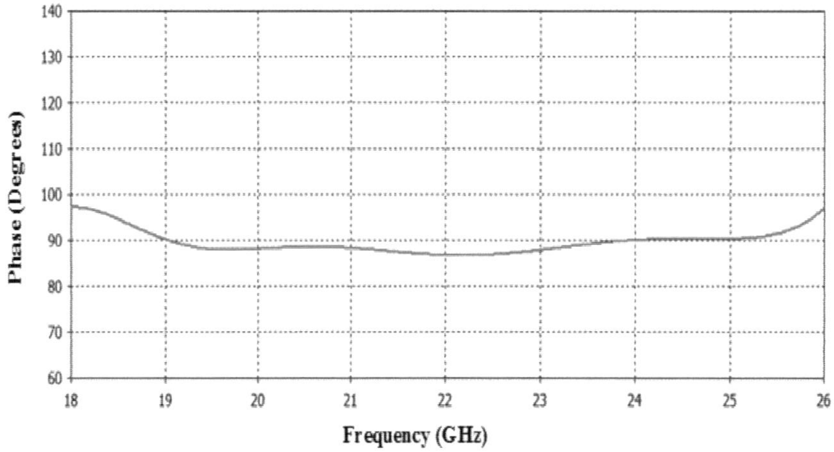

Figure 7. Phase Difference of SIW coupler

5. Conclusion

In this design, the Substrate Integrated Waveguide (SIW) coupler which based on the coupling theory for K-Band applications is analysed. The coupler is designed by using the Computer Simulation Software (CST) 2014 and material substrate Roger 4003C is used. The percentage bandwidth is more than 36.36%. S_{21} and S_{31} achieve the range of the requirement results which are -3 ± 1.5 dB meanwhile S_{11} and S_{41} obtain less than -10 dB. This designed coupler relatively has loss that can be improved by increasing the value of metal and substrate thickness. Coupling will be reduced as $W_{aperture}$ is decreased. Therefore, the simulation results show that the good performance of this designed coupler conceived in 18-26 GHz.

References

[1] B. H. Ahmad, S. S. Sabri and A. R. Othman, *Design of a Compact X-Band Substrate Integrated Waveguide Directional Coupler*, vol. **5**, no. 2, pp. 1905–1911, 2013.
[2] D. Pozar, *Microwave Engineering Fourth Edition*. 2012.
[3] Z. S. Tabatabaeian, *Design Investigation of a Wideband Backward-Wave Directional Coupler Using Neural Network*, pp. 1394–1398, 2016.
[4] A. Kumar, G. Saini, S. Singh, A. Kumar, G. Saini and S. Singh, *Electrical & Electronic Engineering | Review Article A Review On Future Planar Transmission Line Electrical & Electronic Engineering | Review Article*, *Cogent Eng.*, vol. **48**, no. 1, pp. 1–12, 2016.
[5] D. Feature, F*abricate C-Band with Rectangular SIW*, no. May, 2016.
[6] Z. Kordiboroujeni and J. Bornemann, *Designing the width of substrate integrated waveguide structures*, *IEEE Microw. Wirel. Components Lett.*, vol. **23**, no. 10, pp. 518–520, 2013.
[7] Z. Kordiboroujeni, J. Bornemann and P. P. M. So, *Substrate integrated waveguide mode coupler for tracking applications*, *IEEE MTT-S Int. Microw. Symp. Dig.*, vol. **2016–Augus**, pp. 15–17, 2016.
[8] N. A. Smith, *Substrate Integrated Waveguide Circuits and Systems,* no. May, 2010.
[9] F. Qiu, J. Huang, Z. Tang, D. Lei and M. Yao, *Compact Dual-Band Filter Using Miltilayer Dual- Mode Substrate Integrated Waveguide Resonator*, pp. 2–4, 2016.
[10] K. Dong, J. Mo, Y. He, Z. Ma and X. Yang, *Design of a millimeter-wave dual-band bandpass*

filter using SIW dual-mode cavities, 2016 IEEE MTT-S Int. Wirel. Symp. IWS 2016, pp. 2–4, 2016.

[11] Z. Q. Xu, Y. Shi, J. X. Liao and P. Wang, *Substrate integrated waveguide filter with mixed coupled modified trisections*, *Electron. Lett.*, vol. **49**, no. 7, pp. 482–483, 2013.

[12] E. Alfonso, A. U. Zaman, E. Pucci and P.-S. Kildal, *Gap waveguide components for millimetre-wave systems: Couplers, filters, antennas, MMIC packaging, Antennas Propag. (ISAP), 2012 Int. Symp.*, pp. 243–246, 2012.

[13] Z. Dqg, S. Lowhu, D. Rq, M. F. Vhx and H. G. X. Fq, *Millimeter-wave Band Pass Filter Based on Complementary Split Rings and Substrate Integrated Waveguide (SIW) Resonator,* pp. 6–8.

[14] D. Y. Kim, Y. Lim and S. Nam, *Millimeter-wave slot array antenna using SIW and electroforming techniques*, *Asia-Pacific Microw. Conf. Proceedings, APMC*, vol. **3**, pp. 4–6, 2016.

[15] I. Symposium and M. Mashhad, *Development of a Wideband and Inclined Polarized Cavity Backed Slot Antenna Using SIW Structure*, pp. 452–455, 2016.

[16] Z. C. Hao, W. Hong, J. X. Chen, H. X. Zhou and K. Wu, *Single-layer substrate integrated waveguide directional couplers*, vol. **153**, no. 5, 2006.

[17] B. Rahali and M. Feham, *Substrate Integrated Waveguide Power Divider , Circulator And Coupler In [10-15] Ghz Band*, vol. **4,** no. 1, pp. 1–11, 2014.

[18] B. Rahali and M. Feham, *Divider , Circulator and Coupler in [10-15].*

[19] M. Bozzi, A. Georgiadis and K. Wu, *Review of substrate-integrated waveguide circuits and antennas," IET Microwaves, Antennas Propag.* vol. **5**, no. 8, pp. 909–920, 2011.

[20] M. Esquius-Morote, B. Fuchs, J. F. Zurcher, and J. R. Mosig, "A printed transition for matching improvement of SIW horn antennas," *IEEE Trans. Antennas Propag.*, vol. 61, no. 4, pp. 1923–1930, 2013.

[21] F. Taringou, D. Dousset, J. Bornemann and K. Wu, *Broadband CPW feed for millimeter-wave SIW-based antipodal linearly tapered slot antennas*, *IEEE Trans. Antennas Propag.*, vol. **61**, no. 4, pp. 1756–1762, 2013.

[22] J. P. Stang and W. T. Joines, *A Tapered Microstrip Patch Antenna Array for Microwave Breast Imaging*, *IEEE MTT-S Int. Microw. Symp. Dig.*, pp. 1313–1316, 2008.

[23] A. Suntives, *High-Speed Data Transmission Using Substrate Integrated Waveguide-Type Interconnects*, no. February, 2009.

[24] S. S. Sabri, B. H. Ahmad, A. R. Othman and A. Suntives, *High-Speed Data Transmission Using Substrate Integrated Waveguide-Type Interconnects,* no. February, pp. 264–268, 2013.

IOP Conf. Series: Materials Science and Engineering **341** (2018) 012015 doi:10.1088/1757-899X/341/1/012015

Design of 5.8 GHz Integrated Antenna on 180nm Complementary Metal Oxide Semiconductor (CMOS) Technology

A H A Razak, M I A Shamsuddin, M F M Idros, A K Halim, A Ahmad and S A M Al Junid

Electronics Architecture and Application (EArA) Research Group, Faculty of Electrical Engineering (FKE), Universiti Teknologi MARA, 40450 Shah Alam, Selangor

Email: hadi@ieee.org

Abstract. This project discusses the design and simulation performances of integrated loop antenna. Antenna is one of the main parts in any wireless radio frequency integrated circuit (RFIC). Naturally, antenna is the bulk in any RFIC design. Thus, this project aims to implement an integrated antenna on a single chip making the end product more compact. This project targets 5.8 GHz as the operating frequency of the integrated antenna for a transceiver module based on Silterra CMOS 180nm technology. The simulation of the antenna was done by using High Frequency Structure Simulator (HFSS). This software is industrial standard software that been used to simulate all electromagnetic effect including antenna simulation. This software has ability to simulate frequency at range of 100 MHz to 4 THz. The simulation set up in 3 dimension structure with driven terminal. The designed antenna has 1400um of diameter and placed on top metal layer. Loop configuration of the antenna has been chosen as the antenna design. From the configuration, it is able to make the chip more compact. The simulation shows that the antenna has single frequency band at center frequency 5.8 GHz with -48.93dB. The antenna radiation patterns shows, the antenna radiate at omnidirectional. From the simulation result, it could be concluded that the antenna have a good radiation pattern and propagation for wireless communication.

1. Introduction

System-on-Chip (SoC) is an implementation of a system into a chip. It is integrating all components of a computer or electronic system into a single chip. This makes the chip have compact design. Thus, it will reduce the size of the device to become smaller and the device shall able to attach to living things.

The aim of this project is to design an integrated antenna which will allow human to wear in their daily life. The size of the antenna has been set less then 2mm. The guideline of antenna sizing is based on the size of a die and Silterra CMOS 180nm technology. Loop antenna configuration has been chosen as antenna configuration. The configuration criteria are based on integrated circuit level component such as inductor and capacitor. The component structure may need to use same layer structure as antenna. The design geometry is revealed in figure 1.

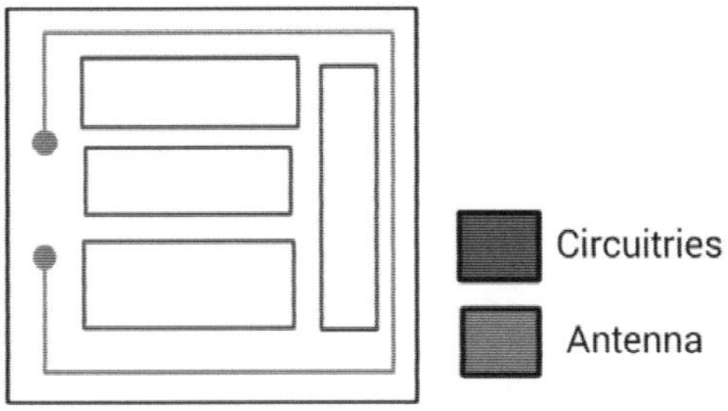

Figure 1. Top view of loop antenna for integration with active devices.

This project designed a 5.8 GHz loop configuration antenna on chip antenna. The frequency of the antenna been chosen based on unlicensed Industrial, Scientific and Medical Band (ISM Band). Below shows ISM band table 1:

Table 1. Range of frequency of ISM Band and it availability [1].

Frequency Range (GHz)		Bandwidth (MHz)	Centre Frequency (GHz)	Availability
2.400	2.500	100	2.450	Worldwide
5.725	5.875	150	5.800	Worldwide
24.00	24.25	250	24.13	Worldwide
61.00	61.50	500	61.25	Subject to local acceptance

The simulation of the antenna was done by using High Frequency Structure Simulator (HFSS). This software is industrial standard software that been used to simulate all electromagnetic effect including antenna simulation. This software has ability to simulate frequency at range of 100 MHz to 4 THz. The simulation set up in 3 dimension structure with driven terminal.

1.1. Theory of antenna

The antenna is the main part of any wireless device, where without a proper design of the antenna, it will affect the transmitted and received data. Antenna is a device to convert an electric signal into a radio signal. The antenna will radiate energy from the signal into electromagnetic waves through the air. In this project, loop antenna will be used as the antenna design. Advantages of using loop antenna design; is that it has the same characteristics as dipole and monopole antenna. It is a simple design and hence less hassle for the designer with the design complexity. It also has a variety of shapes; circular, rectangular and elliptical. Frequency and wavelength can be described as below:

$$\lambda = \frac{c}{f} \tag{1}$$

Where
λ = wavelength of frequency
c = constant speed of light
f = frequency

From the equation (1), the wavelength is inversely proportional to frequency. The highest value of frequency, the smallest wavelength will be produce. Based on the wavelength of the frequency, the size of the antenna will be affected. The size of the antenna is directly proportional to wavelength. Size of full wave antenna is $\lambda/4$.

1.2. Substrate layer of antenna

Substrate layer of antenna is a very important part in designing antenna. Different dielectric material will form different frequency and radiation pattern for the antenna. Therefore, to design on-chip antenna it is a very important step to create layer of the die based on manufactory specification or technology. By follow the manufacturer specification, the output result of the antenna will produce approximate value for the practical design. Below shows the diagram of substrate layer of die and table of layer parameters:

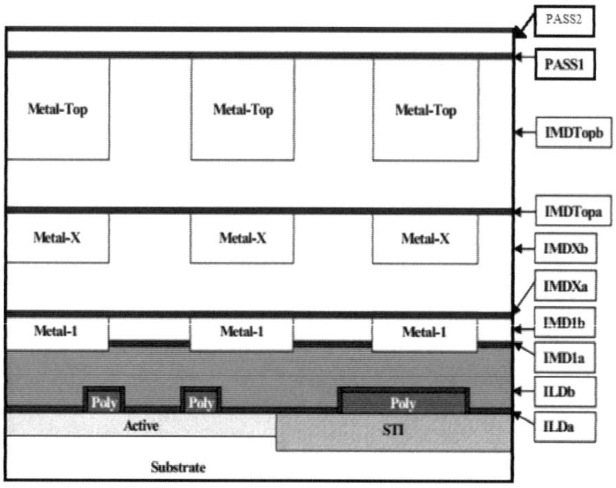

Figure 2. Top view of loop antenna with dimension [2].

Table 2. Substrate layer of CMOS018 Technology [2].

Layer	Thickness (um)	Dielectric Constant
STI	0.309	4.2
OD	0.1079	4.2
POLY	0.2	0
ILDA	0.06	7.5
ILDB	0.6	4.2
METAL1	0.53	0
IMD1a	0.67	3.7
IMD1b	0.18	4.2
METAL2	0.53	0
IMD2a	0.67	3.7
IMD2b	0.18	4.2
METAL3	0.53	0
IMD3a	0.67	3.7
IMD3b	0.18	4.2
METAL4	0.53	0
IMD4a	0.67	3.7
IMD4b	0.18	4.2
METAL5	0.53	0
IMDTOPa	0.55	3.7
IMDTOPb	0.3	4.2
METAL6	2.39	0
PASS1	0.9	4.2
PASS2	0.5	7.5
AIR	40	1

IOP Conf. Series: Materials Science and Engineering **341** (2018) 012015 doi:10.1088/1757-899X/341/1/012015

2. Methodology

In this section, it shows 2 main set up of project. First section it shows the dimension of antenna and lastly simulation set up.

2.1. Dimension of antenna

Right dimension of antenna may produce a good result. The good result is referring to centre frequency of the antenna, radiation pattern of antenna and etc. Below shows dimension of the antenna in micron:

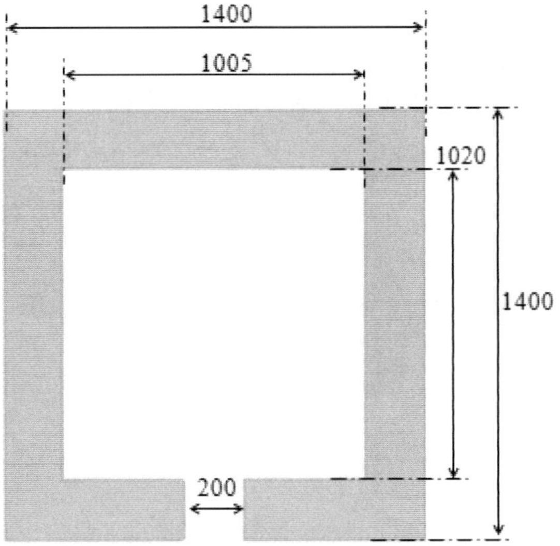

Figure 3. Top view of loop antenna with dimension.

2.2. Simulation set up

In this section shows simulation set up for this project. There are two set up that have been implemented. The set up are terminal driven of the antenna and simulation setup. Below shows the driven terminal set up:

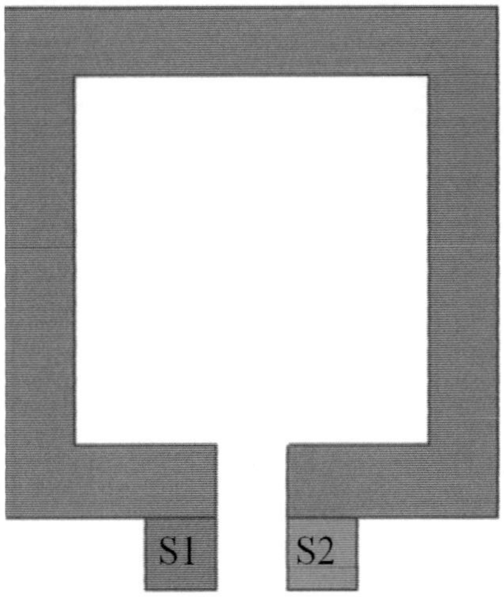

Figure 4. Top view of loop antenna with driven terminal.

IOP Conf. Series: Materials Science and Engineering **341** (2018) 012015 doi:10.1088/1757-899X/341/1/012015

From the figure above, the antenna has been set up as 2 port network system. The simulation had been set up as interpolating frequency sweep from 0.1GHz to 10GHz with 0.1GHz setup. This procedure is done to allocate the frequency range of the antenna.

3. Result and Discussion

Based on simulation set up above, result and findings obtained as shown in figure 5.

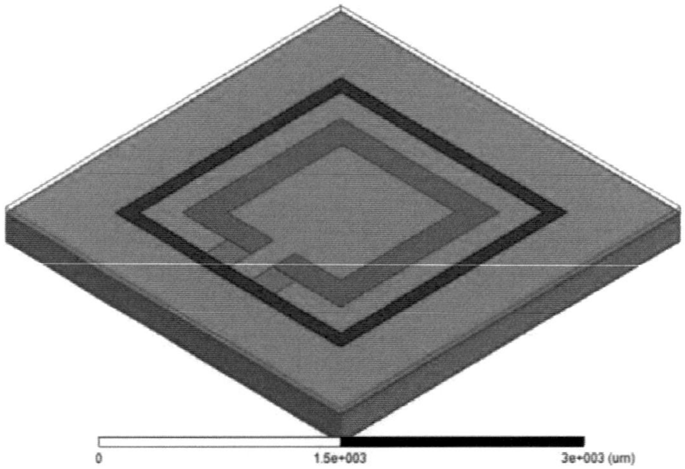

Figure 5. 3D of Integrated Antenna Layout in HFSS.

From the figure above shows antenna model set up in HFSS. The outer ring in the figure is a guide for 2mm size of a die. All result of the simulation for the antenna as shown in figure 6, below:

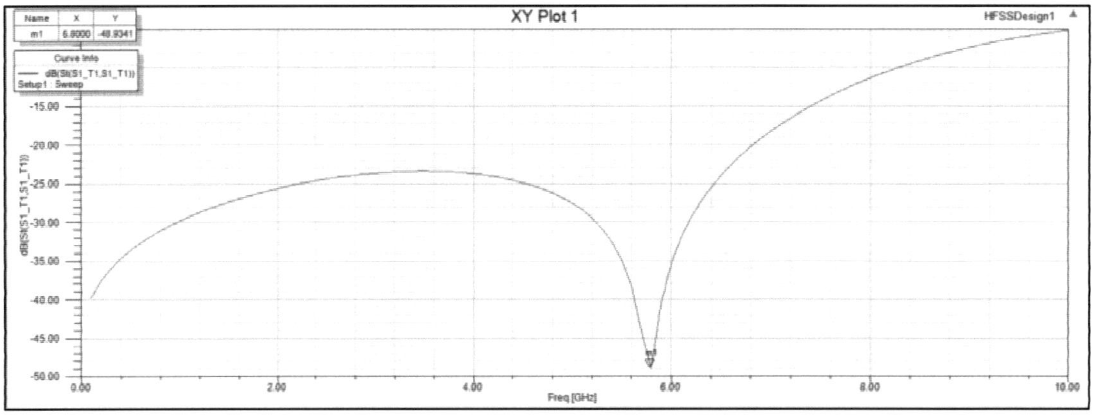

Figure 6. S11 plot Graph of the simulation.

From the figure 6 above shows the frequency drop of the antenna at 5.8 GHz with -48.93dB. From the result the antenna has met the targeted value. Figure 7 shows a graph of total gain of the antenna and radiation pattern:

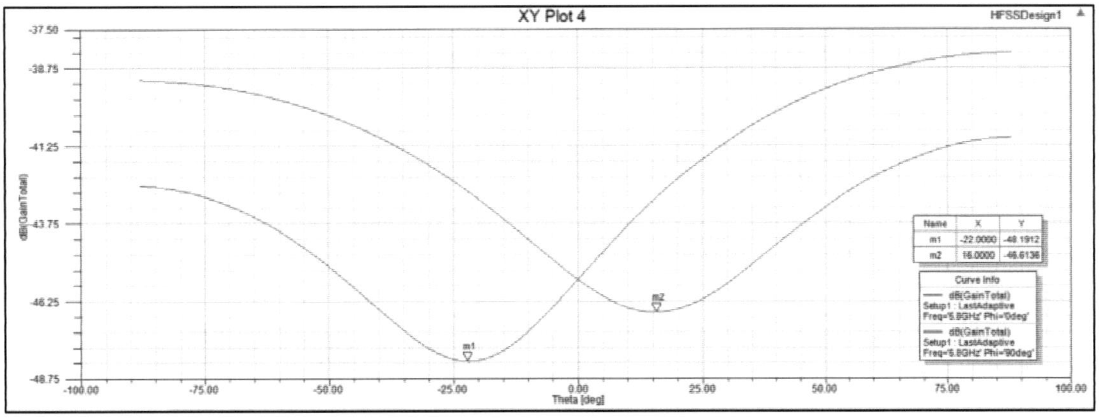

Figure 7. Total Gain of the antenna.

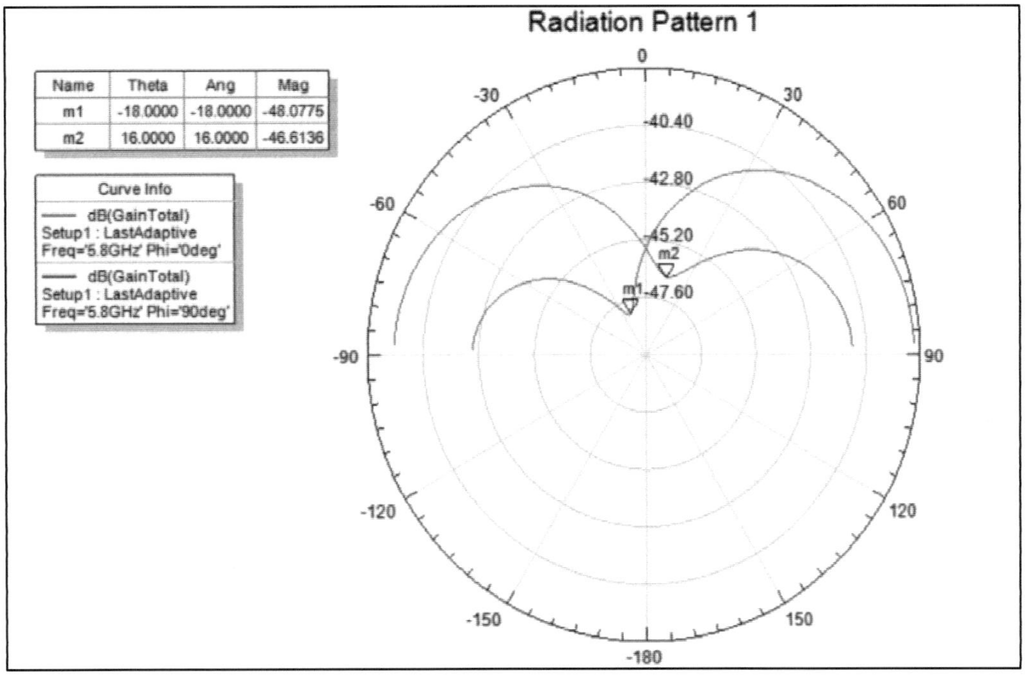

Figure 8. Radiation Pattern.

Figure 7 above shows the total gain of the antenna at phi equal to 0° is -48.19dB, while at phi equal to 90°the value drop at -46.61dB. The value shows different of -1.58dB at phi equal to 0° and 90°. For the radiation pattern graph in figure 8 shows two different degrees of phi 0° and 90° and the lowest magnitude is -48.08 at angle -18.

4. Conclusion

The project presents a design of On-Chip Integrated Antennas for Biomedical Application. Designing integrated antenna into die may reduce the size of device and the size of integrated circuits may have more compact design. The simulation shows that the antenna has single frequency band at center frequency 5.8 GHz with -48.93dB. The antenna radiation patterns shows, the antenna radiate at omnidirectional. From the simulation result, it could be concluded that the antenna has a good radiation pattern and propagation for wireless communication.

5. Acknowledgment

Appreciation to the Universiti Teknologi MARA (UiTM) that provide the academic research training and also Ministry of education for supporting this research through Research Grant Scheme (RAGS) [File No: 600-RMI/RAGS 5/3 (195/2014)]. Not forgotten our thanks to Silterra Technology for access to their CMOS technology and CEDEC that provide a great training session.

References

[1] General Secretariat of ITU, *Radio Regulation*, International Telecommunication Union, Edition of 2012.
[2] Jazril bin Jamil Din and S.Madhanaraj, *CMOS18 Electrical Design Rules*, Silterra Kulim High Tech, Malaysia Edition of 2012.
[3] A. Shamin, *24 GHz Integrated Differential Antenna in Digital Bulk Silicon*, Carleton University, Ottawa, Canada June 2004.
[4] Zhai Mingwei, *Development of Integrated Multi-band Antenna for GSM and GPS Applications*, Nayang Technological University, Nanyang Avenue, Singapore
[5] Ansoft Corporation, *User's Guide – High Frequency Structure Simulator*, Edition of 2005.
[6] Constantine A. Balanis, *Antenna Theory Analysis and Design*, 3rd Edition.
[7] R. Hasse, W.Hunsicker and K. Naishadham, *Analysis and Design of a Partitioned Circular Loop Antenna for Omni-directional Radiation*, Georgia Tech Research Institude, Georgia Institute of technology, Atlanta, Georgia
[8] T. Sasamori, T. Tobana and Y. Isota, *Input Impedance Measurement for Balanced Antenna by S-parameter Method*, Akita Prefectual University, Japan.
[9] M. Kozlov and R. Turner, *A Comparison of Ansoft HFSS and CST Microwave Studio Simulation Software for Multi-channel Coil Design and SAR Estimation at 7T MRI*, Max Planck Institute for Human Cognitive and Brain Sciences, Germany.
[10] Minh Thuy Le, *Design of high gain antenna at 5.8GHz Using Metamaterials Structure*, IMEP-LAHC, Grenoble Inst. of Technol. (GINP), Grenoble, France

IOP Conf. Series: Materials Science and Engineering **341** (2018) 012016 doi:10.1088/1757-899X/341/1/012016

Emotional State Classification in Virtual Reality Using Wearable Electroencephalography

N S Suhaimi[1], J Teo[2], J Mountstephens[3]

[1]Graduate Researcher, Evolutionary Computing Laboratory, Faculty of Computing and Informatics, Universiti Malaysia Sabah, Jalan UMS, 88400, Kota Kinabalu, Sabah, Malaysia

[2]Associate Professor, Faculty of Computing and Informatics, Universiti Malaysia Sabah, Jalan UMS, 88400, Kota Kinabalu, Sabah, Malaysia

[3]Senior Lecturer, Faculty of Engineering, Universiti Malaysia Sabah, Jalan UMS, 88400, Kota Kinabalu, Sabah, Malaysia.

E-Mail: nazmisofian@gmail.com

Abstract. This paper presents the classification of emotions on EEG signals. One of the key issues in this research is the lack of mental classification using VR as the medium to stimulate emotion. The approach towards this research is by using K-nearest neighbor (KNN) and Support Vector Machine (SVM). Firstly, each of the participant will be required to wear the EEG headset and recording their brainwaves when they are immersed inside the VR. The data points are then marked if they showed any physical signs of emotion or by observing the brainwave pattern. Secondly, the data will then be tested and trained with KNN and SVM algorithms. The accuracy achieved from both methods were approximately 82% throughout the brainwave spectrum (α, β, γ, δ, θ). These methods showed promising results and will be further enhanced using other machine learning approaches in VR stimulus.

1. Introduction

Virtual Reality (VR) has been on the rise for every industry especially in the smartphone industry where they want to capitalize this feature. Oculus, HTC Vive and Samsung Gear are some of the notable industries competing in the VR field. Some of the notable features of using VR from applications are such as gaming and 3360 videos that allows people to see and observe what other people have recorded throughout the journey and their surrounding environments.

In many of the demos from the companies that have conducted with VR, the participants would be immersed into the VR and demonstrated body movements and emotions that showed signs of fear such as riding a roller coaster and losing their sense of ground due to cognitive manipulation. However, there were not any application that have the adaptivity to capitalize the mental state of the user that could help the user which would then improve the user's experience inside the VR world.

The mental state of the user which is an ongoing electrical activity of the brain, particularly known as electroencephalography (EEG) which shows the ongoing sensory such as motor movements, emotions and memories. This would provide an indication of the level of engagement for the user immersing into the VR which could be measured and quantified.

There are numerous researchers which have conducted in neuroinformatic to obtain the brainwave signals from humans. However, none has ever approached with the method of classifying mental state of users using VR as the stimulus. Technology has evolved throughout the years and EEG equipment have scaled down and are easily accessible by public to obtain.

In this research, we will be using the Muse headband which was developed by Interaxon for recording of the brainwave activity happening within the user's brain when they are immersed into the VR. The approach for this research will require the user to not have any motion sickness and impaired eyes. With the data obtained from the brainwave, the data will then be filtered with KNN and SVM obtained from the brainwave such as muscle movements and eye twitches using R language data analysis. They will then be tested and trained with Machine Learning Language to identify emotional brainwave patterns to allow adaptive interaction between VR and the current user's mental state.

This review paper will be divided into 5 sections. Firstly, the background and related work for this research paper will be briefed. Secondly, the methodology of this research. Thirdly, the results and data analysis. Fourth, the analysis of the results will be discussed. Lastly, the conclusion and future work will be presented in this section.

2. Background and Related Work

In this section, we will brief through the method and the approach of this research.

2.1. Virtual Reality Immersion

A software was developed and used to obtain the current brain's state of concentration or relaxation in real time [1]. The software uses a plane to be controlled by the user and will move according to the mental state whereby if the user concentrates, the plane will move up, relaxation will move the plane downwards. There were also other simulation such as a driving simulation where the approach of this was to observe the mental state of the user such as boredom and fatigue if they were given a repetitive tasks [2]. Lastly, a group of participants were tasked to solve problems given to them and to differentiate each task for their level of difficulty [3].

2.2. Mental State Classification

Researchers had conducted different approaches to classify the mental state of the brain signals. Methods include electroencephalogram (EOG), electrocorticography (ECoG) and functional magnetic resonance (fMRI). However, electroencephalography (EEG) has a larger network of information stored within the brainwave which could provide responses towards the emotional state of the user [4]. Affective Computing (AC) was developed to identify and classify the emotional state of the user by measuring the behavioural and physiological signals [5].

2.3. KNN and SVM

There are several researches conducting data analysis on EEG with some of the commonly used methods such as K-nearest Neighbor (KNN) and Support Vector Machine (SVM). In some cases, researchers [6, 7, 8, 9] reported that they achieved an average accuracy of 70% and above using the mentioned methods to classify their user's mental state.

2.4. EEG
Most EEG devices that was used in the process of obtaining the brainwave signal were mostly laboratory grade hardware and were very costly in obtaining these devices. These devices are mostly kept by organizations which were difficult to get an appointment and the devices are not portable to be brought out from the premise for experimentation purposes. Therefore, there was a need to approach this problem and Interaxon has developed a portable and small headband known as Muse for consumers to obtain. Most researchers [1, 10, 11, 12, 13] used this convenient headband to obtain their brainwave signals for their research thanks to its low-cost production even though it suffers from the lower quality of signal acquisition.

3. Research Methodology
Ten (2 Female, 8 Male) participants were recruited from various backgrounds aged between 22 through 65 who were residing in Kota Kinabalu, Sabah, Malaysia. The participants had no prior brain damage or mental disorder with little to no motion sickness. Each participant were asked to wear a VR Box headset together with a pair of earphones and a smartphone device along with the Muse Headband. The participant were then asked to play a VR app that was download from the Google App Store.

The participants were asked to explore the VR by observing the surrounding areas of the VR. During the exploration inside the VR, any external stimulus was taken with great care from interfering the participants unless the participant showed any signs of mental fatigue that would be deemed dangerous to continue. The EEG readings were recorded from the Muse headband with the assistance of a third-party app known as Muse Monitor obtained from Google App Store developed by James Clutterback. The positions of where the Muse will collect the data are indicated with the head points from TP9, AF7, AF8, TP10 and reference point at FPz.

The collected data will then be analyzed using both KNN and SVM methods to classify the mental state of the participants and train the models to be able to identify and recognize the brainwave patterns and evaluate its performance on the analysis.

4. Result and Data Analysis
The format of the data recorded was using comma-separated-value (CSV) for use in R programming. The recorded data includes absolute values from different bands with different electrodes performing Fast-Fourier Transform (FFT) from the raw data obtained, raw signal from different electrodes and the horse-shoe indicator (the sensor connectivity to the human skin from the electrode) with marker buttons. The table displaying the data that was obtained from the Muse Monitor application is shown in table 4.0. The original signal from the RAW brainwave of the participant is shown in figure 4.0

Table 4.0 The data that was recorded from Muse Monitor Application

TimeStamp	Delta_TP9	Delta_AF7	Delta_AF8	Delta_TP10	Theta_TP9	Theta_AF7	Theta_AF8	Theta_TP10	Alpha_TP9	Alpha_AF7	Alpha_AF8	Alpha_TP10	Beta_TP9	Beta_AF7	Beta_AF8	Beta_TP10	Gamma_TP9	Gamma_AF7	Gamma_AF8	Gamma_TP10
13:11.3	-0.0694319	1.3304945	0.2931786	0.57081604	-0.65723985	0.55272967	-0.15972473	0.105109446	-0.011432778	0.8100465	0.26195696	0.089846164	0.016325379	0.62165576	0.4616858	0.23926306	0.1295893	0.30562043	0.2387899	-0.046086803
13:11.8	0.9557695	1.0108373	0.2931786	0.77096975	0.81637615	0.08033841	-0.15972473	0.86634845	0.73174727	0.38526773	0.26195696	0.4433261	0.9028859	0.5582183	0.4616858	0.93279964	0.4749549	0.48756295	0.2387899	0.51308364
27/3/2017 20:13																				
13:12.3	0.964833	0.9491199	1.0054566	0.7896134	0.7496924	0.31926966	0.30403918	0.8084942	0.73174727	0.38526773	0.5316054	0.4433261	0.8432555	0.5293639	1.0031792	0.87741077	0.51258916	0.45803526	0.6712591	0.49358433
13:12.9	0.9207533	0.8992336	0.94163257	0.6181794	0.75658625	0.51094383	0.18061322	0.8258473	0.6309069	0.3354203	0.5316054	0.28578928	0.83156735	0.49329025	0.89852756	0.8407546	0.52289003	0.44132134	0.7579186	0.46924487
13:13.4	0.79402405	0.9765179	0.74282765	0.704203	0.2652796	0.48756662	0.062057856	0.32931417	0.31931388	0.086035095	0.41382283	0.14625344	0.59901506	0.4095963	0.85855746	0.638694	0.5863336	0.4027599	0.85242945	0.48146534
27/3/2017 20:13																				
13:13.9	0.79402405	1.4144753	0.73334223	0.704203	0.2652796	0.54014564	-0.04971247	0.32931417	0.31931388	0.19964275	0.105278045	0.14625344	0.59901506	0.40695393	0.67173004	0.638694	0.5863336	0.4708551	0.96142334	0.48146534
13:14.4	0.79402405	1.5353856	0.8114912	0.704203	0.2652796	0.36146268	-0.05138672	0.32931417	0.31931388	0.011417222	0.24282905	0.14625344	0.59901506	0.38864025	0.7578281	0.638694	0.5863336	0.519576	0.96703345	0.48146534
13:14.9	0.43972704	1.4824952	0.7733677	0.001447669	0.37083754	0.5075155	0.034687474	-0.21383089	0.05128072	0.24416283	0.23977302	0.5597331	0.6963234	0.38059396	0.82419086	0.7646912	0.6416874	0.56448543	0.99915016	0.62131375
27/3/2017 20:13																				
13:15.4	0.3388191	0.49085826	0.5607818	0.014346176	0.24300985	0.07326078	-0.13257873	-0.2533908	0.1515054	0.1832922	0.18524596	0.5597429	0.63919014	0.35324204	0.9109961	0.68139845	0.58156496	0.4135686	0.9542114	0.53753406
13:16.0	0.14000924	0.35790095	0.24615596	-0.06734164	0.11472069	0.016930468	-0.046730213	-0.26718387	0.14325468	0.17726634	-0.121232204	0.31999922	0.6316463	0.47692892	0.94509107	0.6319596	0.5150013	0.4255925	0.99559325	0.47927895
13:16.5	0.10842329	0.2194466	0.28298792	0.4480827	-0.1804494	-0.010850434	0.027959526	0.20961058	-0.022236021	0.16024281	-0.057930768	0.42614406	0.4951184	0.56507003	0.9335896	0.45637652	0.3508543	0.46495745	0.89658046	0.33781633
13:17.0	0.37524635	0.29056805	0.4656164	0.8446897	0.46310836	0.13222598	0.20958827	0.87037355	0.687206	0.14672676	-0.002696071	0.75832427	0.50266033	0.5861492	0.86256844	0.4870234	0.32293376	0.5481355	0.7620571	0.352044
13:17.5	0.6655482	0.39187086	0.45371574	0.9330588	0.67284846	0.40588778	0.23015201	1.1625042	0.7484406	0.0912584	-0.041788667	0.7932498	0.65608776	0.5297445	0.67700815	0.5326495	0.4557455	0.5710101	0.48596025	0.42335093
13:18.0	0.9366242	0.40302944	1.0381037	1.0389735	0.7855867	0.47188556	1.0942447	1.1043028	0.75522184	0.046596788	0.6924091	0.79610384	0.69128036	0.4801849	0.6596101	0.5624435	0.4464555	0.5282841	0.79554576	0.50920457
13:18.5	1.0480713	0.5354012	1.014513	1.1044453	0.8785123	0.60009086	0.9560949	1.0083072	0.7485035	0.15634649	0.7783169	0.6969314	0.73763794	0.4610502	0.784582	0.601655	0.44233704	0.5780667	0.79269767	0.5482098
27/3/2017 20:13																				
13:19.0	1.1015865	0.62677586	0.8867008	1.1312946	0.87754303	0.50981534	0.7731689	0.56658856	0.46072045	0.34472477	0.7783169	0.55509496	0.68618125	0.44420484	0.80017334	0.6059915	0.32709512	0.52100354	0.7636472	0.5627566
13:19.6	1.0103414	0.66473293	0.89729315	1.1092775	0.919945	0.45240647	0.39103013	0.32482812	0.3254717	0.35795122	0.8238593	0.4602012	0.6897841	0.39517865	0.8713494	0.6110443	0.36268708	0.43364966	0.7486955	0.48788157
13:20.1	1.7542964	0.63225895	0.91331816	0.41662383	0.46073043	0.22048755	-0.03619597	0.31955835	0.44980353	0.35347316	0.6750391	0.73753166	0.35143498	0.2990514	0.89683646	0.9402809	0.19610524	0.17729181	0.7480159	0.5848965
13:20.6	1.6598529	0.56895125	1.034461	0.530387	0.4909509	0.11855346	0.18646467	0.31377932	0.49386492	0.26868996	0.68562406	0.73753166	0.3693689	0.19778924	0.9408792	0.8394864	0.21610723	0.085568435	0.7183344	0.50970256
13:21.1	1.5033139	0.5501428	0.52231556	0.666	0.46776804	0.06661584	0.33760136	0.2094459	0.5268557	0.08890529	0.34012562	0.6788006	0.37467605	0.10121452	0.8629378	0.78805965	0.18299022	0.112695694	0.534628	0.4797131
13:21.7	1.1738359	0.47820568	0.5452484	0.82848644	0.3020238	-0.00740133	0.25786108	0.10077423	0.52113587	-0.042478092	0.34012562	0.5295681	0.41103753	0.5651584	0.83116704	0.73294556	0.26156828	0.5761903	0.5069249	0.43506464
13:22.2	1.0635668	0.5245007	0.7536052	0.83695036	0.2424155	0.18128613	0.2623465	-0.0762402	0.412449	0.10883577	0.30720064	0.39958614	0.33787736	0.4741143	0.7422255	0.7264142	0.2275692	0.52946824	0.45188746	0.42800388
13:22.7	1.310214	0.9264789	0.8563529	0.81047136	0.69338495	0.25292733	0.26969904	0.12306588	0.87807876	0.2403869	0.09787249	0.6588889	0.33391368	0.43425566	0.63669705	0.74528456	0.26102388	0.49107486	0.4135622	0.4140637
13:23.2	1.2866296	1.1138017	0.9620084	0.63610864	0.74513	0.42644727	0.5578654	0.22949797	0.87807876	0.32244614	0.40650406	0.7016865	0.25533208	0.37446013	0.6254003	0.7294157	0.18714584	0.37149006	0.48146138	0.38034216
13:23.7	1.1823385	1.1391878	1.063599	0.4403802	0.6926266	0.39628342	0.566061	0.22278379	0.87807876	0.28734404	0.40015838	0.705497	0.3791533	0.47194672	0.66853344	0.676359	0.23147552	0.34889227	0.55873245	0.3267802
13:24.2	0.82960635	1.0451733	1.0402081	0.47469652	0.4245246	0.3655958	0.61234725	0.20303105	0.7806957	0.28153387	0.37432286	0.64802897	0.37300935	0.51335484	0.7027497	0.5235694	0.28043458	0.32911548	0.56600153	0.27286842
13:24.7	0.35623246	0.674425	1.0386796	0.54913336	0.17939194	0.1430093	0.4151294	0.18815126	0.45652157	0.08364903	0.50600423	0.4867218	0.34720525	0.47602835	0.63571966	0.45838398	0.3248642	0.31930053	0.50854236	0.20681152
13:25.3	0.25321987	0.6157619	0.8184957	0.65115947	0.17892049	0.25046062	0.35639924	0.28941464	0.46103606	0.16526814	0.12466143	0.4895187	0.22855963	0.3651829	0.68171495	0.4537614	0.36557925	0.20080683	0.43024826	0.2868326
27/3/2017 20:13																				
13:25.8	0.63568753	0.586569	0.85150605	0.71340525	0.12582374	0.4405466	0.33180398	0.32161516	0.5112704	0.28336638	0.14468034	0.580101	0.21236616	0.2910474	0.6722462	0.5404918	0.2839861	0.14355488	0.3774953	0.41695902
13:26.3	0.7557982	1.0502391	0.7619511	0.75334907	0.20346561	1.0275649	0.26293474	0.34568006	0.5088422	0.52197194	0.089678615	0.55364144	0.26186615	0.41822752	0.7360722	0.44589052	0.220372	0.3155573	0.34385642	0.49908987
13:26.8	0.8901597	1.1165732	0.71042454	0.7828593	0.37296674	1.1060264	0.25544629	0.40930545	0.4952633	0.51750666	0.069225565	0.51918614	0.45691824	0.4732093	0.7006678	0.6526717	0.2513044	0.43737924	0.33699366	0.5834428
13:27.4	0.68698853	1.26605	0.6495938	0.33683613	0.4998648	1.2350745	0.096049845	0.28250828	0.36350387	0.50205684	0.05049362	0.12566	0.63985276	0.5293819	0.7787483	0.62386984	0.5052836	0.5646337	0.73016423	0.4709041
13:27.9	0.49389132	0.60841465	0.38258395	0.18734632	0.35082072	0.23761119	0.01862934	0.17189937	0.3909382	-0.11558279	0.015970517	0.28865507	0.57895	0.49505574	0.7175156	0.59819365	0.4905495	0.49590063	0.7223451	0.3991079
13:28.4	0.39579895	0.42634615	0.26636192	-0.14170732	0.26535645	-0.41454354	-0.1199232	0.033760607	0.38924372	-0.1217093	0.07838396	0.3868151	0.53404576	0.44108492	0.7616095	0.6524289	0.5087677	0.400436	0.74668634	0.261175
13:28.9	0.49226463	0.84569365	0.42272034	0.05723698	-0.035227746	-0.15888448	-0.41381466	0.051266197	0.2964567	0.12821414	0.002660208	0.5423596	0.45096353	0.41111714	0.73462677	0.60276914	0.4866308	0.312699	0.6365829	0.29193184
13:29.4	0.764755	1.0634737	0.9574633	0.42273116	-0.08423174	0.042949207	0.14993398	0.22798966	0.35098606	0.14105275	0.5851711	0.555086	0.44965133	0.38476533	0.7222581	0.5546203	0.5169847	0.23706296	0.568459	0.3485501

Figure 4.0 RAW Brainwave Pattern Obtained from the Recording of the Muse Monitor Application

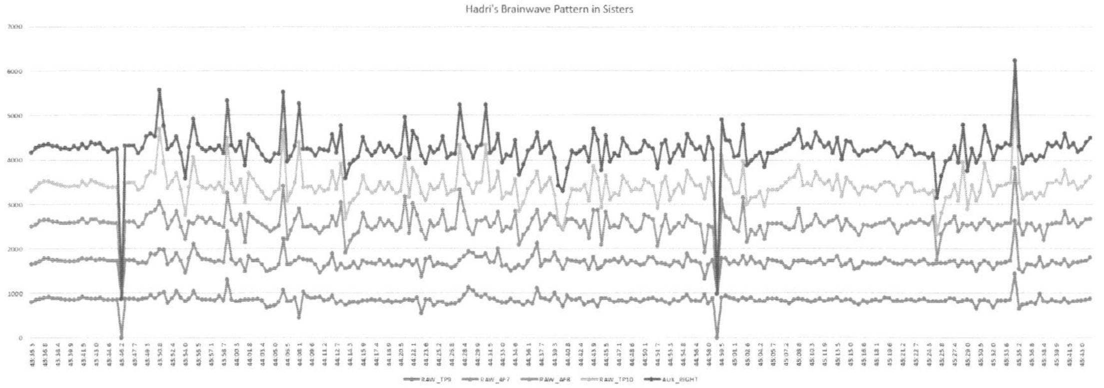

The data is then rearranged to feed the data classifier and was saved in a different filename to avoid any changes to the original data which will be required for future references. The band of interest: alpha, beta, delta, gamma and theta. A length of 20 data points (approximately 1 data points for every 0.5 seconds) was selected before and after the marker was pressed to highlight the significant events that displayed physical emotional changes such as the doll that made a jump scare or any creepy motions that was displayed within the room. Resting state data points was selected from the data points before the subject was subjected to any stimulations.

Table 4.1 Adjustment of data to be fed into the classifier

TimeStamp	Stimulated	Delta_TP9	Delta_AF7	Delta_AF8	Delta_TP10	Theta_TP9	Theta_AF7	Theta_AF8	Theta_TP10	Alpha_TP9	Alpha_AF7	Alpha_AF8	Alpha_TP10	Beta_TP9	Beta_AF7	Beta_AF8	Beta_TP10	Gamma_TP9	Gamma_AF7	Gamma_AF8	Gamma_TP10
14:32.8	0	0.4702636	1.4448537	1.3689204	1.294079	0.5632958	0.9390105	1.0088042	0.73179096	0.5089057	0.75720316	0.5918508	0.6204981	0.61474675	0.95102096	0.8583514	0.9608712	0.7296515	0.831921	1.1078826	0.7191716
14:33.3	0	0.4702636	0.77869886	1.2463939	1.294079	0.5632958	0.49124166	0.90328145	0.73179096	0.5089057	0.27700648	0.5431513	0.6204981	0.61474675	0.89815164	0.8292931	0.9608712	0.7296515	0.7734393	0.90892506	0.7191716
14:33.8	0	1.625126	0.9242136	1.548224	1.6241219	1.189041	0.6580448	0.72545475	0.46780244	1.1491997	0.29524565	1.0775756	0.4447619	1.0837564	0.83298886	0.8792739	0.8336646	1.0748711	0.74763316	0.90910083	0.85254174
14:34.3	0	1.3978115	0.91984195	1.6062238	1.3847667	1.0244961	0.6422009	0.68391055	0.1351585	0.74219716	0.28504544	1.0754951	0.3440304	0.86907005	0.77316535	1.0655247	0.84483993	1.0344296	0.8465832	1.0941479	0.88738036
14:34.8	0	1.3531315	0.92856854	1.6180907	1.3558606	0.9849777	0.6004699	0.7903575	0.13959329	0.74219716	0.3118458	1.0875287	0.36102208	0.8568165	0.601186	1.0818723	0.85503083	1.0367724	0.87012655	1.1761826	0.8859111
14:35.3	0	0.8791667	0.7533727	1.5529832	0.75172496	0.6970887	0.36852586	0.8621382	-0.053613715	0.49266705	0.1461605	0.96448165	0.29640248	0.78084886	0.61907077	1.0522902	0.8935896	0.9898583	0.8197229	1.0474788	0.85195655
14:35.8	0	1.1510863	0.87584645	1.5588522	1.3413274	0.69965804	0.6366123	0.7833951	0.5472783	0.36844844	0.1751968	0.8051373	0.47283944	1.1284982	0.6153265	0.8536888	0.5698306	0.644076	0.6678746	0.8112824	0.5137713
14:36.4	0	1.2008848	0.9418764	1.544988	1.3500347	0.67063016	0.6828532	0.6785856	0.4813813	0.56000227	0.2606252	0.771167	0.48631966	1.1017016	0.6616192	0.78969496	0.58402115	0.70821625	0.5984889	0.71071625	0.4350935
14:36.9	0	1.4646357	1.3486518	1.5135884	1.4766319	0.85891235	0.8403501	0.7355708	0.720746	0.75718707	0.68290174	0.62307227	0.6177444	0.9767871	0.77873945	0.8636657	0.53965396	0.7611048	0.6352788	0.67807937	0.4442541
14:37.4	0	1.4395771	1.3285792	1.427522	1.4159087	0.82185525	0.7626236	0.7388524	0.6773242	0.75334233	0.68290174	0.6194094	0.4983909	0.8517098	0.79612607	0.862851	0.5231401	0.7614218	0.62067515	0.627472	0.38686034
14:37.9	0	1.3187872	1.2385292	1.3369222	1.2590252	0.72234434	0.6865102	0.76039517	0.6425386	0.75026065	0.68577784	0.6204925	0.5985129	0.7689059	0.86551535	0.8191334	0.52088356	0.72769994	0.6294248	0.49333644	0.42099977
14:38.5	0	0.873889	1.173065	1.2248274	0.93232465	0.45134065	0.71147746	0.8536559	0.5808068	0.81822485	0.73737806	0.7847255	0.8356583	0.84656507	0.791993	0.8020532	0.58431244	0.6661651	0.629456	0.5962197	0.42868468
14:39.0	0	0.97810924	1.2046455	1.2189575	1.0106069	0.55769205	0.6724985	0.83900833	0.72368037	0.81822485	0.73946	0.7955265	0.8356583	0.9698971	0.75839156	0.77387977	0.61340326	0.7111795	0.6362762	0.6759256	0.5002387
14:39.5	0	0.93157256	1.1744164	1.1748483	1.0172973	0.58409225	0.6171606	0.79195577	0.7525099	0.81822485	0.7450493	0.7955265	0.8356583	0.7827072	0.8095427	0.7789324	0.6737096	0.5665264	0.58514076	0.88438225	0.52375987
14:40.0	0	0.84057945	0.81887716	0.7981129	0.7028384	0.45415342	0.32368362	0.40198043	0.35603404	0.35082692	0.6127847	0.6147694	0.37024644	0.5781791	0.79833525	0.65135956	0.59078264	0.41386777	0.54584384	0.8844809	0.44304082
14:40.6	0	0.87933296	0.7578865	0.6622417	0.7333302	0.4340467	0.3882871	0.30618864	0.3353136	0.45968774	0.6127847	0.60504055	0.3890681	0.557539	0.7237533	0.66885775	0.5670574	0.42829213	0.45635137	0.8275466	0.40780967
14:41.1	0	1.0454637	0.8331014	0.6551008	0.9001027	0.6005339	0.47189054	0.3047111	0.47320908	0.45968774	0.6186565	0.60504055	0.38763344	0.5272002	0.58217615	0.6866106	0.5366581	0.4343608	0.5017559	0.75984985	0.36348918
14:41.6	1	0.70159304	0.4185297	0.17148803	0.6289742	0.08014066	0.10489965	-0.11189974	0.017270429	0.3136985	0.49378166	0.08814736	0.27391872	0.4271433	0.6040208	0.7779427	0.31212527	0.6257746	0.5610887	0.80127484	0.38474023
14:42.1	1	0.5787262	0.25121227	-0.10331242	0.72609484	0.027486382	-0.024342721	-0.21185662	0.07904645	0.3448082	0.49298206	0.03848197	0.30682275	0.47665673	0.6455309	0.89031416	0.22593047	0.5852257	0.59903413	0.87832445	0.36373797
14:42.6	1	0.5787262	0.42823923	0.04928536	0.72609484	0.027486382	0.13931255	-0.28162143	0.07904645	0.3448082	0.22689122	-0.031622995	0.30682275	0.47665673	0.5932338	0.90363795	0.22593047	0.5852257	0.5496527	0.7672276	0.36373797
14:43.1	1	1.0822525	0.98249733	0.9592365	0.8284461	0.3860691	0.43534598	0.23373556	0.056807037	0.41007808	0.26838684	0.022647485	0.82285035	0.6878854	0.4433266	0.93069863	0.9483117	0.5098578	0.59871775	0.7498791	0.6171594
14:43.7	1	1.4393193	1.1912324	1.1747793	0.77783865	0.46102455	0.57604414	0.42464128	-0.1653786	0.557968	0.34474456	0.32335153	0.9214608	0.8361273	0.43770173	0.8987278	1.1188008	0.66701925	0.6727916	0.78093463	0.83440596
14:44.2	1	1.7861755	1.2799325	1.2790362	0.73067427	0.49784356	0.57463986	0.6090985	0.010717525	0.62472326	0.48941895	0.326127	1.0162629	0.79853964	0.54234064	0.9075257	1.1468498	0.64882433	0.80227494	0.9262733	0.8930584
14:44.7	1	1.8598802	1.1418128	1.1792849	0.68789244	0.47080234	0.6182102	0.61123854	0.44123724	0.61176854	0.66993	0.31499758	1.0643647	0.7863333	0.66570556	0.85547197	1.11162	0.6553729	0.904258	1.0100256	1.0287629
14:45.2	1	2.0301979	0.8087378	0.74872166	0.4152672	0.55097	0.446487	0.32054707	0.48674676	1.0037043	0.66993	0.35388756	1.034475	1.0705516	0.6800462	0.89555836	1.1040215	0.9934134	0.89041936	1.2051115	1.158705
14:45.7	1	1.900638	0.56511235	0.73581016	0.30213407	0.40298066	0.3561981	-0.017102446	0.43309653	0.96531343	0.66993	0.034451697	0.8725567	1.0616974	0.7001183	0.94881356	1.031515	0.97352105	0.8566707	1.181341	1.1889443
14:46.3	1	1.767327	0.4059919	0.57950354	0.088744916	0.27998468	-0.3067676	0.007823396	-0.1329177	0.97227174	0.24246636	0.46413526	0.8197436	0.9551318	0.6386449	0.9310779	0.991842	0.896088	0.6533876	1.1240002	1.0339433
14:46.8	1	1.6711779	0.549719	1.4113514	0.46653745	0.17457798	0.28011402	0.2771673	-0.26282007	0.8790532	0.7956047	0.54495806	0.6789419	0.84076697	0.6208465	0.93305814	0.9346834	0.8144319	0.6559183	0.96942973	0.91109574
14:47.3	1	1.4251059	0.719698	1.7007898	0.63351834	0.18404363	0.46733943	0.548752	-0.15029007	0.70187855	0.7985612	0.6977438	0.35562137	0.742833	0.6850154	0.8677769	0.9618129	0.7914866	0.5652811	0.7641961	0.6174967
14:47.8	1	1.1169457	0.8910682	1.7254672	0.8110374	0.31851402	0.5510014	0.6814199	0.34938404	0.71960074	0.79787433	0.7412413	0.48383212	0.6490186	0.7221512	0.7883418	0.38969973	0.7886514	0.5696487	0.53789514	0.37273636
14:48.3	0	0.7693724	0.8392541	1.5129409	0.6916599	0.072743066	0.41931942	0.62942475	0.27328426	0.616189	0.6324359	0.7528261	0.48957533	0.67783856	0.6829107	0.61901844	0.29299006	0.70529395	0.5051508	0.6318885	0.19164228
14:48.8	0	0.74151266	0.6974636	0.762313	0.67402816	0.040145632	0.31852666	0.3956725	0.27435616	0.5260706	0.16058327	0.5965461	0.58260506	0.69090587	0.5560982	0.6501462	0.33289662	0.67770946	0.67871964	0.7140126	0.19937956
14:49.4	0	0.38846093	0.38270277	0.5937597	0.7675777	-0.13328986	0.23434664	0.24613777	0.19896848	0.2996737	0.16058327	0.56441814	0.4699217	0.74591064	0.51444566	0.6992142	0.41525742	0.58248097	0.6899004	0.7954249	0.07102785
14:49.9	0	0.3042897	0.34998223	0.28573734	-0.18512875	-0.08558332	0.10257322	0.15103732	-1.0321293	0.37153286	0.07407599	0.5482233	0.27802432	0.7192178	0.46852562	0.75162274	0.44704098	0.55787957	0.6182627	0.7511928	0.2778422
14:50.4	0	0.324053	-0.012889571	0.24519175	-0.16917486	-0.12770823	-0.28211254	0.081289805	-0.79611665	0.38182878	0.003562544	0.4991794	0.29347888	0.7354883	0.43452984	0.6869892	0.50046164	0.5281999	0.41717732	0.6612243	0.2951165
14:50.9	0	0.324053	0.16307959	0.24990946	-0.16917486	-0.12770823	-0.30143854	0.010182373	-0.79611665	0.38182878	-0.1912992	0.26634555	0.29347888	0.7354883	0.4483664	0.64305097	0.50046164	0.5281999	0.362505	0.50997597	0.2951165
14:51.4	0	0.19139592	0.18525437	0.25983843	-0.382751	-0.25610107	-0.14517953	0.051692754	-0.7113598	0.38182878	0.045686457	0.4467966	0.29347888	0.7723426	0.55023956	0.65701854	0.56262034	0.3943718	0.46774197	0.6328912	0.29190108
14:52.0	0	0.19139592	0.24005426	0.32582024	-0.382751	-0.25610107	-0.059037272	0.02673632	-0.7113598	0.38182878	0.0889273	0.44906953	0.29347888	0.7723426	0.6664207	0.6749357	0.56262034	0.3943718	0.54120094	0.75580925	0.29190108
14:52.5	0	0.001377275	0.044896938	0.11901	0.033517182	-0.747546	-0.24196804	-0.15024865	-0.40359512	0.24899523	0.092551544	0.43800846	-0.017562956	0.56220824	0.7001049	0.6773646	0.42245826	0.6046164	0.56860995	0.7347477	0.3383584
14:53.0	0	-0.25282705	-0.122516915	-0.005774468	-0.18449455	-0.6119156	-0.4576062	-0.26746324	-0.47097364	0.31602445	0.043808825	0.32291743	0.022585372	0.6006181	0.64823556	0.550971	0.38385087	0.6095483	0.57089394	0.6106175	0.31501457
14:53.5	0	-0.24969108	-0.20126496	-0.24322997	0.005540223	-0.4908654	-0.6957038	-0.41383135	-0.38762432	0.40340912	-0.0346443	0.015253318	0.08133311	0.6092578	0.5021824	0.51278174	0.3923427	0.65760326	0.5570951	0.39842582	0.2731033

However, a few participants were not explicitly displaying their emotions physically during the recording session and it is difficult to establish a marker on the data point in the timestamp. The raw brainwave signal data was then used to analyze and was marked according to the highest peak value of the signal in Y-axis and the approximate time where the participants would be stimulated during the session.

The datasets are then selected for a number of rows to be trained and a number of rows to be tested for the evaluation of the performance with KNN and SVM. Table 4.2 shows the selected datasets for training and table 4.3 shows the selected datasets to be tested for the performance evaluation.

Table 4.2 Datasets which are Selected to be Trained

Datapoint	Delta_TP9	Delta_AF7	Delta_AF8	Delta_TP10	Theta_TP9	Theta_AF7	Theta_AF8	Theta_TP10	Alpha_TP9	Alpha_AF7	Alpha_AF8	Alpha_TP10	Beta_TP9	Beta_AF7	Beta_AF8	Beta_TP10	Gamma_TP9	Gamma_AF7	Gamma_AF8	Gamma_TP10
2	0.208512886	0.084258651	0.20797054	0.092611456	0.229413525	0.217442825	0.297231737	0.169849992	0.280743399	0.091138746	0.306058679	0.15986002	0.014581931	0.111651931	0.13211683	0.064890196	0.098624886	0.16039182	0.121028836	0.10918467
3	0.23738026	0.144569333	0.187950833	0.10066061	0.263048078	0.228231213	0.254577636	0.172780946	0.280743399	0.101053511	0.306058679	0.159275888	0.002190965	0.264112415	0.103500841	0.093794925	0.096824334	0.364264455	0.137351194	0.130435582
4	0.251097929	0.164848246	0.201396692	0.212758874	0.284934584	0.223176611	0.266165075	0.278490678	0.280743399	0.148139516	0.316874857	0.279880544	0.007322977	0.312697572	0.117722854	0.126675699	0.11195183	0.396778283	0.144103562	0.144575743
5	0.305895505	0.172737337	0.199340005	0.286246103	0.270729306	0.184031915	0.234072955	0.439224292	0.231943845	0.134616927	0.294410042	0.301829505	0.027154936	0.359850039	0.099780621	0.173588442	0.168263081	0.45022386	0.154842705	0.116978034
6	0.276096394	0.154763227	0.230638617	0.205521526	0.228471633	0.077135035	0.261842689	0.33327975	0.231943845	0.128846586	0.294410042	0.301829505	0.040046352	0.270207897	0.121076863	0.117745262	0.157144492	0.304714512	0.143464869	0.083330233
7	0.233540726	0.088219493	0.230638617	0.12198862	0.161297577	0.097937018	0.261842689	0.250879507	0.184855905	0	0.294410042	0.2055565	0.047728654	0.007567989	0.121076863	0.094408027	0.163759314	0.028070525	0.143464869	0.061685643
8	0.180284203	0.052912662	0.149571361	0.062212767	0.071796019	0.060214904	0.095923632	0.063862996	0.141492631	0.007807346	0.121589681	0.067057818	0.031181286	0.002046178	0.027090323	0.063561256	0.117318857	0.045658837	0.051710848	0.060617737
9	0.148587974	0.020584594	0.077768672	0.116612499	0.04471592	0.088221559	0.037207461	0.1180399	0.141870477	0.021195327	0.107465388	0.079119062	0.034005075	0.000105284	0.030188683	0.09318878	0.080644932	0.041956741	0.046237116	0.049621269
10	0.086227857	0	0.074801484	0.108359006	0.075378034	0.135433735	0.03775818	0.123293838	0.078638014	0.067170353	0.110101531	0.142939386	0.059808889	0.044224245	0.026361404	0.097934045	0.037840424	0.03833341	0.044167585	0.064580158
11	0.048492538	0.02738118	0.040021273	0.075127314	0.084496803	0.155685002	0.049496557	0.112605604	0.096508683	0.073395577	0.10388202	0.142939386	0.054167357	0.055960705	0.042780334	0.094558346	0	0	0.050951504	0.065437584
12	0.061220647	0.078572892	0.067664486	0.040570163	0.122409437	0.165846737	0.077916838	0.090581861	0.144513249	0.129691176	0.10388202	0.154355878	0.043401838	0.055286742	0.040416142	0.078688255	0.021168092	0.000399299	0.053345554	0.120985394
13	0.077681479	0.198618867	0.113813059	0.096711668	0.085444219	0.139867478	0.015974732	0.111727142	0.13733421	0.117671229	0.094165392	0.12855851	0.009371796	0.047320107	0.060247548	0.089695056	0.09284341	0.010212891	0	0.135311069
14	0.163483135	0.257716709	0.113813059	0.23622013	0.062190887	0.193419898	0.015974732	0.138889978	0.127717933	0.147291492	0.094165392	0.12855851	0.030948708	0.080229936	0.060247548	0.105703154	0.177919329	0.073832012	0	0.158719044
15	0.212895377	0.286200025	0.097520898	0.251381532	0.043594057	0.309208342	0.036702469	0.166290481	0.119300439	0.158595697	0.073197059	0.110019367	0.028089649	0.091671385	0.10167117	0.103056405	0.222591424	0.126675352	0.004536838	0.146963923
16	0.210330614	0.268677161	0.097520898	0.233607504	0.100143561	0.363015529	0.036702469	0.178054195	0.13776579	0.168658245	0.073197059	0.138511405	0.023308543	0.087589609	0.10167117	0.087056582	0.222815739	0.142103821	0.004536838	0.158600689
17	0.1787846	0.298850119	0.141079677	0.09373593	0.191539985	0.378120233	0.162349968	0.210304356	0.235010258	0.171199684	0.160641894	0.237749166	0.052978151	0.073328449	0	0.066625572	0.255921077	0.148110075	0.126361787	0.145024322
18	0.138986316	0.283325709	0.141079677	0.018986108	0.199499688	0.358254947	0.162349968	0.187131701	0.235010258	0.172911657	0.160641894	0.237749166	0.064183938	0.059203739	0	0.071622829	0.231373626	0.130862492	0.126361787	0.144706528
19	0.138986316	0.283325709	0.141079677	0.018986108	0.199499688	0.358254947	0.162349968	0.187131701	0.235010258	0.172911657	0.160641894	0.237749166	0.064183938	0.059203739	0	0.071622829	0.231373626	0.130862492	0.126361787	0.144706528
20	0.822581118	0.897196742	0.159429932	0.921570897	0.797763912	0.877352383	0.189601251	0.877881823	0.916521321	0.9769195	0.260617901	0.914265124	0.747036432	0.877548584	0.240695061	0.883398934	0.773550574	0.870647474	0.281134364	0.869745018
21	0.822572496	0.89712899	0.154155936	0.921566879	0.797433765	0.877238767	0.165945337	0.877556159	0.916521321	0.9769195	0.260617901	0.914265124	0.746614046	0.877520958	0.235734006	0.885456014	0.772656855	0.870586361	0.266978035	0.869484543
22	0.939016387	0.965466629	0.937775556	0.975667991	0.92603551	0.952384173	0.919522749	0.978623615	1	0.999656616	0.999762689	1	0.911689706	0.946777191	0.931304525	0.973945855	0.931686688	0.944140114	0.923253793	0.97622471
23	0.949224554	0.969316833	0.935366223	1	0.939418565	0.956315313	0.918894635	1	1	0.999656616	1	1	0.926882419	0.951726233	0.92868703	1	0.945983292	0.948755075	0.919604599	1
24	0.949224554	0.969316833	0.935366223	1	0.939418565	0.956315313	0.918894635	1	1	0.999656616	1	1	0.926882419	0.951726233	0.92868703	1	0.945983292	0.948755075	0.919604599	1
25	0.949224554	0.969316833	0.935366223	1	0.939418565	0.956315313	0.918894635	1	1	0.999656616	1	1	0.926882419	0.951726233	0.92868703	1	0.945983292	0.948755075	0.919604599	1
26	0.856683526	1	0.807240701	0.857470901	0.84548485	1	0.795351086	0.808616389	0.928037767	1	0.938245786	0.919970988	0.807299337	1	0.770596326	0.806535942	0.830944828	1	0.765485797	0.785591442
27	0.934775379	0.956961433	0.89554351	0.942245078	0.922220857	0.952835068	0.880170985	0.908111465	0.931124589	0.966276411	0.940355377	0.923101872	0.904295263	0.945657244	0.864933425	0.919231729	0.910773706	0.944970237	0.858530536	0.896506328
28	1	0.920276319	1	0.983940944	1	0.913852703	1	0.955413946	0.931124589	0.966276411	0.940355377	0.923101872	1	0.902224022	1	0.97137438	1	0.900620182	1	0.948849949
29	0.89525683	0.521983851	1	0.826276678	0.884535867	0.52244584	1	0.785837218	0.8542427	0.622571753	0.940355377	0.840913716	0.860970826	0.477171263	1	0.787672141	0.864153651	0.459505643	1	0.760487686
30	0.89525683	0.33399487	1	0.493249005	0.884535867	0.422397515	1	0.427778131	0.8542427	0.455886076	0.940355377	0.472227275	0.860970826	0.339174351	1	0.42806214	0.864153651	0.194320731	1	0.37706524
31	0.89525683	0.223383303	1	0.395795935	0.884535867	0.287774489	1	0.279675138	0.8542427	0.248039812	0.940355377	0.301397077	0.860970826	0.247905514	1	0.30480249	0.864153651	0.100810791	1	0.213211465
32	0.89525683	0.223383303	1	0.395795935	0.884535867	0.287774489	1	0.279675138	0.8542427	0.248039812	0.940355377	0.301397077	0.860970826	0.247905514	1	0.30480249	0.864153651	0.100810791	1	0.213211465
33	0.89525683	0.223383303	1	0.395795935	0.884535867	0.287774489	1	0.279675138	0.8542427	0.248039812	0.940355377	0.301397077	0.860970826	0.247905514	1	0.30480249	0.864153651	0.100810791	1	0.213211465
34	0.89525683	0.223383303	1	0.395795935	0.884535867	0.287774489	1	0.279675138	0.8542427	0.248039812	0.940355377	0.301397077	0.860970826	0.247905514	1	0.30480249	0.864153651	0.100810791	1	0.213211465
35	0.89525683	0.135717489	0.394448395	0.155651637	0.884535867	0.231415549	0.312094151	0.175488181	0.8542427	0.183647188	0.230355853	0.159523483	0.860970826	0.138194552	0.427818627	0.24862049	0.864153651	0.094046569	0.508534179	0.280086435
36	0.89525683	0.17367912	0.394448395	0.089319563	0.884535867	0.25767099	0.312094151	0.163170805	0.8542427	0.222892197	0.230355853	0.185872948	0.860970826	0.19439302	0.427818627	0.266384028	0.864153651	0.117503408	0.508534179	0.259247112
37	0.89525683	0.243719202	0.394448395	0.202743218	0.884535867	0.2690922	0.312094151	0.154509534	0.8542427	0.228942748	0.230355853	0.187325688	0.860970826	0.191439981	0.427818627	0.247850299	0.864153651	0.11810676	0.508534179	0.260098311
38	0.89525683	0.293935398	0.394448395	0.310921398	0.884535867	0.294583684	0.312094151	0.165148858	0.8542427	0.182422697	0.230355853	0.152711262	0.860970826	0.225221874	0.427818627	0.249750089	0.864153651	0.135961415	0.508534179	0.257161074

Table 4.3 Dataset which are selected to be Tested for the Accuracy Prediction

Datapoint	Delta_TP9	Delta_AF7	Delta_AF8	Delta_TP10	Theta_TP9	Theta_AF7	Theta_AF8	Theta_TP10	Alpha_TP9	Alpha_AF7	Alpha_AF8	Alpha_TP10	Beta_TP9	Beta_AF7	Beta_AF8	Beta_TP10	Gamma_TP9	Gamma_AF7	Gamma_AF8	Gamma_TP10
126	0	0.257084797	0.394448395	0.250420283	0.170144615	0.386792091	0.312094151	0.353860126	0.194556355	0.175114941	0.230355853	0.115650325	0.142509739	0.319174775	0.427818627	0.161666426	0.243973649	0.342954037	0.508534179	0.213410315
127	0	0.257084797	0.394448395	0.250420283	0.170144615	0.386792091	0.312094151	0.353860126	0.194556355	0.175114941	0.230355853	0.115650325	0.142509739	0.319174775	0.427818627	0.161666426	0.243973649	0.342954037	0.508534179	0.213410315
128	0	0.257084797	0.394448395	0.250420283	0.170144615	0.386792091	0.312094151	0.353860126	0.194556355	0.175114941	0.230355853	0.115650325	0.142509739	0.319174775	0.427818627	0.161666426	0.243973649	0.342954037	0.508534179	0.213410315
129	0	0.257084797	0.394448395	0.250420283	0.170144615	0.386792091	0.312094151	0.353860126	0.194556355	0.175114941	0.230355853	0.115650325	0.142509739	0.319174775	0.427818627	0.161666426	0.243973649	0.342954037	0.508534179	0.213410315
130	0.296456364	0.157516297	0.311861543	0.250420283	0.116373365	0.329866199	0.217691472	0.353860126	0.245032256	0.27879376	0.213932962	0.115650325	0.25306476	0.211892407	0.426535718	0.161666426	0.347415166	0.007583219	0.484871946	0.213410315
131	0.320546681	0.24202893	0.341869406	0.250420283	0.128758794	0.39106283	0.2435533	0.353860126	0.269396023	0.391019381	0.257125359	0.115650325	0.245590878	0.249307222	0.411664621	0.161666426	0.343904934	0.047543616	0.502567417	0.213410315
132	0.321115109	0.24202893	0.343490329	0.250420283	0.143418965	0.39106283	0.252182191	0.353860126	0.259118662	0.391019381	0.257125359	0.115650325	0.268485435	0.249307222	0.416612999	0.161666426	0.340870672	0.047543616	0.503310156	0.213410315
133	0.336017992	0.252582521	0.371061958	0.250420283	0.162158917	0.42100297	0.270192772	0.353860126	0.248896939	0.391019381	0.237456877	0.115650325	0.278205038	0.277393257	0.390789992	0.161666426	0.335010638	0.065526534	0.522254506	0.213410315
134	0.336017992	0.252582521	0.371061958	0.250420283	0.162158917	0.42100297	0.270192772	0.353860126	0.248896939	0.391019381	0.237456877	0.115650325	0.278205038	0.277393257	0.390789992	0.161666426	0.335010638	0.065526534	0.522254506	0.213410315
135	0.336017992	0.252582521	0.371061958	0.250420283	0.162158917	0.42100297	0.270192772	0.353860126	0.248896939	0.391019381	0.237456877	0.115650325	0.278205038	0.277393257	0.390789992	0.161666426	0.335010638	0.065526534	0.522254506	0.213410315
136	0.552242818	0.252582521	0.371061958	0.250420283	0.247592368	0.42100297	0.270192772	0.353860126	0.185278935	0.391019381	0.237456877	0.115650325	0.227712727	0.277393257	0.390789992	0.161666426	0.246651955	0.065526534	0.522254506	0.213410315
137	0.552242818	0.252582521	0.371061958	0.367773121	0.247592368	0.42100297	0.270192772	0.137002681	0.185278935	0.391019381	0.237456877	0.144642054	0.227712727	0.277393257	0.390789992	0.270359718	0.246651955	0.065526534	0.522254506	0.205945269
138	0.552242818	0.252582521	0.371061958	0.367773121	0.247592368	0.42100297	0.270192772	0.137002681	0.185278935	0.391019381	0.237456877	0.144642054	0.227712727	0.277393257	0.390789992	0.270359718	0.246651955	0.065526534	0.522254506	0.205945269
139	0.552242818	0.252582521	0.371061958	0.367773121	0.247592368	0.42100297	0.270192772	0.137002681	0.185278935	0.391019381	0.237456877	0.144642054	0.227712727	0.277393257	0.390789992	0.270359718	0.246651955	0.065526534	0.522254506	0.205945269
140	0.552242818	0.252582521	0.371061958	0.367773121	0.247592368	0.42100297	0.270192772	0.137002681	0.185278935	0.391019381	0.237456877	0.144642054	0.227712727	0.277393257	0.390789992	0.270359718	0.246651955	0.065526534	0.522254506	0.205945269
141	0.552242818	0.252582521	0.371061958	0.367773121	0.247592368	0.42100297	0.270192772	0.137002681	0.185278935	0.391019381	0.237456877	0.144642054	0.227712727	0.277393257	0.390789992	0.270359718	0.246651955	0.065526534	0.522254506	0.205945269
142	0.552242818	0.252582521	0.371061958	0.367773121	0.247592368	0.42100297	0.270192772	0.137002681	0.185278935	0.391019381	0.237456877	0.144642054	0.227712727	0.277393257	0.390789992	0.270359718	0.246651955	0.065526534	0.522254506	0.205945269
143	0.552242818	0.252582521	0.371061958	0.367773121	0.247592368	0.42100297	0.270192772	0.137002681	0.185278935	0.391019381	0.237456877	0.144642054	0.227712727	0.277393257	0.390789992	0.270359718	0.246651955	0.065526534	0.522254506	0.205945269
144	0.552242818	0.252582521	0.371061958	0.367773121	0.247592368	0.42100297	0.270192772	0.137002681	0.185278935	0.391019381	0.237456877	0.144642054	0.227712727	0.277393257	0.390789992	0.270359718	0.246651955	0.065526534	0.522254506	0.205945269
145	0.552242818	0.252582521	0.371061958	0.367773121	0.247592368	0.42100297	0.270192772	0.137002681	0.185278935	0.391019381	0.237456877	0.144642054	0.227712727	0.277393257	0.390789992	0.270359718	0.246651955	0.065526534	0.522254506	0.205945269
146	0.290159091	0.252582521	0.371061958	0.435481032	0.304042775	0.42100297	0.270192772	0.149339543	0.225629133	0.391019381	0.237456877	0.214715407	0.123906028	0.277393257	0.390789992	0.20812098	0.214274338	0.065526534	0.522254506	0.197444175
147	0.331741346	0.252582521	0.371061958	0.435481032	0.344487762	0.42100297	0.270192772	0.149339543	0.239767678	0.391019381	0.237456877	0.214715407	0.09125555	0.277393257	0.390789992	0.20812098	0.197560024	0.065526534	0.522254506	0.197444175
148	0.344408824	0.252582521	0.371061958	0.435481032	0.363463559	0.42100297	0.270192772	0.149339543	0.217392134	0.391019381	0.237456877	0.214715407	0.049059806	0.277393257	0.390789992	0.20812098	0.176683315	0.065526534	0.522254506	0.197444175
149	0.387360181	0.252582521	0.371061958	0.435481032	0.402096704	0.42100297	0.270192772	0.149339543	0.191128935	0.391019381	0.237456877	0.214715407	0	0.277393257	0.390789992	0.20812098	0.158117212	0.065526534	0.522254506	0.197444175
150	0.196176647	0.252582521	0.371061958	0.433910289	0.196147056	0.42100297	0.270192772	0.182165773	0.21444261	0.391019381	0.237456877	0.151413132	0.107848083	0.277393257	0.390789992	0.152079118	0.256420841	0.065526534	0.522254506	0.175249549
151	0.37173857	0.252582521	0.371061958	0.442234385	0.314901368	0.42100297	0.270192772	0.204059599	0.279728067	0.391019381	0.237456877	0.224222219	0.101578354	0.277393257	0.390789992	0.175923248	0.248451351	0.065526534	0.522254506	0.15100294
152	0.37173857	0.252582521	0.371061958	0.433552589	0.314901368	0.42100297	0.270192772	0.211043789	0.279728067	0.391019381	0.237456877	0.23091404	0.101578354	0.277393257	0.390789992	0.189465515	0.248451351	0.065526534	0.522254506	0.153429493
153	0.429848591	0.252582521	0.371061958	0.43319396	0.368653743	0.42100297	0.270192772	0.235611539	0.246880938	0.391019381	0.237456877	0.215019756	0.096400535	0.277393257	0.390789992	0.220510501	0.241791514	0.065526534	0.522254506	0.128668192
154	0.444859622	0.252582521	0.371061958	0.415914687	0.388549953	0.42100297	0.270192772	0.268123233	0.243694909	0.391019381	0.237456877	0.213781182	0.095083409	0.277393257	0.390789992	0.243019494	0.234977175	0.065526534	0.522254506	0.134049977
155	0.444859622	0.252582521	0.371061958	0.415914687	0.388549953	0.42100297	0.270192772	0.268123233	0.243694909	0.391019381	0.237456877	0.213781182	0.095083409	0.277393257	0.390789992	0.243019494	0.234977175	0.065526534	0.522254506	0.134049977
156	0.444859622	0.252582521	0.371061958	0.415914687	0.388549953	0.42100297	0.270192772	0.268123233	0.243694909	0.391019381	0.237456877	0.213781182	0.095083409	0.277393257	0.390789992	0.243019494	0.234977175	0.065526534	0.522254506	0.134049977
157	0.444859622	0.252582521	0.371061958	0.415914687	0.388549953	0.42100297	0.270192772	0.268123233	0.243694909	0.391019381	0.237456877	0.213781182	0.095083409	0.277393257	0.390789992	0.243019494	0.234977175	0.065526534	0.522254506	0.134049977
158	0.444859622	0.252582521	0.371061958	0.415914687	0.388549953	0.42100297	0.270192772	0.268123233	0.243694909	0.391019381	0.237456877	0.213781182	0.095083409	0.277393257	0.390789992	0.243019494	0.234977175	0.065526534	0.522254506	0.134049977
159	0.134873001	0.252654569	0.371061958	0.219188102	0.05165668	0.387536943	0.270192772	0.154867961	0.228602024	0.195213137	0.237456877	0.176246526	0.123173945	0.125789279	0.390789992	0.313233302	0.093692122	0.095759237	0.522254506	0.194193999
160	0.134873001	0.252654569	0.371061958	0.219188102	0.05165668	0.387536943	0.270192772	0.154867961	0.228602024	0.195213137	0.237456877	0.176246526	0.123173945	0.125789279	0.390789992	0.313233302	0.093692122	0.095759237	0.522254506	0.194193999
161	0.1179095	0.240787013	0.371061958	0.172706546	0.036539533	0.326686317	0.270192772	0.134552781	0.227065086	0.183154135	0.237456877	0.132774078	0.130466427	0.115496368	0.390789992	0.321063058	0.091595024	0.10045289	0.522254506	0.191044328
162	0.102181932	0.235289089	0.371061958	0.12131005	0	0.273435319	0.270192772	0.14327849	0.223255422	0.163833178	0.237456877	0.109159055	0.13610657	0.110103098	0.390789992	0.299028356	0.094367129	0.08842903	0.522254506	0.180470976

The data obtained from the Muse Monitor were then trained and tested using KNN and SVM and the accuracy percentage of the algorithm were tabulated as shown in table 4.2.

Table 4.4 The Accuracy of Each Participant using KNN and SVM methods to train and test the observed brainwave patterns.

Participants	KNN(%)	SVM(%)
A	76.54	76.54
B	85.48	85.48
C	82.91	83.42
D	91.29	91.42
E	60.98	60.98
F	69.70	68.94
G	95.94	95.94
H	92.98	92.98
I	86.16	86.16

5. Conclusion

The data analysis showed promising results in classifying he mental state of the participants while immersed under VR. This would provide a foundation for reference point as VR has never been conducted with EEG and would prove useful for future developments.

For future recommendations, the number of participants will be increased to 30 to be able to obtain a better convincing data on the accuracy of the mental classification. There were also a need to change the VR to provide a deeper impact for the participant. However, this will require further reconsideration as the app that was experimented with may traumatize other future participants.

There is also the difficulty of setting up a proper environment for the participant to fully immerse into the VR as not all participant were available to travel to the vicinity of the experiment that was conducted. Instead, we have to arrange logistics to travel to their premise and or using an empty ground area to conduct the experiment while the surrounding environment was dark and quiet. All necessary steps to avoid external stimulus were reassured before the conduct of the research.

Muse headband was not able to fit into most subjects even when the participants were in their teens (female aged 17), when fitted with the rest of the headgears to record the data. A consideration on the head size will be required to evaluate so that future experimentation would ease with the minimal requirement of the head size.

References

[1] Li, Z., Xu, J., & Zhu, T. (2015). Prediction of Brain States of Concentration and Relaxation in Real Time with Portable Electroencephalographs. *arXiv preprint arXiv:1509.07642*.

[2] Tran, Y., Thuraisingham, R., Wijesuriya, N., Craig, A., & Nguyen, H. (2014, August). Using S-transform in EEG analysis for measuring an alert versus mental fatigue state. In *Engineering in Medicine and Biology Society (EMBC), 2014 36th Annual International Conference of the IEEE* (pp. 5880-83). IEEE.

[3] Myrden, A., & Chau, T. (2017). A passive EEG-BCI for single-trial detection of changes in mental state. *IEEE Transactions on Neural Systems and Rehabilitation Engineering*.

[4] Wang, X. W., Nie, D., & Lu, B. L. (2014). Emotional state classification from EEG data using machine learning approach. *Neurocomputing*, *129*, 94-106.

[5] Kim, M. K., Kim, M., Oh, E., & Kim, S. P. (2013). A review on the computational methods for emotional state estimation from the human EEG. *Computational and mathematical methods in medicine*, *2013*.

[6] AlZoubi, O., Calvo, R. A., & Stevens, R. H. (2009, December). Classification of EEG for affect recognition: an adaptive approach. In *Australasian Joint Conference on Artificial Intelligence* (pp. 52-61). Springer Berlin Heidelberg.

[7] Yazdani, A., Ebrahimi, T., & Hoffmann, U. (2009, April). Classification of EEG signals using Dempster Shafer theory and a k-nearest neighbor classifier. In *Neural Engineering, 2009. NER'09. 4th International IEEE/EMBS Conference on* (pp. 327-30). IEEE.

[8] Hosseinifard, B., Moradi, M. H., & Rostami, R. (2013). Classifying depression patients and normal subjects using machine learning techniques and nonlinear features from EEG signal. *Computer methods and programs in biomedicine*, *109*(3), 339-45.

[9] Subasi, A., & Gursoy, M. I. (2010). EEG signal classification using PCA, ICA, LDA and support vector machines. *Expert Systems with Applications*, *37*(12), 8659-66.

[10] Cassani, R., Banville, H., & Falk, T. H. (2015, March). MuLES: An open source EEG acquisition and streaming server for quick and simple prototyping and recording. In *Proceedings of the 20th International Conference on Intelligent User Interfaces Companion* (pp. 9-12). ACM.

[11] Galway, L., McCullagh, P., Lightbody, G., Brennan, C., & Trainor, D. (2015, October). The Potential of the Brain-Computer Interface for Learning: A Technology Review. In *Computer and Information Technology; Ubiquitous Computing and Communications; Dependable, Autonomic and Secure Computing; Pervasive Intelligence and Computing (CIT/IUCC/DASC/PICOM), 2015 IEEE International Conference on* (pp. 1554-59). IEEE.

[12] Karydis, T., Aguiar, F., Foster, S. L., & Mershin, A. (2015, July). Performance characterization of self-calibrating protocols for wearable eeg applications. In *Proceedings of the 8th ACM International Conference on PErvasive Technologies Related to Assistive Environments* (p. 38). ACM.

[13] Mann, S., Janzen, R., Wu, H., Lu, M. H., & Guleria, N. (2015, October). Bright ideas: A wearable interactive" Inventometer"(brainwave-based idea display). In *Games Entertainment Media Conference (GEM), 2015 IEEE* (pp. 1-8). IEEE.A reference

IOP Conf. Series: Materials Science and Engineering **341** (2018) 012017 doi:10.1088/1757-899X/341/1/012017

Hydrogenated nanostructure boron doped amorphous carbon films by DC bias

A Ishak[1,2, a)], K Dayana[3], I Saurdi[1,2,], M F Malek[1] and M Rusop[1]

[1]NANO-SciTech Centre, Institute of Science, Universiti Teknologi MARA, 40450 Shah Alam, Selangor, Malaysia

[2]Faculty of Electrical Engineering, UiTM Sarawak Kampus Kota Samarahan Jalan Meranek, Sarawak, Malaysia

[3]Faculty of Electrical Engineering, Universiti Teknologi MARA (UiTM),Cawangan Johor, Kampus Pasir Gudang, Jalan Purnama, Bandar Seri Alam, 81750 Masai, Johor, Malaysia

[a)]ishak@sarawak.uitm.edu.my

Abstract. Hydrogenated nanostructure-boron doped amorphous carbon thin film carbon was deposited at different negative bias using custom-made deposition bias assisted-CVD. Solid of boron and palm oil were used as dopant and carbon source, respectively. The hydrogenated nanostructure amorphous films were characterized by Field emission scanning electron microscopy, Fourier transform infrared spectroscopy, Raman spectroscopy, while the photo-response studies of thin film is done by I-V measurement under light measurement. The results showed the carbon film were in nanostructure with hydrogen and boron might be incorporated in the film. The Raman spectra observed the increase of upward shift of D and G peaks as negative bias increased which related to the structural change as boron incorporated in carbon network. These structural changes were further correlated with photo-response study and the results obtained are discussed and compared.

1. Introduction

Allotropes carbons have been prepared by multiple deposition techniques using different carbon precursor, palm oil for example has been successfully used for synthesize carbon nano-tubes by chemical vapor deposition technique. Palm oil contained carbon (67), hydrogen (127) and oxygen (8) to form the chemical binding of $C_{67}H_{127}O_8$ and contained highest number of carbon compared to other carbon precursor used so far. It is reported that various properties of carbon films not only depend on their methods and conditions of deposition but also on the starting precursor materials [1,2].

Many deposition parameters have been introduced so far with the target to improve the quality of thin film such as temperature, dopant, deposition time, flow rate of carrier gas, position of substrate, etc. It was reported, DC bias voltage helped for improve quality of films through bombarding effect [3-5]. Among deposition parameters, negative bias applied to the substrates could significantly change film properties due to enhancement of adatom mobility and the effect of ion bombardment [3-5]. It determines the kinetic bombardment energies of the ions arriving on the substrate and can significantly affect the chemical composition, structure and residual stresses in the as-deposited films.

The hydrogenated amorphous carbon (a-C:H) deposited from many types of precursors and methods showed a good potential use in amorphous carbon solar cells. However, it was reported as weakly p-type in nature despite of very complex structure, high density of defect and those factors restricted to dope efficiently and the main barrier for its application in various electronic devices. In order solve that problem, the control of doping could reduce the existing of defect and modified the electronic properties [1,2].

Palm oil, a natural source is known so far as the highest carbon contents and may have both sp^2 and sp^3 carbon. Application of light energy conversion for solar cell applications was reported using amorphous palm oil carbon film [1,2]. However, detail structural properties of palm oil carbon films related with photo-response were rarely discussed. In fact, there have been reports [3-5] about the influences of the substrate bias voltage on the structure and properties of carbon and as-deposited metal oxide films. However, there is no many literatures reporting on the effect of DC negative bias on hydrogenated nanostructure a-C:B films using this preparation.

2. Methodology

The a-C:B film was prepared by custom-made bias-assisted-CVD using the glass substrates (thickness: 1mm) and n-Si (100) (thickness 325 + 25 μm, resistivity 1-10 Ω cm). Substrates were cleaned with acetone (C_5H_{6O}) followed by methanol (CH_3OH) for 15 min in ultrasonic cleaner (power Sonic 405), respectively and the glass substrates were then rinse with deionizer water (DI) water for 15 min. Moreover, excess oxide layers of n-type silicon substrates were continued by the etching process with diluted hydrofluoric acid (10%) solution for about 3 min before rinsing in DI water. Substrates were then blown with nitrogen gas. The cleaned of substrates were finally placed inside the furnace chamber

The deposition temperature was set at 350°C for 1h deposition. A liquid of palm oil precursor was heated in the bottle outside the chamber at around 150°C by using hot platter (Stuart CB162). The vaporized of palm oil was pressured into the furnace chamber using air pumps (model GA8000) where the flow rate of vaporized palm oil is controlled by adjustable valve. The amount of vaporized palm oil, carrier gas argon was set to be constant at 100 mL/min, 200 mL/min, respectively by using AALBORG flow meter. For doping process, approximately 1 g of boron was placed on the aluminium foil above the metal plate heater in the chamber.

Fourier transform infrared spectroscopy, JEOL JSM-7600F field emission scanning electron microscope (FESEM), Current-voltage measurement (I-V) (Bukuh Keiki EP200), and Raman spectroscopy are used to characterize the existing of hydrogen, particle size, structural bonding and electrical properties, respectively.

3. Results and discussion

Figure 1 show FTIR spectra of the carbon films deposited at different negative bias. The absorption bands in the spectra were assigned based on comparison with the standard patterns reported in the literature [6-9].The FTIR spectra were measured in the wave number range of 1000-3000 cm^{-1} to determine the vibration modes of deposited films. Different types of band appearing in the films which is, C=C, CO_2, and C-H formed. The band appearing at the wave number in the range of 1050-1300 cm^{-1} is corresponding to C-O bond while in the range of 1500-1600 cm^{-1} is corresponding to C=C bond (the sp^2 bonding) [6-9]. The band situated in the range of 2950-3060 cm^{-1} is attributed to C-H stretching vibration of sp^2-C, while for sp^3-C in the range 2850-2945 cm^{-1} range [6-9]. It was clearly observed the bonding of hydrogen with carbon films deposited at -10 V to -50 V, the films exhibit ‘characteristic C-H related absorption’ in the range 2800-3000 cm^{-1} and therefore we called it as hydrogenated boron doped amorphous carbon thin film.

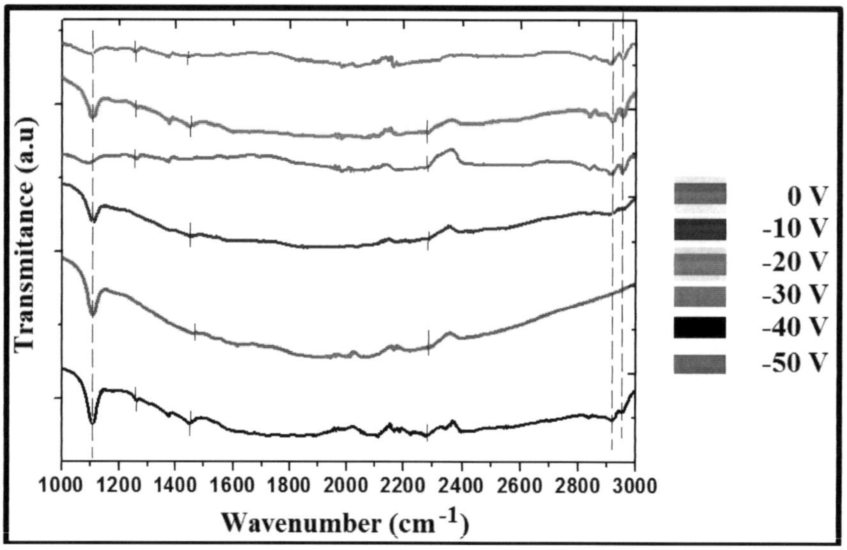

Figure 1. FTIR spectra of hydrgenerated a-C:B films deposited at difference negative bias

However, the carbon film at 0 V is found almost no C-H peak but we predict small number of hydrogen existed. Since the C-H bond signal is found very weak for 0 V, so it is assumed the attachment of H bond is less. It was predicted; the passivated of hydrogen in carbon thin film might reduce defect and therefore can increase conductivity. In contrast, C-H stretching peaks indicates at -10 to -50 V and might be successfully incorporated into carbon network thus might be passivated dangling bonds. The high C-H stretching peaks might be the role of negative bias to attract more opposite ion, in this case positive ion hydrogen to binding with the carbon network. In contrast, the existing peak of CO_2 and no hydrogen bond in FTIR spectrum in carbon network might contribute for defect in carbon network and therefore might suggest the low conductivity carbon film at 0V.

Figure 2 shows the FESEM images of hydrogenated a-C:B films deposited at different negative bias. The surface morphology images were captured by the JEOL JSM-7600F field emission scanning electron microscope (FESEM) between 100kx to 250kx magnification, using 5kV. It clearly observed, the images were really in amorphous nature, since no sharp images were determined. FESEM images consist of irregularly scattered ball-like particles in between 90 nm to less than 10 nm. The sample at 0 V (by only using thermal process only) shows the nanostructure particles with average size in between 51 nm to 27.5 nm. A gradual refinement of the nanostructure particles sizes slightly observed as bias voltage at -10 V (Figure 2 (II)).The nanostructure particles sizes were almost constant at -20 V (Fig. 2 (III)) and deteriorated at -30 V (Figure 2 (IV)) to -50 V (Figure 2 (V)). This suggest that, negative bias from -10 to -50 V give less significant effect on the particles sizes of hydrogenated nanostructure of a-C:B films. The agglomerated particles contain C-O binding might produce defect and C-H binding might passivated the dangling bond as supported by FTIR analysis. It is observed that, the lowest size of particles with less agglomerated form was found at -20 V.

The possibility mechanism of this process can be explained as follow; during the deposition, the palm oil precursor decomposed in a variety of forms (molecule, ion, etc.) and they are moving randomly in the chamber. Similarly, boron powder heated by thermal process released boron ion, molecule. The negative bias attracts more positive ions boron and carbon beside other elements to the substrate by carrier argon. This effect will fill the voids between the grains, break down the large particles growth and create more nucleation sites, thereby promoting the particles refinement [3-5].

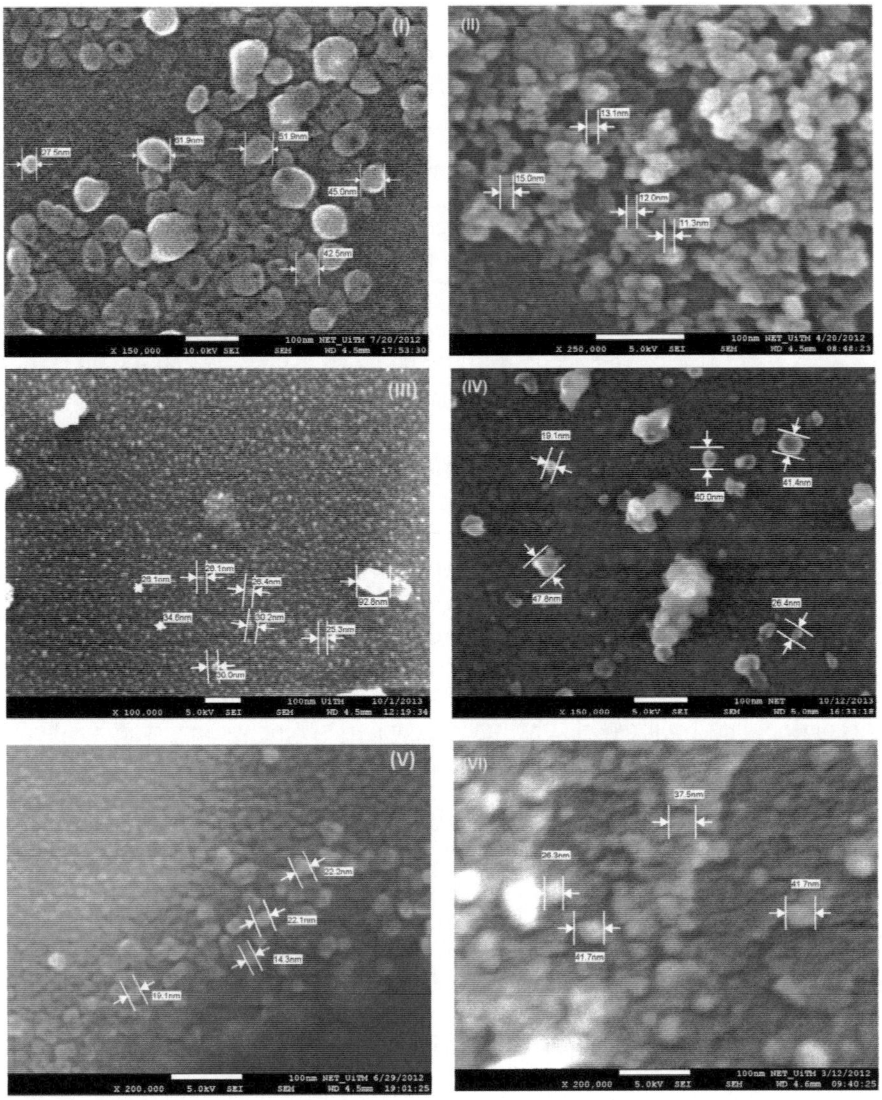

Figure 2. The FESEM images of hydrogenated nanostructure a-C:B films deposited at different negative bias

An intense positive-ion bombardment on the growing film surface is occurred when the optimum condition achieved. In contrast, at less optimum condition only thermal process (0 V), the ions energy is less effective to penetrate into the growing surface and most of the ions are only trapped on the growing surface, resulting in the formation of the large particles size.

The study on nanostructure materials was successfully done by different preparation technique for crystalline material such as zinc oxide (ZnO), compound semiconductor, silicon material, etc. [7, 8]. The result showed nanostructure film has improved efficiency in dye sensitized solar cell [7,8], increased sensitivity and reliability of humidity sensor[205], and increase high surface area in silicon solar cell [7,8]. However, there is less report on the nanostructure of a-C alone or with other dopants in the literatures. K. Dayana *et al* [9] reported that nanostructure iodine doped (a-C:I) film was achieved particle size of nanostructure of about 45 to 65 nm by thermal CVD using camphor precursor. In addition, D.B. Mahadik et al [10] reported a nanostructure carbon thin film by thermal CVD process about 1000°C using turpentine oil achieved by SEM images showed the particles size of 80 to 9 nm. In technological point of view, the result prepared by this technique achieved the size of particles nanostructure a-C:B lower than in the literatures [9,10]. The result showed that, the nanostructure a-C film can be achieved not only by conventional techniques using camphor and turpentine precursors but also can be prepared by this technique.

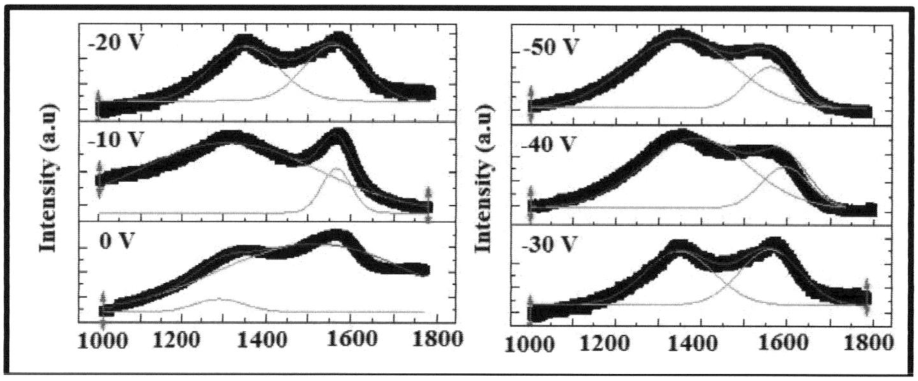

Figure 3. The Raman spectra of nanostructure a-C:B films at different negative bias

Figure 3 shows the Raman spectra nanostructure a-C:B films deposited at different negative bias . The G and D bands observed from 0 to -50 V are relatively broad indicating that the crystallite size of the carbon film is small [12]. The peak around 1550 cm^{-1} shows the amorphous nature of the a-C films [6,9,12]. This same phenomenon has been reported by other groups [6,9,12] but using other methods, precursors, and dopants sources [6,9,12] and also supported by FESEM images as discussed previously. At no bias applied (0 V), the D band position is low compared to a-C:B film deposited with negative bias voltage and slightly increase thereafter until -20 V. The D band of hydrogenated nanostructure a-C:B film for the negative bias between -20 V to -30 V remains almost unchanged after which it has changed obviously from 0 V to -20V. Thus, the structure of the films remains almost unaltered upon incorporation of B between -20 V and -30 V. Both, Raman D and G peaks are upshifted with increase of negative bias from 0 V to -50 V. The increase of D band and G in Raman spectra for graphitization of hydrogenated nanostructure a-C:B film might be correlated with the microstructural change of carbon due boron incorporated in carbon network and might due to C-H binding as correlated with the FTIR spectra might improve the hydrogenated nanostructure a-C:B film quality. Furthermore, the increasing of D band peak height in Raman spectra reported by Pardanoud et al [13] was correlated with the H content in graphene and therefore might be correlated with H incorporated in nanostructure a-C:B film.

This phenomenon might be indicated the component of the sp^2 bonds increased with the increased of boron incorporation with carbon network. The upward shifting of Raman G peak together with the increase of Raman D peak position and intensity might reflect to the progress of "graphitization" because Raman D peak is reported to be assigned to sp^2-bonded C which seems to be aromatic in structure [14-16]. Under this refinement process, the sp^2/sp^3 ratio may also be decreased to a certain value as supported by the upshifting of the D-position. The same phenomenon was agreed by others [6,12] when D-position increased. The changing of size particle in FESEM images and C-H binding from FTIR spectra might be related to the increasing of D-position by the structural disorder. The main factor affecting bands position, width and intensity is the clustering of sp^2 phase. The G band position measures topological disorder, which arises from the size and shape distribution of sp^2 [15] clusters.

The changing of D and G position might be to a change in the clustering of sp^2 phase of nanostructure a-C:B film. Although there is no evident to justify the clustering of sp^2 phase of hydrogenerated nanostructure a-C:B film changed, but changing of D position might be an indicator for changing of clustering of sp^2 phase.

Table 1. Raman spectra parameters of a-C:B films deposited at difference negative bias

Negative bias voltage (V)	$\frac{Pos\ (G)}{(cm^{-1})}$	$\frac{Width\ (G)}{(cm^{-1})}$	$\frac{Pos\ (D)}{(cm^{-1})}$	$\frac{Width\ (D)}{(cm^{-1})}$	$\frac{I(D)}{I(G)}$
0	1582.436	448.861	1332.504	212.642	0.447
10	1587.133	182.979	1319.889	642.474	0.778
20	1560.283	182.978	1344.504	194.825	0.969
30	1560.283	182.979	1344.503	194.825	1.069
40	1620.337	137.796	1410.199	319.879	1.785
50	1552.452	140.871	1340.450	334.438	1.829

The refinement of a-C:B films particle size by negative bias of FESEM images might be correlated with a dense of boron and carbon peaks together with other elements which might contribute for the change of clustering sp^2 phase and therefore attribute for a change in D and G in Raman spectra. The change in bond hybridisation (sp^2/sp^3) and density of the particles as revealed from EEL analysis was reported more sensitive to starting precursor [17]. According to the literatures [6,9,12] an increase in sp^2-bonding resulted in a G peak shifted to a higher wave number and the upward shifting of the Raman D and G peak positions indicated the reduction of bond angle disorder and growth of graphitic domains. Thus, the result implies that the hydrogenated nanostructure a-C:B films prepared from palm oil precursor are dominated by sp^2 rather than sp^3 as negative bias increased from 0 V to -50 V.

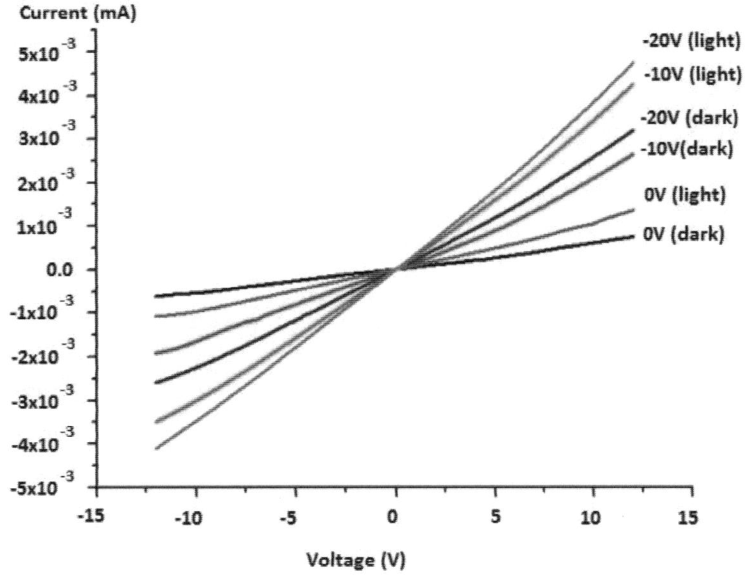

Figure 4. The current-voltage relationships of hydrgenerated nanostructure boron doped amorphous carbon under dark and light measurement

Figure 4 illustrates the current voltage (I-V) characteristic of a-C:B films under dark and illumination. All a-C:B films do not show rectifying behavior either in dark or under illumination. The reason is that the work function of Aurum is larger than a-C:B and might be due to low resistivity of a-C:B which attributed to the existence of sp^2-rich interfacial layer [18, 19].

In a dark measurement (refer Figure 4), the conductivity of a-C thin film deposited by negative bias compared with zero bias ($2.048 x 10^{-7}$) is gradually increases from $4.53 x 10^{-6}$ to $4.878 x 10^{-6}$ Scm^{-1}. The result shows with the help of negative bias increase the conductivity of hydrogenated a-C:B film. This phenomenon might be correlated with the increase of true dopant of boron in the thin film. Similar result was found by W. Chan et al [20] conductivity increase with increasing boron concentration in reliable boron-doping. The electrical properties could be related with the changing density of crystal defects, such as the vacancy, grain boundary and film surface, carrier concentration and mobility of the electron [21]. It is reported that the surface morphology (grain size), carrier concentration and carrier mobility also play an important role in the electrical conductivity [10, 22]. Deposition by negative bias showed higher conductivity compared to thermal process only. This might be increased carrier concentration due to ability of negative bias attract positive boron, increase true doping boron incorporated in the film and this might explained why conductivity increased as negative bias increased. In addition, the electron mobility of the a-C:B film might be increased due to the film in nanostructure and ability hydrogen passivated in film as discussed previously.

Table 2. Cconductivity and photoresponse of a-C:B films by applied negative bias

Applied DC bias voltage (V)	Conductivity (σ)		photoresponse
	Under dark	**Under illumination**	($\sigma_{illumination}/\sigma_{dark}$)
0	$2.0438 x 10^{-7}$	$3.126 x 10^{-6}$	1.531
-10	$4.536 x 10^{-6}$	$8.407 x 10^{-6}$	1.856
-20	$4.878 x 10^{-6}$	$9.746 x 10^{-6}$	1.998

Table 2 shows photo-responses of a-C:B films with a function of DC applied bias. Increment in conductivity values under illumination indicate photoresponse characteristic for thin films. Photoconductivity of thin films is important in addressing the suitability of thin films to be evaluated in photovoltaic solar cell applications. The photoresponse of a-C:B film ranging from 0 V to -20 V is 1.531, 1.856, and 1.998, respectively. It observes that a-C:B films have the overall range photo-response below 2 with the highest one 1.998 at -20V. Although photo-responses are low, all hydrogenated nanostructure a-C:B films are rarely reported by using this preparation. The film shows the ability to absorb light by producing current flow in the external circuit. Current flow is the results of the photoexcited carriers drift before recombination [23]. For hydrogenated a-C:B films under illumination, the ohmic form is similar with hydrogenated nanostructure a-C:B films under dark but higher than under dark measurement is due to the higher generation of excess carriers. We predict, negative bias as discussed previously gives more excess of holes and thereby generates more excess carriers. The nanostructure of a-C:B film drift electron-hole pair in the load rather and thus slow recombination time . The hydrogen terminated the dangling bond reduced less defect in hydrogenated nanostructured a-C:B film and thus prevent recombination excess of carrier in the film surface.

4. Conclusion

The hydrogenated nanostructured a-C:B films were successfully prepared by bias assisted-CVD. FTIR and FESEM images confirmed the carbon is C-H binding and in nanostructure, respectively. The FESEM images at 0 V showed particles size was less than 45 nm. Furthermore, a more uniform of particles size less than 25 nm was achieved at -20 V. Raman spectra showed the a-C:B films deposited by different negative bias were amorphous in nature. Raman analyses showed a change of structural properties was correlated with incorporated of carbon, boron, hydrogen and other elements in carbon network. Hydrogen was believed to be incorporated with dangling bonds near the surface as correlated with vibration peaks at 28850 cm^{-1} of FTIR spectra. The overall of photo-response is below 2 with the highest one 1.998 at -20V. Current flow is the results of the photoexcited carriers drift before recombination. For hydrogenated a-C:B films under illumination, the ohmic form is similar with hydrogenated nanostructure a-C:B films under dark but higher than under dark measurement is due to the higher generation of excess carriers. The nanostructure of a-C:B film drift electron-hole pair in the load rather than recombine and thus reduce recombination time. The hydrogen terminated the dangling bond reduced defect in hydrogenated nanostructure a-C:B film and thus slower recombination of excess carrier in the film surface.

Acknowledgments
The authors thank to Ministry of Higher Education (MOHE) Malaysia, UiTM Kota Samarahan Sarawak for the scholarship, NANO-ElecTronic Centre, Faculty of Electrical Engineering, Universiti Teknologi MARA, NANO-SciTech Centre, Institute of Science, Universiti Teknologi MARA, and Research Management Institute (RMI) Universiti Teknologi MARA (UiTM) for the facilities.

References

[1] Ishak A, Dayana K, Mamat M, Malek M and Rusop M 2015 *Optik*, **126** 1610-1612.
[2] Annuar I, Rouhi J and Rusop M 2015 *IEICE Elec. Expr.* **12** 20140937-20140937.
[3] Mallikarjuna A, Reddy, Sivasankar Reddy A and Sreedhara Reddy P 2011 *Mater. Chem. and Phys.* **125** 434-439.
[4] Zhang H, Zhang Z, Li J and Cai S 2007 *Electrochemistry comm.* **9** 605-609.
[5] Buijnsters J G,Camero M, Vázquez L, Agulló-Rueda F, Gómez-Aleixandre C and Albella J 2007 **Vacuum** **81** 1412-1415.
[6] Podder J, Rusop M, Soga T and Jimbo T 2005 *Diam. and rel. mater.* **14** 1799-1804.
[7] Gharbi R, Fathallah M, Alzaied N, Tresso E and Tagliaferro A 2008 *Mater. Sci. and Eng. C* **28** 795-798.
[8] Zhang Q and Cao G 2011Nano *Today* **6** 91-109.
[9] Liu Y, Lai T, Li H, Wang Y, Mei Z, Liang H, et al. 2012 *Small* **8** 1392-1397.
[10] Kamaruzaman D, Ahmad N, Annuar I and Rusop M 2013 *Japan. J. of Appl. Phys.* **52** 11NL02.
[11] Mahadik D, Shinde S, Bhosale C and Rajpure K 2011 *J. of Alloys and Comp.* 509 1418-1423.
[12] Rusop M, Soga T and Jimbo T, 2005 *Thin solid films* **482** 280-286.
[13] Pardanaud C, Martin C, Roubin P, Giacometti G, Hopf C, Schwarz-Selinger T, et al. 2013 *Diam. and Rel. Mater.* **34** 100-104.
[14] Jones D and Stewart A 1982 *Philosophical Mag. B* **46** 423-434.
[15] Ferrari A C and Basko D M 2013 *Nature nanotechnology* **8** 235-246.
[16] Burgess J S, Acharya C K, Lizarazo J, Yancey N, Flowers B, Kwon G, et al. 2008 *Carbon* **46** 1711-1717.
[17] Urbonaite S, Hälldahl L and Svensson G 2008 *Carbon* **46** 1942-1947.
[18] Clough F, Milne W, Kleinsorge B, Robertson J, Amaratunga G and Roy B 1996 *Elect. Lett.* **32** 498-499.
[19] Tian X, Rusop M, Hayashi Y, Soga T, Jimbo T, and Umeno M 2003 *Solar energy mater. and solar cells* **77** 105-112.
[20] Chan W, Allen T, Gaspari F, Szurmak J, Sagnes E and Zukotynsk S 1997 *MRS Proceedings.*
[21] Huang L and Meng 2007 *Mater. Sci. and Eng.: B* **137** 310-314.
[22] Kumari L, Prasad V and Subramanyam S 2003 *Carbon* **41**1841-1846.
[23] Suemasu T, Saito T, Toh K, Okada A and Khan M A 2011 *Thin sol. Films* **519** 8501-8504.

IOP Conf. Series: Materials Science and Engineering **341** (2018) 012018 doi:10.1088/1757-899X/341/1/012018

Modified hyperbolic sine model for titanium dioxide-based memristive thin films

Raudah Abu Bakar[1]*, NurSyahirah Kamarozaman[1], Wan FazlidaHanim Abdullah[2], and Sukreen Hana Herman[1, 2]

[1]NANO-Electronic Centre (NET), Faculty of Electrical Engineering, UniversitiTeknologi MARA, Shah Alam, Selangor Darul Ehsan, Malaysia .
[2]Integrated Sensors Research Group, UniversitiTeknologi MARA, Shah Alam, Selangor Darul Ehsan, Malaysia.

E-mail: raudah088@salam.uitm.edu.my

Abstract. Since the emergence of memristor as the newest fundamental circuit elements, studies on memristormodeling have been evolved. To date, the developed models were based on the linear model, linear ionic drift model using different window functions, tunnelling barrier model and hyperbolic-sine function based model. Although using hyperbolic-sine function model could predict the memristor electrical properties, the model was not well fitted to the experimental data. In order to improve the performance of the hyperbolic-sine function model, the state variable equation was modified. On the one hand, the addition of window function cannot provide an improved fitting. By multiplying the Yakopcic's state variable model to Chang's model on the other hand resulted in the closer agreement with the TiO_2 thin film experimental data. The percentage error was approximately 2.15%.

1. Introduction

In the past few years, memristor or memory resistor has been recognized as the fourth fundamental passive circuit element [1]. This newest device not only could relate the fluxes and charges relationship but also has the ability to remember its current state even without the external applied voltages[1-2]. Because of these, memristor can be incorporated in many applications including non-volatile memory device, neuromorphic computation system, artificial intelligence, analog computing and others [3-5].

To date, the memristive behavior a pinched-hysteresis current-voltage (I-V) characteristicshas been discovered in many transition metal oxide materials such as titanium dioxide (TiO_2), zinc oxide (ZnO), nickel oxide (NiO), tungsten oxide (WO_x) and tantalum oxide (TaO_x) [6-12]. Theoretically, the observed distinctive memristive properties in these materials are due to either the migration of oxygen vacancies or the formation of filamentary conducting path within the oxide layer [13].

Recently, a lot of studies on memristor modeling have been carried out to predict the performance of metal oxide-based memristive device and to allow successful memristor implementation in various applications [14-23]. Other than linear ionic drift model [14], models on hyperbolic sine function have been proposed to approximate the I-V relationship of metal-insulator-metal (MIM) configuration [21-23]. Unlike linear memristor model, this model suggests an exponential relationship between the current and applied voltage through a thin potential barrier [24]. Although using a hyperbolic sine function in modeling the I-V relationship in MIM device satisfies both electronic measurements and material composition [25], only few of them were developed based on the experimental data. One of them is Chang *et al.* model [23]. In Chang's model, both Schottky and tunneling effects were considered in the memristor I-V relationship. The state variable derivative on the other hand was included with the diffusion term.

Since the state variable provides the change in the resistance, modeling this function is therefore very important. Thus, in order to provide a well fitted model to the experimental data, the Chang's state variable function was modified by adding the previously reported window functions [14-17]. Other than that, the inclusion of Yakopcic's state variable function [25] was also examined. The model parameters were initially correlated to a sol-gel-derived spin-coated TiO_2 thin film annealed at 250°C for 20minute and optimized. Various window functions were then added and the memristive behavior was analyzed. The characterization data were obtained from [26]. The simulation process was conducted using Linear Technology Simulation Program with Integrated Circuit Emphasis (LTspice).

2. Hyperbolic sine function model

The I-V relationship and state variable function for Chang's model [23] are described by the following equations:

$$I(t) = (1 - x(t))\alpha\left[1 - e^{\beta V(t)}\right] + x(t)\gamma \sinh(\delta V(t)) \quad (1)$$

$$\frac{dx}{dt} = \lambda[\eta_1 \sinh(\eta_2 V(t))] \quad (2)$$

As given in (1), the I-V relationship is the combination of the Schottky barrier effect between the oxide layer and bottom electrode and the tunnelling effect through the MIM junction. The Schottky conduction becomes dominant if the state variable, $x(t)$ is zero and vice versa when $x(t)$ is 1. The state variable function, on the other hand, is the function of fitting parameters λ, η_1 and η_2 which are used to shape the dynamics of the state variable equation [27]. The state variable function is then modified by adding an ion diffusion term to (2) to account for the natural oxygen vacancy (V_{ox}) diffusion [27]. The new equation is defined as:

$$\frac{dx}{dt} = \lambda\left[\eta_1 \sinh(\eta_2 V(t)) - \frac{x(t)}{\tau}\right] \quad (3)$$

whereτ is a time constant on the order of a few seconds.

3. Window function

In this study, the window functions developed by Strukov [14], Joglekar [15], Biolek [16] and Prodromakis[17] were considered to examine the effects of window function inclusion to state variable derivative function. Table 1 summarized the window function used in this study. The window function was initially introduced by Strukov*et al.* in order to compensate the boundary problem in the linear ionic drift model. This can be achieved by multiplying the state-variable derivative (*dx/dt*) with window function (*f(x)*) as in (4).The window function will not only ensure the proper state-variable shaping near the boundary [0, D] but also capable to include the nonlinear behaviour effects close to the boundary [25, 28].

$$\frac{dx}{dt} = ki(t)f(x), \qquad k = \frac{\mu_v R_{ON}}{D} \quad (4)$$

Table 1. Window functions described by Strukov [14], Joglekar [15], Biolek [16], and Prodromakis [17]

Models	Window Function (f(w))	References	
Strukov	$f(x) = x(1-x)$	[14]	(5)
Joglekar	$f(x) = 1-(2x-1)^{2p}$	[15]	(6)
Biolek	$f(x) = 1-(x-stp(i))^{2p}$	[16]	(7)
Prodromakis	$f(x) = j(1-(x-0.5)^2+0.75)^p$	[17]	(8)

4. Results and discussions

Prior to the modification of state variable equation, the Chang's model was correlated to the I-V characteristic of TiO_2 thin films prepared by spin coating sol-gel method. The thin films were fabricated using titanium (IV) isopropoxide ($Ti[OCH(CH_3)_2]_4$), glacial acetic acid (CH_3COOH), Triton X-100 ($C_{34}H_{62}O_{11}$), absolute ethanol (C_2H_5OH) and deionized water, which acted as the precursor, stabilizer, surfactant, and solvent. The coated TiO_2 layers were then subjected to annealing process at 250°C for 20 min to enhance the films crystallinity. The I-V characteristic of the memristive device structure consisted of TiO_2 thin film sandwiched between platinum (Pt) electrode and indium-doped tin oxide (ITO) coated glass substrate (Pt/TiO_2/ITO glass) was then characterized by a two-point probe measurement method using a Keithley 4200 semiconductor characteristic system. The detailed fabrication method and results are reported in [26].Figure 1 shows the annealed TiO_2 thin film I-V characteristic.As apparent in Figure 1, the TiO_2 thin film exhibits a pinched hysteresis loop. From the I-V characteristic, the low and high state resistances, $R_{ON}$and R_{OFF} were calculated as 51 and 137 Ω respectively.

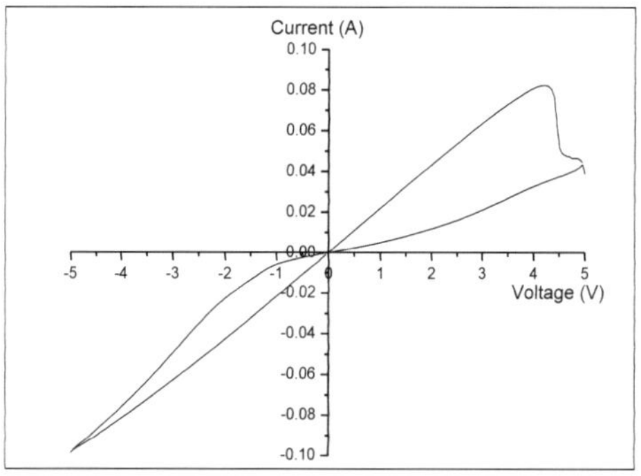

Figure 1. I-V Characteristic of TiO_2 thin film

The Chang's SPICE model was simulated using the circuit shown in Figure 2a. Figure 2b shows the comparison between the simulated I-V characteristic results and the experimental data for TiO_2 thin film. Using the fitting parameters $\alpha = 3.195$, $\beta = 6.505\times10^{-3}$, $\gamma = 2.45\times10^{-1}$, $\delta = 3.25\times10^{-2}$, $x_{max} = 0.95$, $x_{min} = 0.05$, $drift\_bit = 150$, $\lambda = 0.06$, $\eta_1 = 12.5\times10^{-6}$, $\eta_2 = 4$ and $\tau = 1$ resulted in an asymmetrical I-V characteristic with a percentage error of 4.75%.

IOP Conf. Series: Materials Science and Engineering **341** (2018) 012018 doi:10.1088/1757-899X/341/1/012018

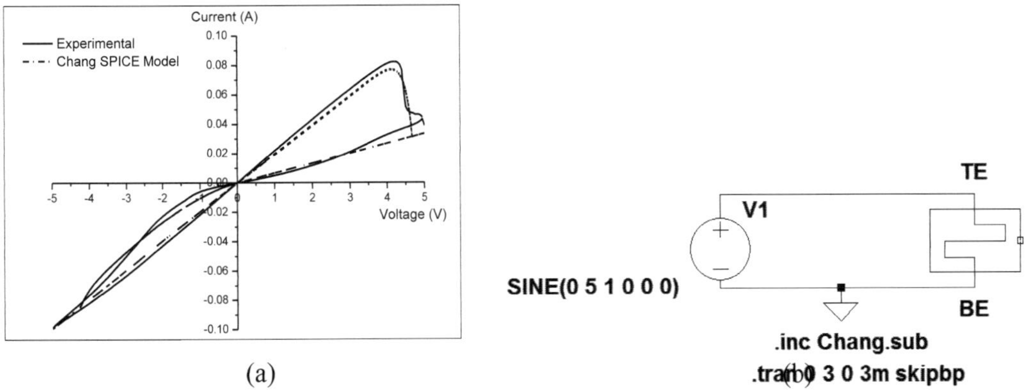

(a) (b)

Figure 2. (a) Schematic circuit for parameters optimization in LTspice (b) Simulation result for TiO_2 thin film when modeled using Chang's hyperbolic sine function model for a sinusoidal input of 5 V

Figure 3 a-d show the simulation results for the combination of Chang's model and various window function types. As can be observed from the I-V characteristics, the addition of window function to the state variable derivative function leads to the un-fitted memristive behaviour especially during the applied negative voltages. The improvement on the positive voltages side however can be obtained by the inclusion of both Joglekar and Biolek window functions. Considering to Joglekar's window function in (6), the improved characteristic is achieved because of the addition of control parameter, p to their window function to control the linearity and non-linearity of the drift phenomenon [15]. Biolek on the hand, managed to address the boundary problem when the doped and un-doped layers reach either end of the device by introducing switching window function (*stp(i)*) [16]. Thus provide better correlated I-V characteristics than Joglekar.

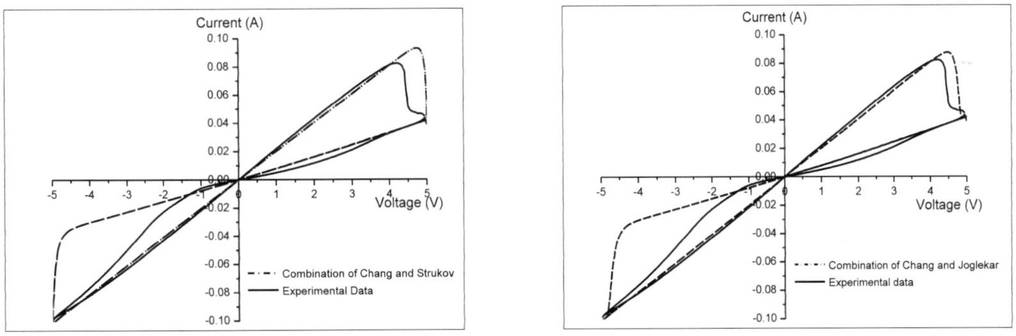

(a) (b)

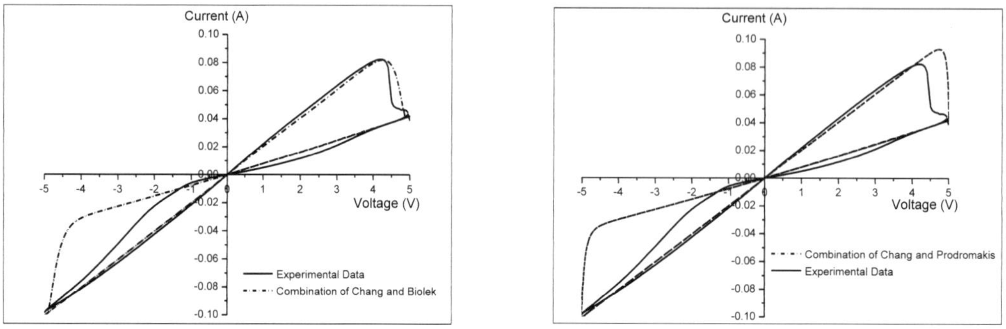

(c) (d)

Figure 3. Simulation result for the combination of Chang's hyperbolic sine function model with (a) Stukov (b) Joglekar (c) Biolek and (d) Prodromakis window functions

Since the I-V characteristics of TiO_2 thin film is asymmetrical, the function that is capable to model a switching state variable depending on the input voltage polarity is therefore essential. One of the memristor models which could provide this capability is the model created by Yakopcic [25]. In his model, the state variable *f(x)* is modeled by dividing the state variable motion into two different regions as given in (9) and (10),

$$f(x) = \begin{matrix} e^{-\alpha_p(x-x_p)} w_p(x, x_p), \\ 1, \end{matrix} \quad (9)$$

$$f(x) = \begin{matrix} e^{\alpha_n(x+x_n-1)} w_n(x, x_n), \\ 1, \end{matrix} \quad (10)$$

where α_p, α_n, x_p and x_n are the parameters that determine where the state variable motion is no longer linear and the degree to which the state variable motion is dampened [25] while w_p and w_n are the window function which is given by the following formula

$$w_p(x, x_p) = \frac{x_p - x}{1 - x_p} + 1 \quad (11)$$

$$w_n(x, x_n) = \frac{x}{1 - x_n} \quad (12)$$

As can be observed in Figure 4, the incorporation of Yakopcic's state variable model to Chang's model resulted in the improved correlated TiO2 thin film I-V characteristic. The percentage error was calculated to be approximately 2.15%.

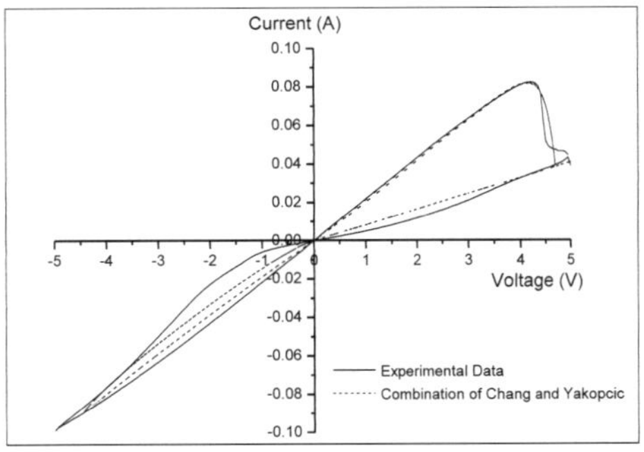

Figure 4.Simulation result for the combination of Chang's hyperbolic sine function model with window function

5. Conclusion

This study is carried out to examine the modified hyperbolic sine function model for titanium dioxide-based memristive thin film. The modification was performed by adding the various types of window functions and Yakopcic's state variable function to Chang's hyperbolic sine function model. The inclusion of window function however resulted in an un-fitted I-V characteristics. A better correlated characteristic was achieved when incorporating the Yakopcic's model.

References

[1] L. Chua 1971*IEEE Trans. on Circuit Theory* **18** 507
[2] Williams R-S 2009 *IEEE Spectrum* **45** 28
[3] Hamdioui S, Xie L, Nguyen H A D, Taouil M, Bertels K, Corporaal H, Jiao H, Catthoor F, Wouters D, Eike L and Lunteren J V 2015 *Design, Automation & Test in Euro. Conf. Exhibit.* p 1718
[4] Potrebic M and Tosic D 2015 *Radioeng.***24** 409
[5] Tetzlaff R, Schmidt T 2012*IEEE Int. Symp. Circ Sys.* 1590
[6] Kasim S M M, Shaari N A A, Bakar R A and Herman S H. 2015 *App. Mechan. Mats.***749** 308
[7] Shaari N AA, Kasim SMM, Sauki NSM and Herman SH. 2015 *4th Int. Conf. Electron Device Syst. Apps.* 99 p 1
[8] Tripathi S K, Kaur R and Rani M 2014 *Solid State Phenomena***222** 67
[9] Lee S, Park J-B, Lee M-J and Boland J J 2016 *AIP Advances***6** 125010
[10] Pergament A, Stefanovich G, Malinenko V and Velichko A 2015 *Adv. Condensed Matt. Phys.***654840** 1
[11] Mohammad B, Jaoude M A, Kumar V, Al Homouz D M, Nahla H A, Al-Qutayri M and Christoforou N 2016 *Nanotechnol. Rev.*
[12] Fauzi F B, Othman R, Mohamed M A, Herman S H, Ahmad A A Z and Ani M H 2015 *Japan Ins. Metals Mats.***8** 1302
[13] Sawa A 2008 *Mats. Today***11** 28
[14] Strukov DB, Snider GS, Stewart D R, and Williams R S 2008*Nature***453** 80
[15] Joglekar Y N and Wolf S J 2009 *Euro. J. Phys.***30** 661
[16] Biolek Z, Biolek D and Biolková V 2009 *Radioengin..* **18** 210
[17] Prodromakis T, Peh BP, Papavassiliou C and Toumazou C 2011 *IEEE Trans. Elect. Devices* **58** 3099
[18] Kvantinsky S, Friedman E G, Kolodny A and Weiser U C 2013 *IEEE Trans. Cct. Systs – I: Regular Papers* **60** 211

[19] Abdalla H and Pickett MD 2011 *IEEE Int. Symp. Cct. Syst. (ISCAS)* 1832
[20] Laiho M, Lehtonen E, Russel A and Dudek P 2010*Neuromorphic Engin.*
[21] Eshraghian K, Kavehei O, Cho K R, Chappel JM, Iqbal A, Al-Sarawi S F and Abbott D 2012 *Proc. IEEE* 100 1991
[22] Yakopcic C, Taha T M, Subramanyam G and Pino R E 2013 *Int. Joint Conf. Neur Net.* 1
[23] Chang T, Jo SH, Kim KH, Sheridan P, Gaba S, and Lu W 2011 *Appl Phys. A*, **102** 857
[24] Merkel C.E Thermal profiling in CMOS/memristor hybrid architectures. Master thesis, Rochester Institute of Technology, Rochester, New York, 2011.
[25] Yakopcic C M S Memristor device modeling and circuit design for read out integrated circuits, memory architectures and nueromorphic systems. PhD thesis, University of Dayton, Dayton, Ohio. 2014.
[26] Kamarozaman N S, Mohamed Soder M F, Musa M Z, Bakar R A, Abdullah W F H, Herman S H and Rusop M 2014 *Adv-Maters. Research***925** 125
[27] Chang T. Tungsten oxide memristive devices for neuromorphic applications. PhD thesis, The University of Michigan, Michigan. 2012.
[28] Radwan A G and Fouda ME 2015 Memristor: models, types and applications *On the Mathematical Modeling of Memristor, Memcapacitor and Meminductor*, 1st ed. (Switzerland: Springer Int. Publishing)pp 13-23

MATLAB Simulation of Photovoltaic and Photovoltaic/Thermal Systems Performance

Farah H. M. Nasir, Yusnira Husaini

Faculty of Electrical Engineering, Universiti Teknologi MARA, Shah Alam, Selangor, Malaysia

yusni458@salam.uitm.edu.my

Abstract. The efficiency of the photovoltaic reduces when the photovoltaic cell temperature increased due to solar irradiance. One solution is come up with the cooling system photovoltaic system. This combination is forming the photovoltaic-thermal (PV/T) system. Not only will it generate electricity also heat at the same time. The aim of this research is to focus on the modeling and simulation of photovoltaic (PV) and photovoltaic-thermal (PV/T) electrical performance by using single-diode equivalent circuit model. Both PV and PV/T models are developed in Matlab/Simulink. By providing the cooling system in PV/T, the efficiency of the system can be increased by decreasing the PV cell temperature. The maximum thermal, electrical and total efficiency values of PV/T in the present research are 35.18%, 15.56% and 50.74% at solar irradiance of 400 W/m2, mass flow rate of 0.05kgs-1 and inlet temperature of 25 °C respectively has been obtained. The photovoltaic-thermal shows that the higher efficiency performance compared to the photovoltaic system.

1. Introduction

Renewable energy is defined as the energy that is generated from natural resources that can continually be recharge. There are various types of renewable energy, one most important are solar energy. The photovoltaic system is one of the renewable electric power generations. It can generate direct current electricity without contamination and environmental impact when exposed to solar radiation. To improve the performance of the photovoltaic system, it was brought to recover the heat dissipated by convection and conduction, this system use of photovoltaic-thermal in order to operate thermal and electrical performance for both occupied the same space.

The photovoltaic or known as PV cells or solar cells, are the semiconductor pn-junction that consists of p-type and n-type [1]. When sunlight is exposed, photons with energy greater than band gap energy is absorbed to create an electron-hole pair to the incident irradiation is produced [2]. While, the photovoltaic-thermal (PV/T) is a combination of a photovoltaic panel that integrates to a solar thermal collector to removes waste heat from the Photovoltaic module for water heating and cooling system [3]. The photovoltaic-thermal (PV/T) system studied in this research is Solar PV, a rectangle tunnels, an absorber, and an insulator. The rectangle tunnel was placed below PV cells as the heat conductor to increase the efficiency of thermal [3] . The purpose of the absorber plate is to cool the PV module and to improve its electrical performance at the same time to collect the thermal energy produced. The absorber plate was covered with PV cells and hence constituted the combined photovoltaic/thermal (PV/T) absorber.

PV and PVT system have been investigated and discussed by many researchers. For the photovoltaic module, many researchers used simplest the equivalent circuit PV cell to characterize the I-V and P-V characterization [1], [2], [5]–[8]. The software that uses to model the simulation of equivalent circuit PV module are Matlab/Simulink and PSIM software. While for thermal modeling, has been developed in COMSOL Multiphysics [9]-[10]. A block diagram of the stage by stage model for Simulink modeling of PV module was developed by other researchers [5]-[7]. Two important parameters which are solar irradiation and temperature are discussed of PV simulation study [6]-[7].

From the other researchers, the PV results show the I-V and P-V output characteristics with varying irradiance at the constant temperatures and constant irradiance with varying temperature [5]–[7]. Effects of an increase in solar irradiance, the open circuit voltage also increased. While, effects of varying temperature increase, the open circuit voltage drop and short circuit current will increase [5]–[7]. Besides that, the PV results in effects of varying series resistance, R_S and shunt resistance, R_{SH}. The value of R_{SH} must higher than the value of R_S to reduce the power output of PV cell [8]. The output power reduces when the temperature increase and efficiency increase depends on the weather conditions [1].

Thermal/Electrical of a PV module to enhanced cooling system has been studied by earlier researchers to investigate the electrical and thermal performance [11]. The result shows that the performance PV/T with cooling influenced of increasing irradiance (200-1000 W/m^2) at constant temperature (25°C) and increasing temperature (0-50°C) at absorbed radiation of 800W/m^2.

Another research has been carried out by evaluation Single-pass Photovoltaic-Thermal (PV/T) air collector with absorber is to increase its electrical efficiency and to produce heat energy [4]. The efficiency will increase with the increase of mass flow rate. At a mass flow rate of 0.0754 kg sec-1, electrical and thermal efficiency can achieve 10.06% and 75.16% respectively shows that the PV/T system has the higher performance in efficiency of thermal and electrical.

The analysis of a practical single diode equivalent circuit PV model was implemented by using Matlab/Simulink. I-V and P-V output characteristics with a constant temperature (25°C) and solar irradiation (1000 W/m^2) were observed for the PV module. Then, another observation was done for different surrounding conditions by the varying the solar irradiation (200 W/m^2 and 600 W/m^2) and temperature (50°C and 75°C).

In this project, the aim is to develop the model for photovoltaic (PV) and photovoltaic-thermal (PV/T) using Script Editor Matlab. The main objective is to investigate I-V and P-V output characteristics by using the equivalent circuit for the PV cell in order to calculate the total efficiency. In addition to that is to investigate the PV/T and characterize their thermal and electrical performances.

2. Methodology

2.1. Photovoltaic system modeling

The equivalent circuit of PV cell as shown in figure 3 where Rs is the series resistance, Rsh is the shunt resistance, Iph is the photocurrent source which is dependent on the irradiation and temperature and D is the p-n diode. Voltage, V and current, I represent the output voltage and current respectively.

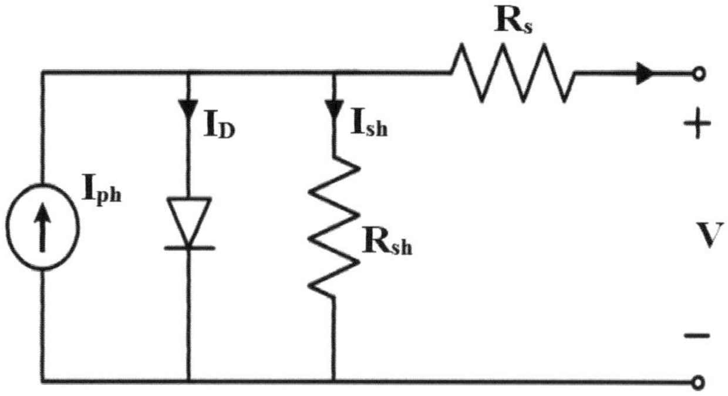

Figure 1 A practical single diode equivalent circuit of PV cell

The mathematical equations of equivalent circuit PV cell shown in figure 1 can be expressed as follow (1)-(5) [1], [2], [5]–[9].

$$I = Iph - Id \tag{1}$$

$$Id = Is\ (\exp(\frac{q(V + IRs)}{NsAkT} + \frac{V + IRs}{Rsh})) \tag{2}$$

$$Iph = Isc + ki(T - 298) \times (\frac{G}{1000}) \tag{3}$$

Voltage V is open circuit voltage, $V_{oc.}$ The generated photo-current I_{ph} is the light generated current at specific temperature and irradiance condition (25°C and 1000 W/m^2). While I_s represents the reverse current. The electron charge q is a constant at 1.602×10^{-19}. In this model, the series and shunt resistance, Rs and R_{sh} are to represent the real internal voltage drop voltage drops and internal losses due to flow of current and any leakage current to the ground when diode is in reverse biased [4]. The value of shunt resistance, R_{sh} is very large than the value of series resistance, R_s. Boltzmann constant, k is 1.38×10^{-23} J/K. The ideality factor A is assumed to be 1 and N_s are the number of cells connected in series. I_{sc} is the short circuit current, ki is the temperature coefficient the short circuit current (0.017), T is the surface temperature of PV cell and G is the solar radiation (W/m^2). The saturation current, I_o varies with temperature, which is given by:

$$Io = Irs \times (\frac{Tn}{T})^3 \exp(\frac{qEg}{Ak})(\frac{1}{Tn} - \frac{1}{T}) \tag{4}$$

where, the reverse saturation current, I_{rs};

$$Irs = \frac{Isc}{\exp\left(\frac{qVoc}{NskAT}\right) - 1} \tag{5}$$

In (4) and (5), Voc represents the open circuit voltage, E_g is the energy band gap for silicon (E_g=1.12) and T_n is the reference temperature of PV cell. The reference values are generally provided by manufacturers of PV cells for specified operating condition (Standard Test Conditions, STC).

2.2. Photovoltaic system modelling

Heat transfer and thermal capacity are the important parameters of thermal because heat transferred to the PV/T by convection and removed from the PV module and PV/T. The performance of PV/T can be depicted by the combination of efficiency expression. It comprised of the electrical efficiency and thermal efficiency. The total of the efficiency is used to evaluate the performance of the system.

$$\eta\_total = \eta\_th + \eta\_elec \tag{6}$$

The electrical efficiency is expressed in (7), where η_{Tref} as the electrical efficiency at references conditions is 10%, T_{pv} is the PV module surface temperature and β_{ref} as thermal coefficient of PV module are 0.003 [12].

$$\eta\_elec = \eta\_Tref\,[1 - \beta(T\_pw - T\_ref\,)] \quad (7)$$

The total amount of energy and the output thermal energy extracted by cooling water can be determined by the given equation in (8) and (9). Solar irradiation, Q is the energy in, $\dot{m}$ is the mass flow rate and Cp is the specific heat capacity. T_{outlet} and T_{inlet} are the outlet and inlet temperature respectively. The thermal efficiency, η_{th} can be expressed by equation (10).

$$E_{in} = Q.A \quad (8)$$

$$E_{water} = \dot{m}C_p(T_{outlet} - T_{inlet}) \quad (9)$$

$$\eta_{th} = Ewater/Ein \quad (10)$$

Hence, the quantity of the total input energy converted to electrical energy (E_{pv}) is obtained by the total amount energy multiplying by electrical efficiency.

$$E_{pv} = \eta_{elec}E_{in} \quad (11)$$

By assuming that there are no energy losses, the total efficiency of the PV/T can be calculated by equation (12):

$$\eta_{total} = (Ewater + Epv)/Ein \quad (12)$$

2.3. Comparison of PV and PV/T systems modeling

The models of photovoltaic and thermal systems are implemented by using Matlab/Simulink. The inputs for photovoltaic cell are the surface temperature and solar irradiation, while the output calculated and shown in the I-V and P-V characteristics. KC200GT is taken as the reference module for the simulation and parameters from the manufacturer's datasheet measured under standard test condition as shown in Table 1 [13].

Table 1 Parameter of KC200GT for Photovoltaic module

Parameter	Variable	Value
Open circuit Voltage	V_{OC}	*32.9 V*
Voltage at MPP	V_{MPP}	*26.3V*
Short circuit current	I_{SC}	*8.21A*
Current at MPP	I_{MPP}	*7.61A*
Voltage temperature coefficient	A_V	*-0.123 V/K*

Current temperature coefficient	A_I	*0.0032 A/K*
Maximum power	P_{MAX}	*200.14 W*

3. Results and Discussion

The model of the photovoltaic was implemented using Matlab script editor. The parameters of the model are evaluated using the equations listed in the methodology part. Figure 2– 7 illustrates the dependence of I-V and P-V characteristics on temperature and irradiance for a KC200GT model. Figure 4 and 5 display the I-V and P-V output characteristics of PV module with the constant temperature of 25°C and vary the irradiance from 200 W/m^2, 600 W/m^2 and 1000 W/m^2 respectively.

From figure 2, the PV cell current is dependent on the solar irradiation. The open circuit voltage increase and also current and power output increases when the solar irradiation increased from 200 W/m^2 to 1000 W/m^2 at the constant temperature. This is in accordance to other papers that also proving that by increasing irradiance parameter, the output of current and voltage outputs also increase.

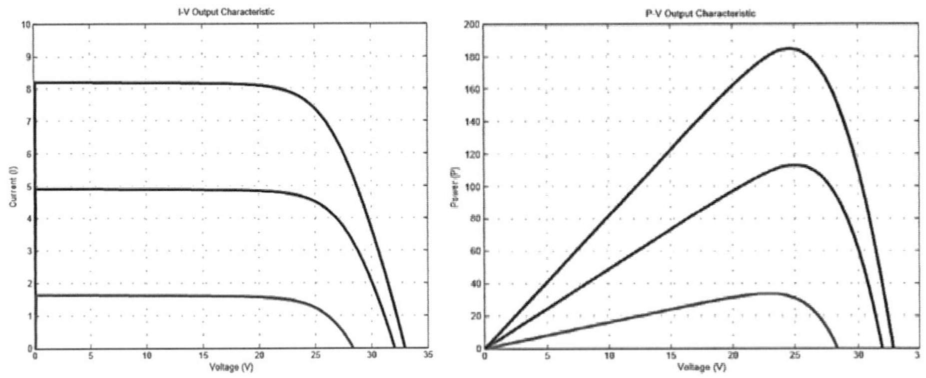

(a) (b)

Figure 2 I-V and P-V characteristics with a constant temperature and varying irradiance

Figure 3 display the I-V and P-V output characteristics of PV module with a constant irradiance of 1000 W/m^2 with varying temperature at 25°C, 50°C and 75°C respectively. From the figures, the open circuit voltage and power output decreases with rise of cell temperature at a constant irradiation. This results are validating the other researchers findings [6]-[7], where by varying the PV cell temperature, the output voltage drops at a faster rate while maintaining at the output current.

Figure 6 shows the efficiency of PV module effects by PV cell temperature. The results show that when the temperature is increased, the efficiency will drop slightly. When the PV cell temperature is increased from 25°C to 75°C, the electrical efficiency dropped from 17.8% to 15.5%. This condition is normal and the main reason of needs in the cooling system research for PV system.

Figure 7 and 8 are showing that the electrical, thermal and total efficiency by the effect of the mass flow rate at solar irradiation 400 W/m^2 and 600 W/m^2 respectively. The results shown that the efficiency (photovoltaic, thermal and total) was increasing with respect to solar radiation. The electrical efficiency varies from 9.82% to 15.5% at solar radiation of 400 W/m^2 and inlet temperature of 30°C. The thermal efficiency varies from 21.10% to 35.18%. The combined efficiency varies from 40.92% to 46.86%. The results in Figure 8 show that the electrical efficiency varies from 9.56% to 11.46% at solar irradiation of

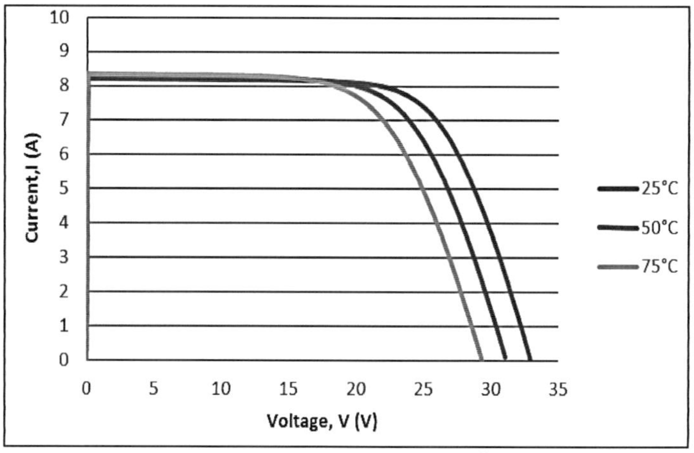

Figure 3 I-V characteristics with 1000 W/m^2 irradiance across the PV surface temperature T at 25, 50 and 75 degree Celsius.

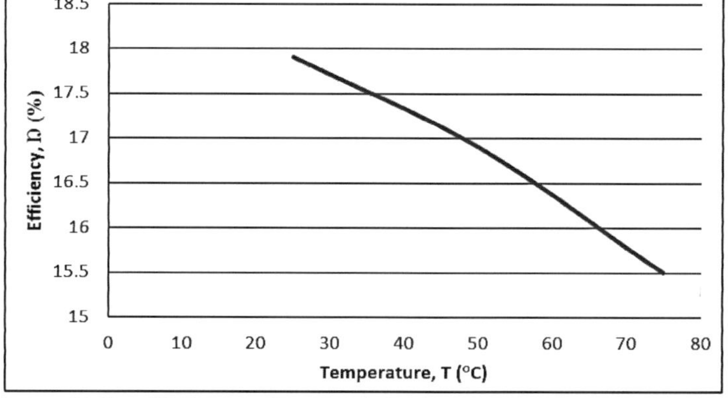

Figure 4 The efficiency of PV module decreases with the increase of PV surface cell temperature.

600 W/m^2. The thermal efficiency varies from 9.89% to 48.5% and combined efficiency of PV/T varies from 18.98% to 58.9%. Figure 5 and 6 also show the comparison of two different values for solar irradiation (400 W/m^2 and 600 W/m^2). The solar irradiation increase by increasing the effect of mass flow rate, but the efficiencies (electrical, thermal and total) is slightly decreased.

The Thermal efficiency reaches steady stage when the mass flow rate over 0.05 kg s-1, the thermal efficiency can increase about 20%. From the equation (9) and (10), the output thermal energy will increase

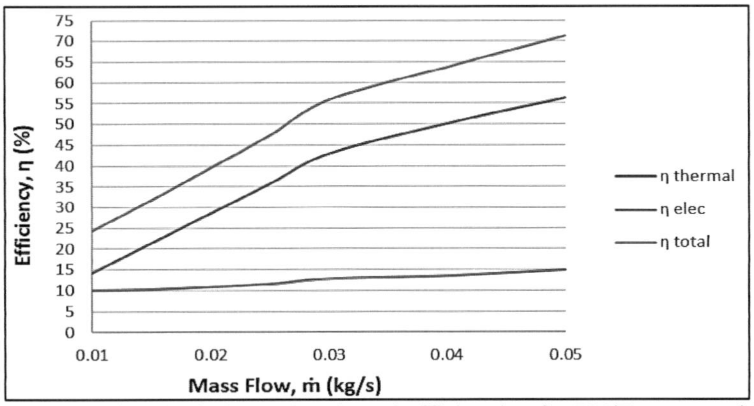

Figure 5 Electrical, thermal and total efficiencies for PV/T system with the effect of the mass flow rate at solar irradiation of 400 W/m^2

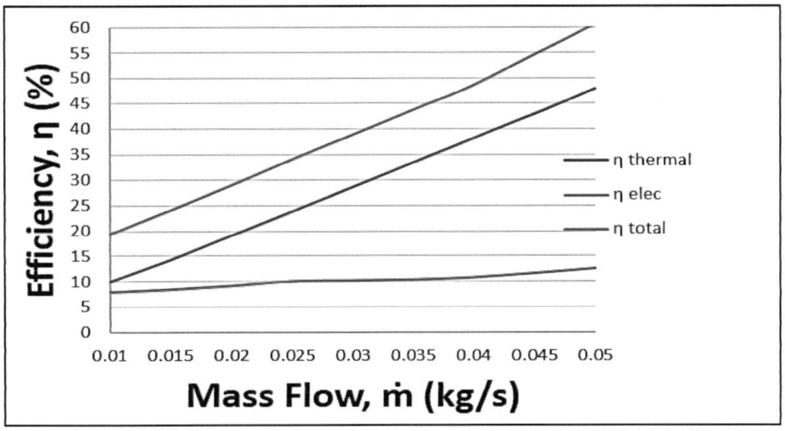

Figure 6 Electrical, thermal and total efficiencies for PV/T system with the effect of the mass flow rate at solar irradiation of 600 W/m2

and also input energy will increase. This is due to the fact that when the mass flow increases, the heat extracted from the air. The more the amount of heat extracted by the water, the lower the PV/T temperature and increases the electrical and thermal efficiency.

4. Conclusion

The I-V and P-V output characteristics are studied and shown with effect of irradiance and PV cell temperature. At a constant solar irradiance, the PV cell temperature has become an important parameter to distinguish the output current, voltage and power. Fortunately, with a cooling system, PV cell temperature can be at constant value. In this condition, an increase of solar irradiance can benefit the system by having an increasing output of current, voltage and power output. Thus, it is showing that the performance of PV cell can be increased and optimized by the integration of thermal module into the system. The maximum of thermal, electrical and total efficiency values of PV/T in the present research at the mass flow rate of 0.05 kgs^{-1} and solar irradiance (400 W/m^2) are 35.18%, 15.56% and 50.74% respectively.

References

[1] S. Rustemli and F. Dincer 2011 Elektron. ir Elektrotechnika **3** pp 35–40
[2] D. Bonkoungou, Z. Koalaga, and D. Njomo 2013 Int. J. Emerg. Technol. Adv. Eng. **3** pp 493–502
[3] Karthick S, and RL HC 2014 Indonesian Journal of Electrical Engineering and Computer Science **12** pp. 8467–8474
[4] G. L. Jin, A. Ibrahim, Y. K. Chean, R. Daghigh, H. Ruslan, S. Mat, M. Y. Othman, and K. Sopian 2010 Am. J. Appl. Sci. **7** pp. 277–282
[5] N. Pandiarajan and R. Muthu 2011 *Proc. 1st Int. Conf. Electr. Energy Syst. ICEES 2011* pp. 258–263
[6] T. Salmi, M. Bouzguenda, A. Gastli, and A. Masmoudi 2012 Int. J. Renew. ENERGY Res., **2** 6
[7] M. Abdulkadir, a. S. Samosir, and a. H. M. Yatim 2012 ARPN J. Eng. Appl. Sci. **7** pp. 616–623
[8] S. Nema, R. K. Nema, and G. Agnihotri 2010 Int. J. ENERGY Environ. **1** pp. 487–500
[9] B. Alsayid and J. Jallad 2012 J. Res. Rev. **2** pp. 1327–1331
[10] A. Elnozahy, A. K. A. Rahman, A. H. H. Ali, M. Abdel-salam, and S. Ookawara, 2015 **4** pp. 1–7
[11] G. JIN, H. Ruslan, S. Mat, M. Y. Othman, A. Zaharim, and K. Sopian 2010 *9th WSEAS International Conference on System Science and Simulation in Engineering, ICOSSSE'10*
[12] Fontenault, B. J. and Gutierrez-Miravete, E. 2002 *Proc. Comsol Conference, Boston, MA*
[13] J. Park, H. Kim, Y. Cho, and C. Shin 2014 **73** pp. 147–155

Effects of Electrical Insulation Breakdown Voltage And Partial Discharge

F S Bahrim, N F A Rahman*, H C M Haris, N A Salim
Faculty of Electrical Engineering, Universiti Teknologi MARA
40450 Shah Alam, Selangor. MALAYSIA
*farahaida@salam.uitm.edu.my

Abstract. During the last few decades, development of new materials using composite materials has been of much interest. The Cross-linked Polyethylene (XLPE) which is insulated power cables has been widely used. This paper describes the theoretical analysis, fundamental experiments and application experiments for the XLPE cable insulation. The composite that has been tested is a commercial XLPE and Polypropylene with 30% fiber glass. The results of breakdown strength and partial discharge (PD) behavior described the insulating performance of the composite.

Keywords*—Insulation breakdown, Cross-linked Polyethylene (XLPE), partial discharge, high voltage.*

1. Introduction

For the last decade, natural rubber has been replaced by synthetic rubbers and plastics as cable insulation. The Cross-linked Polyethylene or commonly known as XLPE is a form of cross-linked polyethylene. XLPE is a thermo set material, which does not deform at higher operating temperature. It is used from 400 V up to transmission voltage [1]. According to previous researches, long term exposure of the insulator surfaces to environmental and operational stress causes several changes on the composition, surface morphology, and reduce water repellency. The excellent resistance to thermal deformation and excellent property of XLPE cable enables it to carry high electric current under normal (90ºC), emergency(130ºC) and short circuit (250ºC) conditions.

However, even though XLPE has great electrical properties, a breakdown in power cable still cannot be avoided due to certain cases [2]. When XLPE is exposed to an extended period of stress, the breakdown strength of XLPE cable will decrease from its original value. Thus, it leads to cable breakdown and it will act as a conductor [3]. By adding an addictive or filler to the XLPE material can increase the breakdown strength and other electrical properties of the cable itself [4]. Thus, in this paper, the XLPE breakdown voltage and partial discharge behavior and how to prolong the breakdown voltage of XLPE are investigated by adding XLPE and Polypropylene with 30% fiberglass filler. Polypropylene with 30% fiberglass is chosen to be a filler due to several advantages: cheaper in market value, more environmentally friendly and has more sustainable construction [5]. Besides, it exhibits high mechanical and electrical performances but low notched strength and resisting stress cracking. Fiberglass is a reinforcement agent that is mostly used in polypropylene based composites, as it has good balance between properties and cost. The mixture between polypropylene and fiberglass has resulted significantly improved tensile and flexural strengths over regular glass reinforced with polypropylene. It also improves dimensional stability, rigidity and the strength. Polypropylene with 30% chemically-coupled glass reinforcement increases the tensile strength over the non-reinforced polypropylene of 80% while a 50% improvement over conventional glass-reinforced grade. Thus, these two types of materials have been mixed industrially with 30% of fiberglass to ensure the material is stronger in terms of electrical properties.

The significance of this study is to investigate the best insulation with its high insulating performance. The objectives of this study are to study the insulating performance, to collect and analyze data of partial discharge and breakdown voltage, and to compare the partial discharge and breakdown voltage behavior between pure XLPE and modified XLPE (with Polypropylene 30% fiberglass filler). The experiment was conducted based on the breakdown voltage according to IEC 60156, IEC 60502, and IEC 60840 standards. Observation, collection, and evaluation of the results of the dielectric properties of pure XLPE and XLPE with added Polypropylene with 30% fiber glass are also included.

2. Methodology

Testing and experimental work is carried out at Nano Laboratory, Polymers Laboratory and High Voltage Laboratory, Faculty of Electrical Engineering and Faculty of Applied Sciences, Universiti Teknologi MARA Shah Alam. A sample of 160-gram XLPE and several samples of polypropylene with 30% fiberglass filler are measured precisely according to their percentages which are 3%, 5%, 7% and 9% as shown in Table 1.

TABLE 1 : List of Sample

Sample	Description
S1	100% XLPE
S2	100% XLPE + 3% Polypropylene 30% Fiber Glass
S3	100% XLPE + 5% Polypropylene 30% Fiber Glass
S4	100% XLPE + 7% Polypropylene 30% Fiber Glass
S5	100% XLPE + 9% Polypropylene 30% Fiber Glass

Fig. 1 shows the flow chart of sampling process which includes collecting information, sampling preparation, extrusion process, heating cooling and compressing process, testing and finally, collecting and analyzing data. Samples are cut with a diameter of 5 cm and 2 mm thickness. This shape is referred to the standard test (CIGRE METHOD II) as shown in Fig. 2.

The experiment was conducted using the High Voltage AC (HVAC) as shown in Fig. 3 and 4. Fig. 3 shows the circuit diagram with earthed point and Fig. 4 shows the testing circuit. Setting value of this experiment is stated in Table 2. Applied voltage is set from 5 kV and up to 100 kV, and each sample is tested until they are fully damaged. For safety purpose, the final voltage is set lower than the limit voltage. The duration of the step is 20 seconds.

TABLE 2: Parameter setting MDSWIN

Limit Voltage (kV)	100.00
Final Voltage (kV)	80.00
Starting Voltage (kV)	5.00
Step Voltage (kV)	1.00
Rate of Voltage rise (kV/s)	3.00
Tolerance of step volt (%)	2.00
No of cycle	20
Duration of Step (s)	00:00:20
Break Duration (s)	00:00:05

3. Result and Discussion

Table 3 shows the results of voltage breakdown across the 5 samples of insulation material that has been tested. In addition, Fig. 5 shows the graph of voltage breakdown against number of test according to the percentage of filler. The graph in Fig. 5 shows the relationship between percentage of filler and the value of breakdown voltage. The breakdown voltage increases as the percentage of filler increases. The results show that XLPE with added polypropylene with 30% fiberglass performs better in insulation behavior compared to the pure XLPE.

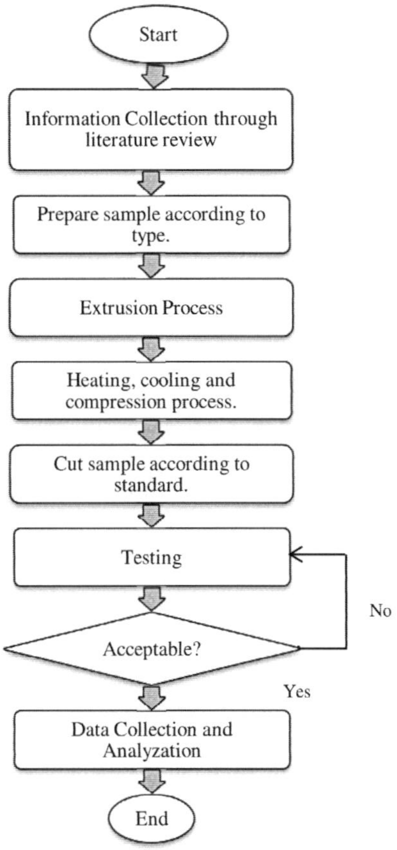

Fig.1 Flow chart of sampling process

Fig. 2 CIGRE METHOD II test

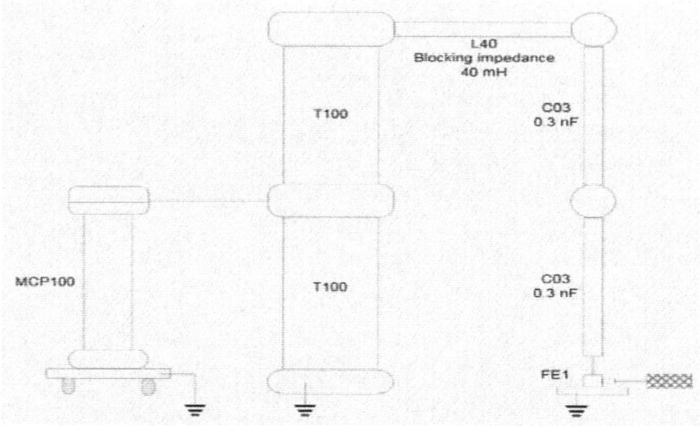

Fig.3 Testing circuit with earthed points

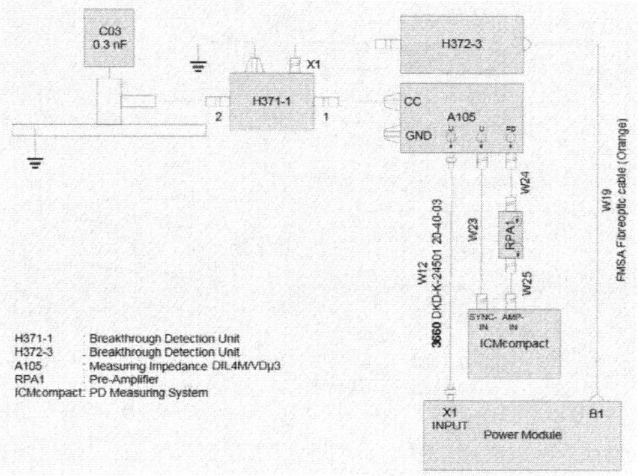

Fig.4 Measuring circuit

TABLE 3: Breakdown voltage for each sample

SAMPLE		S1	S2	S3	S4	S5
BREAKDOWN VOLTAGE(kV)	T1	23.68	26.32	28.43	29.96	24.20
	T2	21.54	26.15	23.40	22.20	24.20
	T3	18.33	23.09	19.83	17.65	24.3 0
	T4	16.98	17.66	13.63	7.39	24.20
	T5	16.21	14.82	8.40	4.39	23.30
PARTIAL DISCHARGE		0 - 100	0 - 300	0 - 400	0- 450	0 - 500

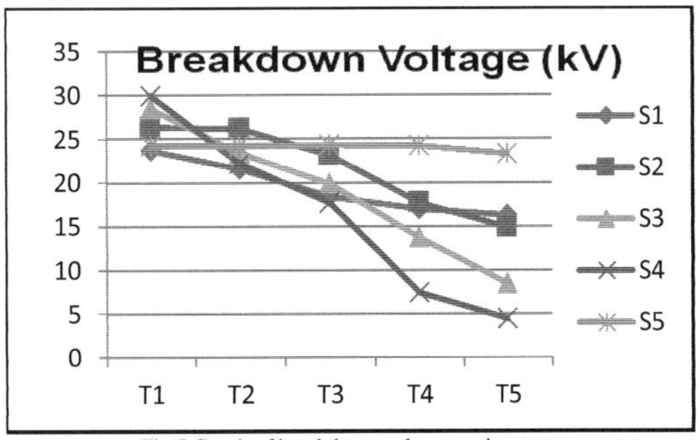

Fig.5 Graph of breakdown voltage against test.

Before the breakdown occurs, the corona sound starts to appear at higher voltage and spark occurs during breakdown. At this moment, the spark (partial discharge) acts as a bridge between two conducting electrodes in insulation system [7, 8]. This sparking can be seen in Fig. 6. The electrode strikes straight to the sample and focus on one point. Tests are repeated for several times on the same part to test on the insulation behavior of the insulation. Once the breakdown occurred, the material may or may not behave as an insulation because the breach has been altering the structure molecule.

Fig. 6 Sparking during breakdown

The insulators are fully broken down at the last stage of testing. During this stage, the sample has the lowest tracking resistance and when voltage is applied, it starts to conduct electric current. It is then leads the insulation to generate heat and causing the surface to become dry.

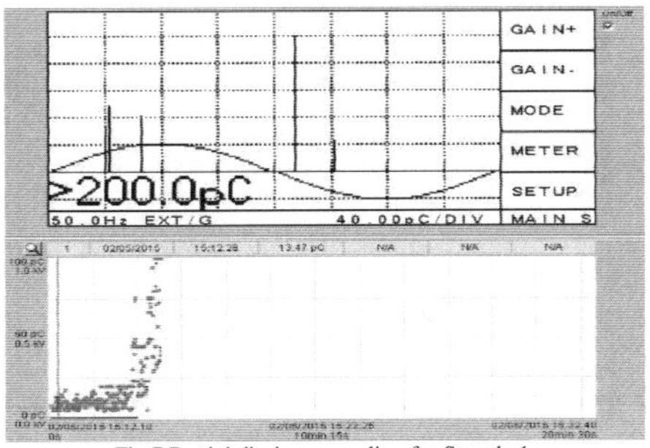

Fig.7 Partial discharge reading for Sample 1

Fig. 7 shows the development of tree length, fractal dimensions, and partial discharge characteristics of a representative branch tree and time for Sample 1. There are 3 regions of tree growth. The tree regions which are A, B and C are clearly defined by the development of the curve in figure above. In region A, the initial tree extends rapidly with a branch structure in the first several seconds. The average rating of growth during this time is 10 pC. The partial discharge activity in region A is intensive with discharge magnitude up to 20 pC. In region B, the branches are growing and a new part of void is formed, thus extending the region. Then, after several minutes with no growth or discharge activity, a little pine branch is formed at the tip of the main channel with the shortest distance from the electrode. The fractal of the tree increases as the time increases in this region. The B-C transition occurs when the pine branch exceeds the original leading branches. In region C, the pine branch grows rapidly away from the electrode. The average growth rate in region C is 75 pC and increases gradually as the trees are away from electrode. The fractal dimension also continues to increase with the tree growth up to 100 pC, thus leading to the breakdown.

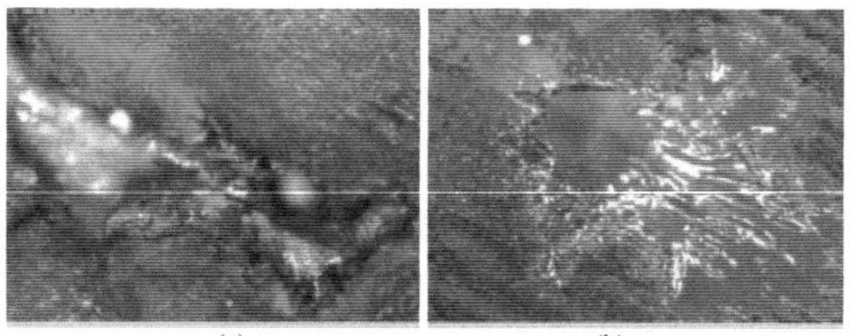

(a) (b)

Fig. 8 Surface of Sample 1

Fig. 8(a) and 8(b) show the real surface degradation of Sample 1 which consists of hollow channels resulting from the decomposition of material after performing several series of breakdown test. First figure shows the upper-surface condition and second figure shows the back-surface condition. By comparing Sample 1 and Sample 4, Sample 4 has higher voltage breakdown of 29.96 kV and higher value of partial discharge. It shows that Sample 4 has longer life time.

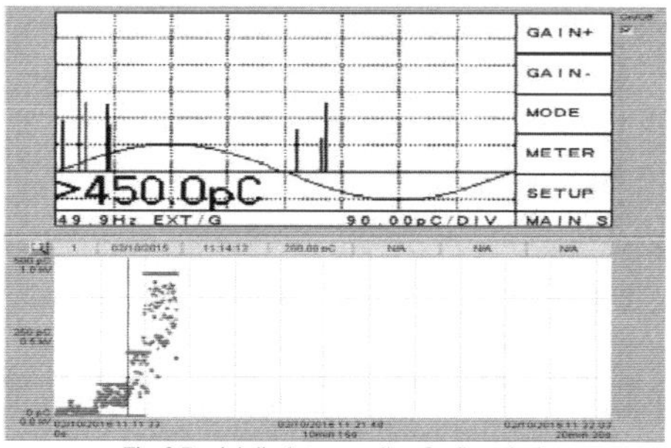

Fig. 9 Partial discharge reading for Sample 4

The tree propagation rate in region A is 120 pC, 450 pC in region B and up to 500 pC in region C as shown in Fig. 9. Fig. 10(a) and 10(b) shows the real surface damage due to the electric stress on Sample 4 after performing several series of breakdown test.

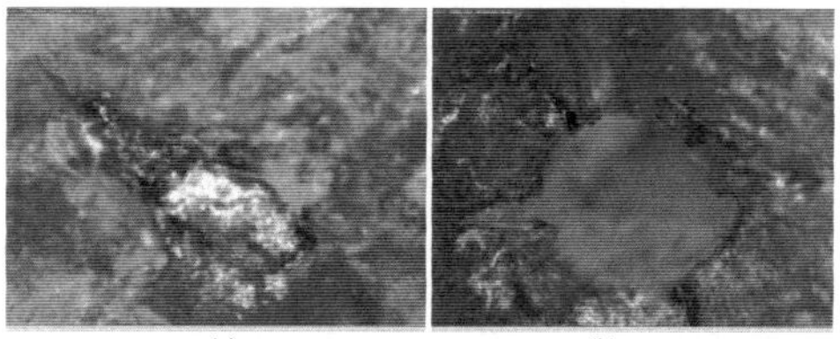

(a) (b)

Fig. 10 Surface of Sample 4

When breakdown occurs in voids, the degradation phenomena will happen, forming positive ions and electrons and they will have sufficient energy. When positive ions and electrons reach the void surface, they break the chemical bond. In each discharge, heat is dissipated in the cavities and carbonize the voids and this will cause erosion of insulator. An active discharge product formed during breakdown causes the chemical degradation All of these effecting results in a gradual erosion of the material and consequent reduction in the thickness of insulation lead to breakdown. Compared to the 100% XLPE, the rate of tree growth for Sample 4 is slower that leads to the increasing time of insulation breakdown. As a result, it shows that the breakdown strength, time and partial discharge increases with the increase in percentage of filler.

Next is Sample 5. Sample 5 is a little bit different. The breakdown voltage of this sample performs at constant value as shown on the graph in Fig. 11 below and has higher number of cycle of test compared to other samples. The sample is not fully broken down and the insulation behavior of this mixture cannot be identified. This problem is related to several reasons.

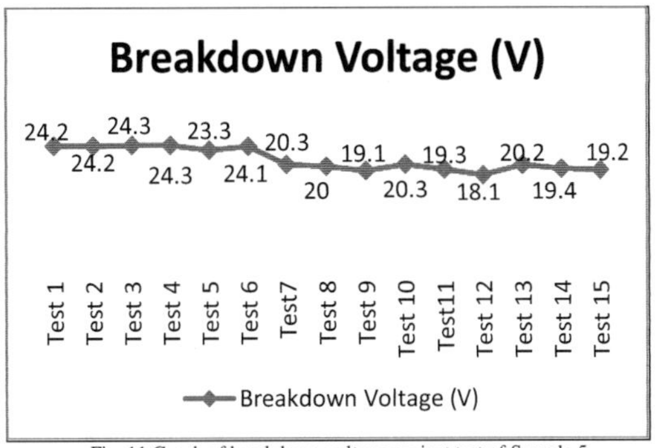

Fig. 11 Graph of breakdown voltage against test of Sample 5

Firstly, there are too many voids and unwanted particles in the insulation that may lead to tracking and treeing phenomena. Partial discharge of Sample 5 is unstable as the initiation and growth of tree is a random process and very complex. Parameters like material morphology, electric field, voltage and temperature play key roles in determining tree growth. This treeing growth is the key failure of the experiment on Sample 5. Water tree growth is marked on the surface of the sample. This can be seen in Fig. 12(a) and 12(b). Voids and unwanted particles in insulation cable can be minimized by a proper process of extrusion and compressing. The insulation needs to be dried in an oven for a night after the extrusion process. Then it must be compressed with hot pressing machine for the next day to make sure that there is no moisture in the insulation that causes void, bubble, etc.

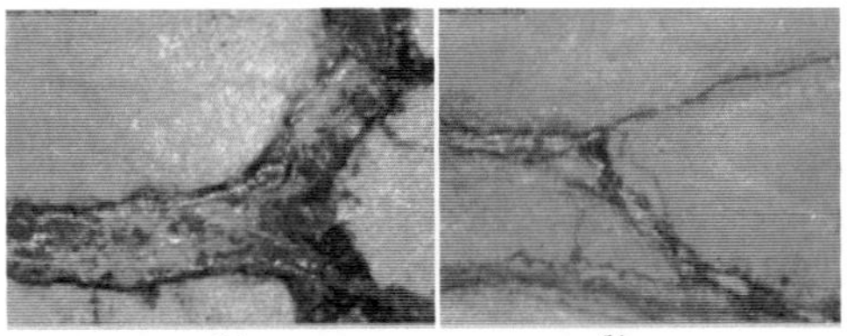

(a) (b)

Fig. 12: Surface of Sample 5

This will prolong the sparking process that causes carbonization at the region of sparking and it will act as permanent conducting channels. The electric field stress of the conducting region increases as the process acts effectively. More and more conducting channels are produced on Sample 5. Thus, the carbonized tracks bridge the gap between the electrodes and finally leads to flash over or system failure. The flash over voltage is defined as the voltage at which current flashes from electrode to electrode or ground with the formation of a sustained arc. This is shown in Fig. 13 where the voltage disperses to the other part of the insulation that has the higher potential of partial discharge phenomenon to occur.

Fig. 13 Flash over during breakdown

Breakdown results on the chemical decomposition and erosion of the insulation material. It is noticeable surface carbonization but these degradations do not affect its flashover properties. However, the insulation properties of Sample 5 failed to be observed as the flash over voltage phenomena occurred.

Next, the silicon rubber that fills the gap between the specimen jigs are not the suitable type or thickness. This silicon rubber can be torn off or can be burned. Therefore, choosing the right silicon rubber is very important procedure. Lastly, the tree grew in the sample because of the presence of moisture. This moisture degrades the performance of the sample. For future study, samples need to be placed or stored at appropriate places.

4. Conclusion

The experiments have revealed that by adding polypropylene with 30% fiber glass with XLPE has improved the breakdown characteristic and performance of insulation compared to pure XLPE breakdown performance. At higher voltages, the partial discharge is high and leads to the formation of tree growth. The trees has produced conducting branches then shows the cross-over from a branch to branch and thus, resulting in increasing fractal dimension, degradation and breakdown phenomena. The most suitable insulation for power cable is XLPE with added 9% polypropylene 30% fiber glass as it can be concluded that the higher percentage of filler increases the insulation performance even though the test for Sample 5 is not a success compared to other samples.

Acknowledgement

The authors would like to thank the Research Management Institute (RMI), Universiti Teknologi MARA, Malaysia and the Faculty of Electrical Engineering, Shah Alam, Malaysia through research grant 600-RMI/DANA 5/3 LESTARI (77/2014) for the financial support of this research.

References

[1] M. Z. H. Makmud, Y. Z. Arief, Aulia, and M. U. Wahit, "Ageing and degradation mechanism of linear low density polyethylene-natural rubber composites due to partial discharge," in *Power and Energy (PECon), 2012 IEEE International Conference on*, 2012, pp. 985-989.

[2] Xiangrong Chen, Yang Xu, Xiaolong Cao, S. J. Dodd and L. A. Dissado "Effect of Tree Channel Conductivity on Electrical Tree Shape and Breakdown in XLPE Cable Insulation Samples" *IEEE Transactions on Dielectrics and Electrical Insulation Vol. 18, No. 3; June 2011*

[3] R. Kawashima, Y. Murakami, H. Shiota, T. Tsurimoto, and M. Nagao, "Influence of filler added into one side of insulation layers on electrical treeing characteristics," in *Electrical Insulation and Dielectric Phenomena (CEIDP), 2013 IEEE Conference on*, 2013, pp. 518-521.

[4] M. Abou-Dakka, A. Bulinski, and S. S. Bamji, "Effect of additives on the performance of cross-linked polyethylene subjected to long term single and periodically reversed polarity DC voltage," *Dielectrics and Electrical Insulation, IEEE Transactions on,* vol. 20, pp. 654-663, 2013.

[5] Shozo Sekioka, "Discussion of CIGRE Current-Dependent Grounding Resistance Model", in 2013 *International Symposium on Lightning Protection (XII SIPDA), Belo Horizonte, Brazil, October 7-11, 2013.*

[6] S. Mohamed Ghousel, S. Venkatesh2, R.Rajesh3, and S. Natarajan4 "Effects of SiO2 and TiO2 Nano Fillers in Enhancing the Insulation Breakdown Strength of Epoxy Nano Composite Dielectric under Divergent Electric Fields" *International Journal on electrical Engineering and Informatics - Volume 5, Number 4, December 2013*

[7] Ling Zhang1, Yuanxiang Zhou1, Xiaoyang Cui, Yanchao Sha1, TrungHieu Le1, "Effect of Nanoparticle Surface Modification on Breakdown and Space Charge Behavior" *IEEE Transactions on Dielectrics and Electrical Insulation Vol. 21, No. 4; August 2014.*

[8] B. X. Du, Z. L. Ma, Y. Gaoand T. Han "Effect of Ambient Temperature on Electrical Treeing Characteristics in Silicone Rubber" *IEEE Transactions on Dielectrics and Electrical Insulation Vol. 18, No. 2; April 2011*

IOP Conf. Series: Materials Science and Engineering **341** (2018) 012021 doi:10.1088/1757-899X/341/1/012021

Optimization of Artificial Neural Network using Evolutionary Programming for Prediction of Cascading Collapse Occurrence due to the Hidden Failure Effect

N. H. Idris[1], N. A. Salim[1], M. M. Othman[1] Z. M. Yasin[1]

[1]Faculty of Electrical Engineering, Universiti Teknologi MARA, 40450, Shah Alam Selangor, Malaysia

norhazwaniidris@yahoo.com

Abstract. This paper presents the Evolutionary Programming (EP) which proposed to optimize the training parameters for Artificial Neural Network (ANN) in predicting cascading collapse occurrence due to the effect of protection system hidden failure. The data has been collected from the probability of hidden failure model simulation from the historical data. The training parameters of multilayer-feedforward with backpropagation has been optimized with objective function to minimize the Mean Square Error (MSE). The optimal training parameters consists of the momentum rate, learning rate and number of neurons in first hidden layer and second hidden layer is selected in EP-ANN. The IEEE 14 bus system has been tested as a case study to validate the propose technique. The results show the reliable prediction of performance validated through MSE and Correlation Coefficient (R).

1. Introduction

Cascading failure is a complicated and complex phenomena which occurs causes increase the stress of power system due to the sequence of line outages [1]. Normally, cascading failure is happen in power grids when one of the elements fails completely or unsuccessfully function and moves its load to nearby elements in the system. One of the main reasons for blackouts in power transmission grids is due to the cascading failure. These blackouts effect on the industrial growth, social and economic of the country [2]. Cascading failures may occurs due to loss of generating units at a station, voltage instability, the change of environment and weather, the software and the failure in protection system. As reported by [3], the most severe of disturbances in power grids convoluted hidden failure of a protection system. Two-third of the most serious disturbances that occurred on the year 1988 and 1984 involved the malfunction or hidden failure in protection system [4].

The severity of the occurrence has inspired to understand and analyze past blackout events and propose the new intelligent technique for prediction model to prevent massive blackouts. Artificial Neural Network (ANN) model is well known in prediction tasks because this model is nonlinear system modeling which capable by learning ability using collected data. ANN provides many advantages compared to the systems of conventional computational. Robustness, ability to memorize and rapidity are the most keen and interesting characteristics of ANN [5]. The architecture of an ANN depends on the type of activation functions which consists of bias and weights among neurons and number of neurons in training parameters.

The training parameters of the ANN model such as learning algorithm, the type of activation functions and number of hidden layer were determined heuristically. Therefore, the overall training process for training parameters became very tedious and consume longer time [6]. This paper proposed with aim to optimize the momentum rate, learning rate and number of neurons in first and second hidden layer during training process for predicting cascading collapse occurrence due to the effects of hidden failure. With a hybrid of Evolutionary Programming (EP) and ANN, the training process becomes adaptable and faster because EP is well-known with fast convergence technique. EP has come out as a useful optimization tool for solving nonlinear programming problems.

2. Methodology

In this paper, a new contribution which is intelligent technique is suggested to predict cascading collapse occurrence due the effects of hidden failure model in steady state condition with hybrid method which is EP is employed to optimize training parameters.

2.1. Data Collection

The proposed models evaluates system performance probabilistically and predicted the historical data from the hidden failure simulation. The historical database is obtained through the cascading events occur in several years and being simulated in the hidden failure model. The probability of tripping for each line has different value depends on the increment of the load. The load is increased to 50% increment from the line bus system. The probability of the hidden failure has been trained and tested in the ANN model for prediction purposed. IEEE 14 bus system is used as a case study.

2.2. ANN-Based Prediction Model

ANN model has been chosen for prediction cascading collapse occurrence with two layer feedforward neural network. ANN with backpropagation was chosen because it can capture a variety of pattern accurately and can be used to perform nonlinear statistical modelling. ANN is an intelligence processing architecture that is inspired from biological nervous system similar with the brain that function for processing the information. ANN is one type of Artificial Intelligence (AI) that attempts to initiate the way a human brain works. It is used in various engineering field and demonstrated remarkable success.

The inputs consists of three variables which are random number generated with constrain 0 to 1, probability of hidden failure and exposed lines. All the data were separated with 80:20 rata for training and testing process. Training and testing data were normalized based on the type of the transfer function [5]. In this study the transfer function used is sigmoid and purely linear transfer function with *Levenberg-Marquardt* learning algorithm to improve the learning speed and avoid the local minimum. This approach will give result in Mean Absolute Percentage Error (MAPE) and Mean Square Error (MSE) as in equation (1) and (2) according to the highest correlation coefficient (R).

$$MAPE = \frac{100}{T}\sum\frac{|Pa - Pp|}{Pa} \tag{1}$$

Where:
P_a The actual the transmission line tripping
P_p The predicted the transmission line tripping
T The total number of the data

$$MSE = \frac{1}{N}\sum_{i=1}^{N}(X'_i - X_i)^2 \tag{2}$$

Where:
N The total number of prediction
X_i The actual vector

X'_i The vector of N prediction

2.3. EP-ANN Model

Despite the ANN had been advantageous in prediction of cascading collapse occurrence, the trial and error method could causes the network to acquire a local optimal solution rather than a global optimal solution. As a result, hybrid model called EP-ANN combination is proposed for solving the optimal parameters during ANN training process that is used to predict the cascading collapse occurrence due to the effect of hidden failure model. The training parameters such as number of neurons in the hidden layer, the learning rate and momentum rate were optimized during the evolution of EP process.

The algorithm of EP consists of initialization of population, fitness computation, mutation, combination, selection, and convergence test. EP has been employed to optimize the three parameters with objective function to minimize the MSE during the training. The proposed EP-ANN model is summarized in the following steps below:

2.3.1. Step 1

Initialize the population with generate 20 random number that correspond to the number of neurons hidden layer1 and hidden layer 2, the learning rate and momentum rate denoted with x_1, x_2, x_3 and x_4. x_1 and x_2 is produced within a range from positive integer 1 to 20 while x_3 and x_4 is generated from 0 to 1 value. After forms the initial candidates of the optimal solutions, this process finally known as parent [6].

2.3.2. Step 2

Fitness process is computed with the set of random number by training the parent in the ANN model called from the step 1. The computation in fitness value is according to the equation (2) which is the MSE value. The correlation coefficient, R and MAPE is also evaluated to check the validity of the error. MAPE should less than 10% error.

2.3.3. Step 3

Mutation process is performed to breed new offspring from the parent. The Gaussian mutation technique as in equation (3) is used to breed offspring.

$$x_{i+m.j} = x_{i.j} + N(0, \beta\,(x_{jmax} - x_{jmin})\,(\frac{fi}{fmax})) \tag{3}$$

Where:

$x_{i+m.j}$ Offspring
$x_{i.j}$ Parents
N Gaussian mutation
β Search step
x_{jmax} Maximum parents
x_{jmin} Minimum parents
fi Fitness 'ith'
$fmax$ Maximum fitness

2.3.4. Step 4

Fitness process is repeated same in the step 2 for each offspring.

2.3.5. Step 5

The process of combination occurred whereas the parents combined with offspring. After that, the population is ranked with the combination of parents and offspring. It is ranked according to the ascending order based on the fitness value. The selection process is performed. The highest MSE will be ranked at the bottom while the lowest MSE will be ranked at the top of the population.

2.3.6. Step 6
Convergence test is evaluated based on the stopping criterion which is the difference between the maximum fitness and minimum fitness should be less than 0.001. If the stopping criterion is violated, step 3 is repeated. Otherwise, the final parents in this step 6 is denoted as the final solutions.

The flowchart [7] of the whole steps is briefed in section 2.4 as shown below:

2.4. Flowchart

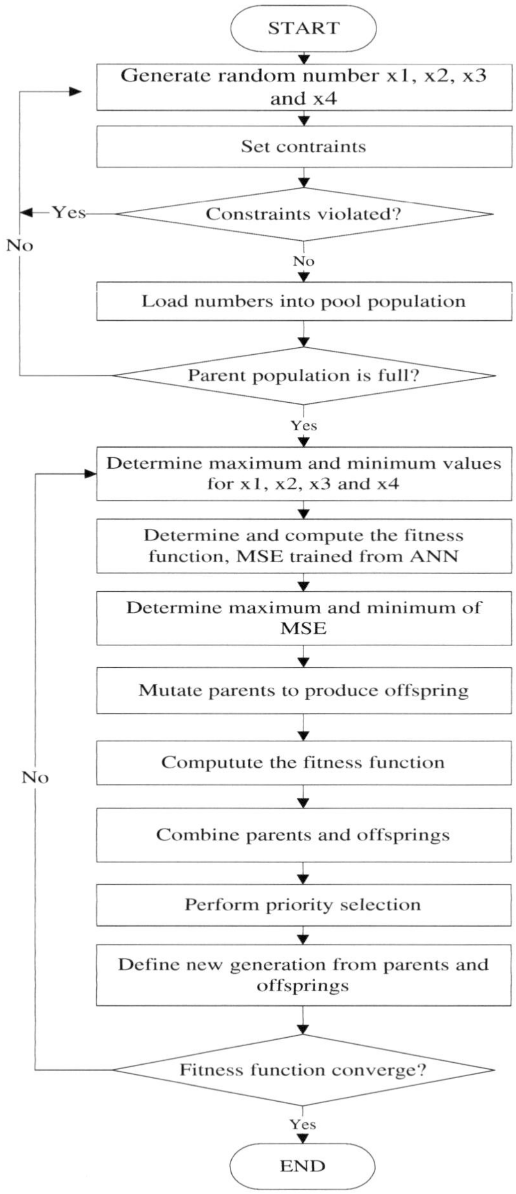

Figure 1. Flowchart of ANN-EP algorithm

3. Results and Discussion
The prediction of the cascading collapse occurrence due to the effects of hidden failure of a protection system is compared with the ANN and EP-ANN model. The IEEE 14 bus system is used as a case

study as shown as in Figure 2. The performance of the proposed EP-ANN model is tabulated in Table 1. Comparison of ANN model and EP-ANN model is tabulated in Table 2. The total load for this test system is 259MW. This test system consists of 5 generators, 14 buses, 20 transmission lines and 11 loads.

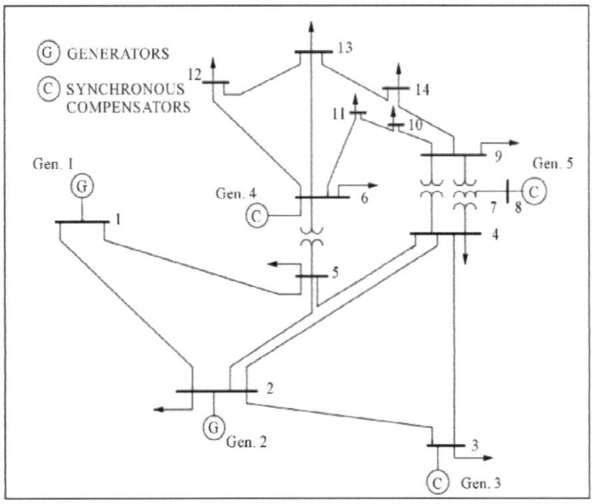

Figure 2. Single line diagram for IEEE 14 bus system

Table 1. ANN-EP Optimization of ANN Parameters and Configuration

Item	Model
Network Configuration	Logsig, logsig, purelin
Momentum Rate	0.01
Learning Rate	0.01
Number of Nodes in Hidden Layer 1	1
Number of Nodes in Hidden Layer 2	1
Training Algorithm	Lavenberg-Marquadt (lm)
Goal	0.001
Correlation Coefficient, R (Testing)	0.95817
Correlation Coefficient, R (Training)	0.98134
Training Number Pattern	888
Testing Number Pattern	222
MSE Training	0.0179
MSE Testing	0.0001
MAPE	0.68839%

The training process is conducted with 888 data patterns while testing process is conducted with 222 data patterns. The objective function is to minimize the MSE with optimal number of training parameters. The error is set to 0.001 with 1000 iterative updates (epochs). The training algorithm used in this algorihm is *Levenberg-Marquardt* with network configuration *logsig-logsig-purelin*. From table 1, the optimal MSE obtained is 0.0001. The best value for the number of nodes in hidden layer 1, x_1 and hidden layer 2, x_2 is 1, while the learning rate, x_3 and momentum rate, x_4 is 0.01 respectively.

The process stop at 9th iteration. The optimal MSE obtained is 0.0001 which is the best value after undergo the fitness computation with minimizing the error. During the optimization process, the 20 sets random numbers is generated and combined with the parents and offspring with 40 sets of random number. The selection process will select the best value from the best 40 sets of random numbers with the lowest fitness value from parents and offspring. The selection process yield the optimal MSE and also the ANN parameters respectively.

The R value obtained for training and testing are 0.98134 and 0.95817 which close to unity.

Table 2. Comparison Results of ANN-EP and ANN Model

Item	EP-ANN	ANN
Network Configuration	*logsig, logsig, purelin*	*logsig, logsig, purelin*
Momentum Rate	0.01	0.5
Learning Rate	0.01	0.7
Number of Nodes in Hidden Layer 1	1	8
Number of Nodes in Hidden Layer 2	1	2
Training Algorithm	*Lavenberg-Marquadt (lm)*	*Lavenberg-Marquadt (lm)*
Goal	0.001	0.001
Correlation Coefficient, R (Testing)	0.95817	0.96123
Correlation Coefficient, R (Training)	0.98134	0.99045
MSE (Testing)	0.0001	0.0073
MSE (Training)	0.0179	0.0265
Training Number Pattern	888	888
Testing Number Pattern	222	222
MAPE	0.68839%	0.6019%

In Table 2, the number of nodes in the first hidden layer and second hidden layer for EP-ANN is 1 while for ANN model the first and second hidden layer are 8 and 2. The details of the performance of EP-ANN had been conducted by evaluating the MSE and R for testing and training process. The error is reduced during the optimization process. The fitness is converged in 9th iteration which is the optimal solution as shown in Figure 3.

After 9th iteration, the error obtained is 0.0001 which is obviously reduce compared to the ANN model which is 0.0073. The optimal R value is 0.95817 for EP-ANN while the value of R in ANN model is 0.96123. Even though there is slightly different value for R, the objective function has been achieved in ordered to minimize the error. The best value for R is more than 0.95 to validate the performance is reliable and good.

The performance of MSE and R is illustrated in a bar chart with training and testing process and it is compared with ANN and EP-ANN model. For EP-ANN model, the MSE performance give 0.0179 for training and 0.0001 for testing process, while for ANN model, the MSE shown is 0.0265 for training while 0.0073 for testing process.

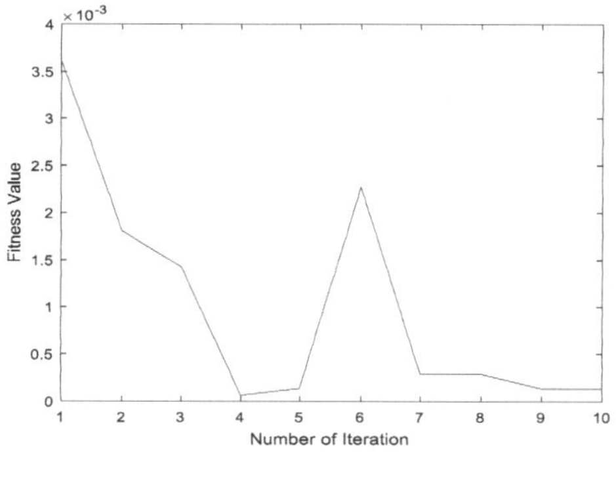

Figure 3. Fitness value with different number of iteration

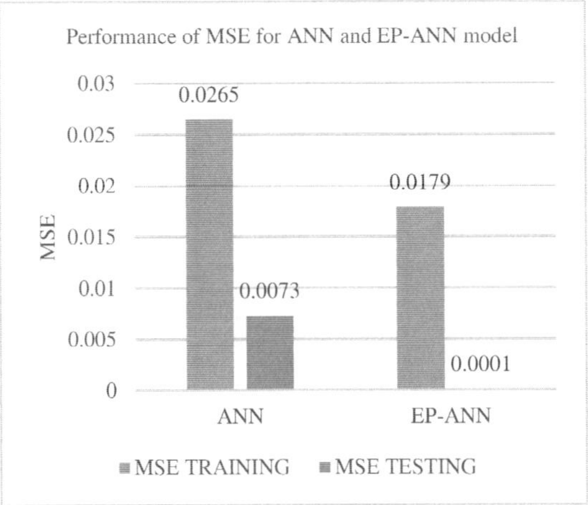

Figure 4. Performance of MSE for ANN and EP-ANN model

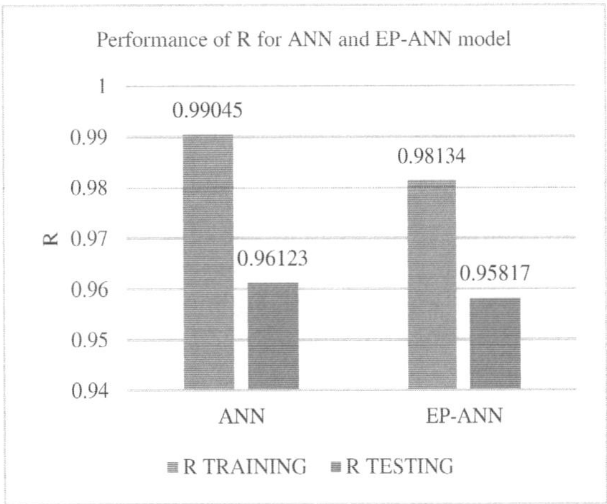

Figure 5. Performance of R for ANN and EP-ANN model

4. Conclusion

This paper has presented the novel of EP-ANN hybrid model for predicting the cascading collapse occurrence due to the effect of hidden failure in a protection system. The performance of MSE shows that the EP-ANN is proved for reliable model and fast convergence contrast to the performance of ANN model. EP-ANN model successfully determine the optimal number of hidden layer 1 and 2, learning rate and momentum rate with minimal error, and correlation coefficient close to unity. In short, the EP-ANN model is highly acceptable model for predicting the occurrence of cascading collapse in conjunction with the protection system hidden failure.

Acknowledgment

The authors would like to thank the Research Management Institute (RMI), Universiti Teknologi MARA, Malaysia and the Ministry of Higher Education (MOHE), Malaysia through research grant 600-RMI/GIP 5/3 (0036/2016) for the financial support of this research.

References

[1] P. Dey, M. Parimi, A. Yerudkar, and S. R. Wagh, "Real-Time Estimation of Propagation of Cascade Failure using Branching Process," *Int. Conf. Power Eng. Energy Electr. Drives*, vol. 2015–Septe, pp. 629–634, 2015.

[2] S. Gupta, R. Kambli, S. Wagh, and F. Kazi, "Support-Vector-Machine Based Proactive Cascade Prediction in Smart Grid using Probabilistic Framework," *IEEE Trans. Ind. Electron.*, vol. 62, no. 4, pp. 2478–2486, 2015.

[3] N. A. Salim, M. M. Othman, I. Musirin, and M. S. Serwan, "Risk Assessment of Cascading Collapse Considering the Effect of Hidden Failure," *PECon 2012 - 2012 IEEE Int. Conf. Power Energy*, no. December, pp. 778–783, 2012.

[4] N. A. Salim, M. M. Othman, I. Musirin, and M. S. Serwan, "Identifying Severe Loading Condition During the Events of Cascading Outage Considering the Effects of Protection System Hidden Failure," *Proc. 2013 IEEE 7th Int. Power Eng. Optim. Conf. PEOCO 2013*, no. June, pp. 375–379, 2013.

[5] Z. M. Yasin, "Quantum-Inspired Evolutionary Programming- Artificial Neural Network for Prediction of Undervoltage Load Shedding," pp. 583–588, 2013.

[6] S. I. Sulaiman, T. Khawa, and A. Rahman, "Fast Evolutionary Programming-Based Hybrid Multi- Layer Feedforward Neural Network for Predicting Grid-Connected Photovoltaic System Output," pp. 44–47, 2012.

[7] S. I. Sulaiman, T. K. A. Rahman, S. Member, I. Musirin, and S. Shaari, "Optimizing Three-layer Neural Network Model for Grid-connected Photovoltaic System Output Prediction 1," no. July, pp. 338–343, 2009.
[8] S. I. Sulaiman, "Optimization of an Artificial Neural Network using Firefly Algorithm for Modeling AC Power from a Photovoltaic System," vol. 3, pp. 2013–2016, 2015.
[9] S. I. Sulaiman, T. K. A. Rahman, and I. Musirin, "Optimizing One-hidden Layer Neural Network Design Using Evolutionary Programming," pp. 288–293, 2009.
[10] N. A. Salim, M. M. Othman, I. Musirin, and M. S. Serwan, "Cascading Collapse Assessment Considering Hidden Failure," *Proc. - 1st Int. Conf. Informatics Comput. Intell. ICI 2011*, pp. 318–323, 2011.
[11] K. Bell, Y. Chen, B. Chowdhury, I. Dobson, P. Hines, M. Papic, S. Miller, and P. Zhang, "Risk Assessment of Cascading Outages : Methodologies and Challenges," vol. 27, no. 2, pp. 631–641, 2012.
[12] J. Chen, J. S. Thorp, and I. Dobson, "Cascading dynamics and mitigation assessment in power system disturbances via a hidden failure model," *Int. J. Electr. Power Energy Syst.*, vol. 27, no. 4, pp. 318–326, 2005.

THE IMPACT OF MINIMUM ENERGY PERFORMANCE STANDARDS (MEPS) REGULATION ON ELECTRICITY SAVING IN MALAYSIA

***Siti Fatihah Salleh*[1], *Mohd Eqwan Roslan*[2], *Aishah Mohd Isa*[3], *Mohd Faizal Basri Nair*[1] *and Siti Syafiqah Salleh*[1]**

[1]Institute of Energy Policy and Research (IEPRe), Universiti Tenaga Nasional
[2]Department of Mechanical Engineering, Universiti Tenaga Nasional
[3]Department of Electrical Power Engineering, Universiti Tenaga Nasional

Abstract. One of Malaysia's key strategies to promote efficient energy use in the country is to implement the minimum energy performance standards (MEPS) through the Electricity Regulations (Amendment) 2013. Five selected electrical appliances (refrigerator, air conditioner, television, domestic fans and lamp fittings) must comply with MEPS requirement in order to be sold in Malaysian market. Manufacturers, importers or distributors are issued Certificate of Approval (COA) if products are MEPS-compliant. In 2015, 1,215 COAs were issued but the number of MEPS products in the market is unknown. This work collects sales data from major manufacturers to estimate the annual sales of MEPS appliances and the cumulative electricity consumption and electricity saving. It was found that most products sold have 3-star rating and above. By year 2015, total cumulative electricity savings gained from MEPS implementation is 3,645 GWh, with air conditioner being the highest contributor (30%). In the future, it is recommended that more MEPS products and related incentives be introduced to further improve efficiency of energy use in Malaysia.

1. INTRODUCTION

Energy efficiency has long been identified as one of the key measure to meet our global energy needs. In an effort to promote efficient use of energy in Malaysia, the Electricity Supply Act was enacted in 1990 in order to provide for the regulation of the electricity supply industry and to serve for the establishment of comprehensive control on any electrical installation, plant and equipment. Despite industrial sector being the major consumer of electricity in this country, residential sector continues to consume a significant portion of the final electricity consumption by over one fifth of the total electricity consumption in 2014 [1]. In order to encourage efficient domestic use of electricity, a Voluntary Energy Efficient Appliances and Equipment standards accompanied by a labelling program were implemented from 2009 to 2012. Later in 2013, Malaysia embarked on a regulatory agenda, by adopting the internationally well-known Minimum energy performance standards (MEPS) policy which dictates the minimum energy performance certain household products should meet in order to be marketed [2]. MEPS policy has been adopted in more than 70 countries, with significant economic and environmental benefits [3,4].

The enforcement of Malaysian Standard (MS) to comply with the MEPS requirement falls on five primary household appliances which are air conditioner, refrigerator, fan, television and lamp fitting.

With the regulation set in place, these five domestic appliances will be issued with a Certificate of Approval (COA) by the Energy Commission Malaysia. In order to be issued with a COA, these appliances must satisfy both the safety and performance requirements and must be affixed with an Energy Rating Label (2 to 5 star rating) apart from the safety label. The star rating allows the consumers to compare the energy efficiency and estimate the running costs of an appliance before they buy. More stars simply means more efficient, when compared to products of the same capacity.

After three years of MEPS implementation, the existing standards need to be reviewed in order to learn from best practices. This is the driver for the current research work, on which the main objectives were to evaluate the impacts of MEPS regulation on the domestic appliances market and overall electricity saving. In order to achieve the research goals, sales data of MEPS-compliant products were collected from participating major electrical appliances companies and analyzed systematically.

2. METHODOLOGY

Sales data were collected from 25 major importers, manufacturers, and distributors of electrical appliances in the local market. The collected sales data were then aggregated and analyzed according to appliance category and star rating. Except for refrigerator, the annual electricity consumption (GWh) was calculated based on the individual equipment wattage and average daily consumption (hour). As for refrigerator, the annual electricity consumption was taken from the equipment performance test report, provided by the national development agency, Standard and Industrial Research Institute of Malaysia (SIRIM). The annual electricity saving gained from the use of each MEPS-compliant appliances was also calculated based on SIRIM's guideline, accomplished by comparing the equipment's annual electricity consumption with the lowest 2-stars rating model of that equipment's type.

3. RESULTS AND DISCUSSION

3.1. Sales Volume of MEPS-compliant Household Appliances

The sales volume of MEPS-compliant appliances sold from 2013 to 2015 are shown in Figure 1. For air conditioner and refrigerator, a steady increase on the sales can be observed, starting from over 0.5 million and almost 0.2 million units sold respectively in 2013, to 0.65 million and almost 0.4 million units sold respectively in 2015. Meanwhile, for fan and television, the sales volumes fluctuate within these 3 years. Lamp fitting stands out from the rest, as the highest MEPS appliance category ever sold for the last three years, with the latest record of 6.4 million units sold in 2015.

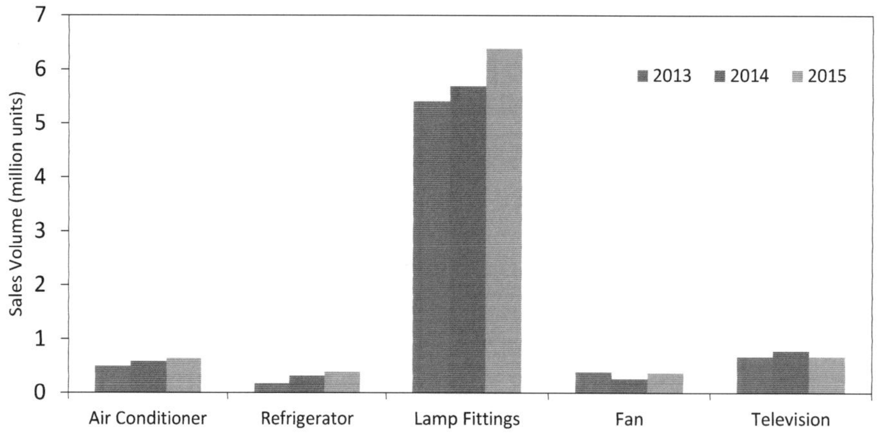

FIGURE 1: Sales volume of MEPS-compliant appliances sold from 2013 to 2015

Most of the units sold were of 5-star rated for television and fan. In the meantime, over 60% of air conditioners sold are 3 stars and below but the reverse is true for refrigerator. It is likely that the difference in consumer preference to purchase lower rating air-conditioners is influenced by the large price point difference between higher star to lower star rating air conditioners. On the other hand, the price point difference may be lower for televisions, fans and refrigerators. Further study and survey will have to be conducted in order to fully understand the underlying factor for this situation. Unlike the other appliances, lamp fitting does not have any star rating since its performance is rated based on luminous efficacy measured in lumens per watt.

3.2. Annual Electricity consumption of MEPS-compliant Household Appliances

Figure 2 shows the annual electricity consumption of MEPS-compliant appliances from 2013 to 2015. Air conditioner demonstrates the highest annual electricity consumption for all year, reaching 2,050 GWh in 2015. This is explainable because the power input for air conditioner is among the highest compared to other types of home appliances, ranging from 0.6 to 21 kW per unit. The hot tropical weather of Malaysia causes its citizens to use air conditioner 8 hours per day in average throughout the year. The second highest annual electricity consumption in 2015 is by lamp fittings (259 GWh), followed by television (192 GWh), refrigerator (156 GWh), and fan (62 GWh). All of the household appliances electricity consumption increase each year, except for fan and television, mainly due to the fluctuation of its sales in this research database. The total annual electricity consumption combining all MEPS-compliant appliances however increases about 10% every year, reaching 2,220 GWh in 2013, 2,510 GWh in 2014 and 2,720 GWh in 2015.

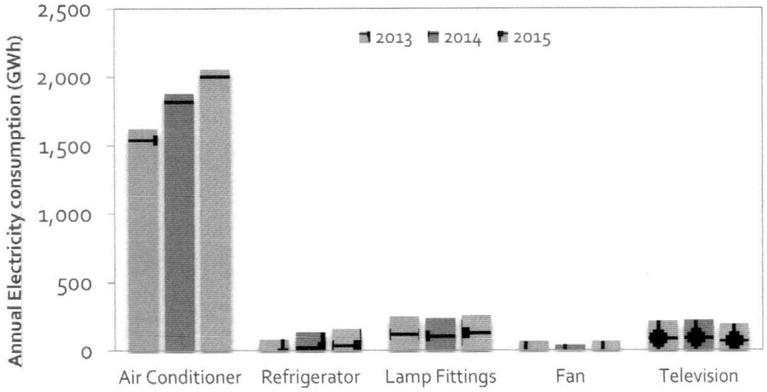

FIGURE 2: Annual electricity consumption (GWh) of MEPS- compliant appliances from 2013 to 2015

3.3. Annual Electricity saving of MEPS-compliant Household Appliances

Figure 3 shows the annual electricity saving of MEPS-compliant appliances. In 2015, the highest electricity saving is demonstrated by air conditioner (485 GWh), followed by lamp fitting (442 GWh), television (336 GWh), refrigerator (70 GWh), and fan (62 GWh). The annual electricity saving by air conditioner almost doubled from 210 GWh in 2013 to 401 GWh in 2014. Meanwhile, the annual electricity savings increase steadily for refrigerator and lamp fittings. However, the same trend of annual electricity saving is not demonstrated by fan and television. Again, this is due to the limitation of the research database. Like electricity consumption, the total annual electricity saving achieved increases every year, from 983 GWh in 2013, to 1,270 GWh in 2014 and 1,390 GWh in 2015.

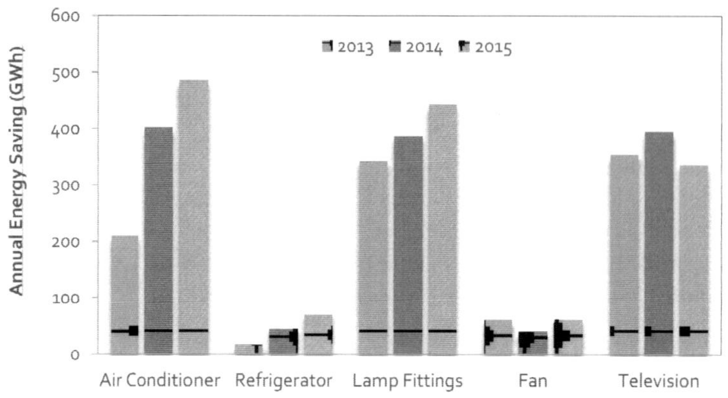

FIGURE 3: Annual electricity saving (GWh) of MEPS- compliant appliances from 2013 to 2015

Based on the analysis on the total amounts of electricity consumption and saving of MEPS-compliant appliances, air conditioner has been found as the highest contributor to both amounts computed compared to the rest of appliances. There are two main types of air conditioner being sold in Malaysia; the inverter and non-inverter type. Only about 5% of the air conditioners sold in 2013 are of the inverter type as shown in Figure 4. However, the number doubled up from 5% to 10% in 2014, and further up to 15% in 2015. Over 78% of the non-inverter type air conditioner sold are 3-star and below. On the other hand, over 96% of the inverter type air conditioner sold are 5-star.

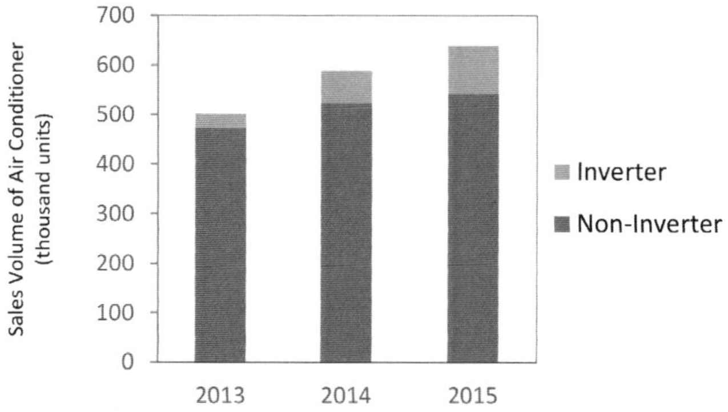

FIGURE 4: Sales volume of air conditioners sold from 2013 to 2015

Figure 5 and Figure 6 show the average electricity consumption and electricity saving by these two types of air conditioners per each unit sold. The inverter types are greatly more efficient in terms of energy use and brought in significant electricity saving compared to the non-inverters or regular ones. This can be contributed to the fact that the inverter type air conditioners are equipped with an innovative technology that allows them to adjust their motor speed and capacity of the compressor. Therefore, the power demand varies based on the periodic cooling requirement of the room. On the other hand, regular air conditioners will have to run at the peak power requirement every time the compressor is running.

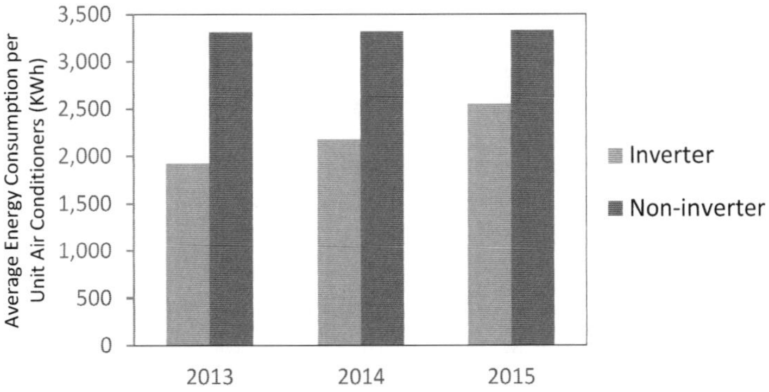

FIGURE 5: Average electricity consumption per unit air conditioners sold from 2013 to 2015

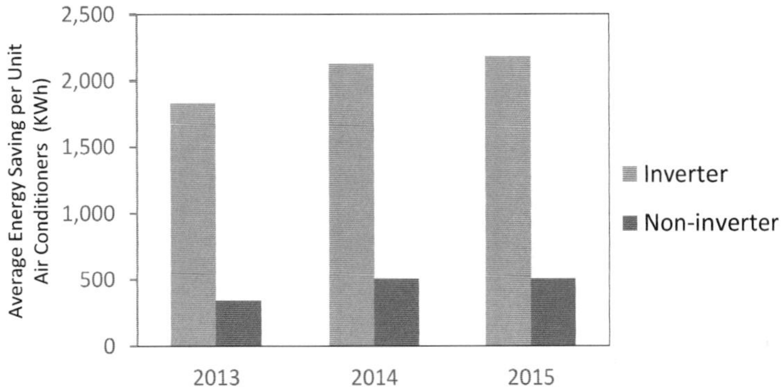

FIGURE 6: Average electricity saving per unit air-conditioners sold from 2013 to 2015

4. CONCLUSION AND RECOMMENDATION

The 2013 MEPS regulation has achieved its target to increase the efficiency of household appliances in Malaysia. The electricity consumption, albeit increasing every year due to the increased in population and other factors, has also been accompanied by increasing electricity saving due to adoption of more and more highly efficient products with 3 and above star ratings. The limited existing research database on the sales volume of fan and television, along with increasing trend of more efficient inverter type- air conditioners sales in the local market suggest that there is still a huge potential for electricity saving in the future. More awareness campaign and tax incentive should be allocated to manufacturers of highly energy efficient appliances such as the inverter type- air conditioners in order to induce a healthy market transformation towards more energy efficient products.

5. ACKNOWLEDGEMENT
The authors would like to acknowledge Suruhanjaya Tenaga for generously funding the current work through the Chair in Energy Economics research fund. The authors would also like to thank the Director of Institute of Energy Policy and Research (IEPRe) and Chair for their suggestions and advice in the formation of this paper.

References
[1] Ministry Coordinator of Strategic Sectors. National Energy Balance 2014:218.
[2] Worrell E, Price L. Methodology for predicting market transformation due to implementation of energy efficiency standards and labels. Energy Conversion Management. 2004;45:1785–93.
[3] Molenbroek E, Smith M, Surmeli N, Ecofys SS, Waide W, et al. Savings and benefits of global regulations for energy efficient products: Final Report. European Commission. 2015.
[4] Armand M, Tarascon J. Achievements of appliance energy efficiency standards and labelling programs: A Global Assessment. International Energy Agency 2015;7:2–7.

IOP Conf. Series: Materials Science and Engineering **341** (2018) 012023 doi:10.1088/1757-899X/341/1/012023

Effect of Listening to the Al-Quran on Heart Sound

N F Daud and Z Sharif

Faculty of Electrical Engineering
University Teknologi MARA,
40450 Shah Alam, Selangor
zaiton623@yahoo.com

Abstract. This paper investigates the effect on the heart sounds upon listening to the chosen verses of the Al Quran. A signal of the heart sounds is extracted using Thinklabs Phonocardiography software and then the frequency components are extracted using MATLAB 7.11.0. Frequency components during diastolic are compared for two sessions; before and during listening sessions. Diastolic is a period where the chamber of the heart is filled with the blood when the heart muscle is in a relaxed condition. From this study, it is found that the frequency of the heart sound during listening to Al-Quran is lower than the one before listening to Al-Quran. This indicates that, the state of calmness can be achieved by listening to this selected verses of the Al-Quran.

1. Introduction

Scientific studies have shown that listening or reciting Quranic verses have many benefits and is essential for psychological health [1,-3,7, 11,12] and under certain circumstances supersedes listening to classical music [2]. Study by Zulkurnaini [2] showed that, listening to Al-Quran can give more relaxation effects as compared to listening classical music where they demonstrated that 12.67% improvement in alpha band before and after listening to the Quran while only 9.96% improvement in study of classical music. Another research stated that by reading, listening and watching the text of Quran can reduce stress and tension of our life and also can achieved mental, spiritual and physical relaxation [3]. Furthermore, a study by Siti Awa Abu Bakar [1] reported that 30 minute of the Quran recitation for mechanically ventilated patients in ICU promotes the psycho-spiritual comfort and reduce the patients' heart rate. Zulkurnaini [2] also reported that if human brain is exposed to one of the verses in the Quran resulted in the increasing in the alpha band than listening to classical music. Thus improve the relaxation and alert condition. These are among the scientific studies done to show that listening and reciting Quran can give benefits to human; Muslims or non-muslims..

The degree of one's health can be measured through several parameters. Among them are from EEG signals which give information on the state of the brain through the measurement of the alpha band, ECG will give information on the heart rate and so on so forth. By knowing this information, the state of one's health condition can be determined. In this study, the information on the diastolic measurements from the heart sound analysis will be used as the parameter to determine the correlation between the relaxation and health of a person due to listening to the selected verses of the Quran.
Diastolic measurements are normally used by the physician to determine the relationship between high diastolic pressure and heart failure. High diastolic pressure can cause the heart muscle to stiffen [4]. Diastole is the time when the heart muscle is in a relaxed condition and it occurs in S2 and the S1.

2. Heart Sounds

The heart sounds contain the two normal sounds which is the first heart sound (S1) and the second heart sound (S2) which are referred as lub dub' sounds. Besides, third (S3) and fourth heart sound (S4) can also exist under certain pathological conditions [5]. These rhythmic set of S3 and S4 are often called gallops which are associated with diastolic filling and low frequency sounds.
Cardiac cycle is the chronological succession of electrical and mechanical issues that pass during the heart in a single round namely referred as diastole and systole. Diastole attributes to the part of relaxation where the ventricles are relaxed and blood comes into them from the atria while during the ventricles is contracting is the point of systole [6]. Systole on contrary is assigned to the contraction. Figure 1 shows the occurrence of diastole and diastole in the heart sound signal.

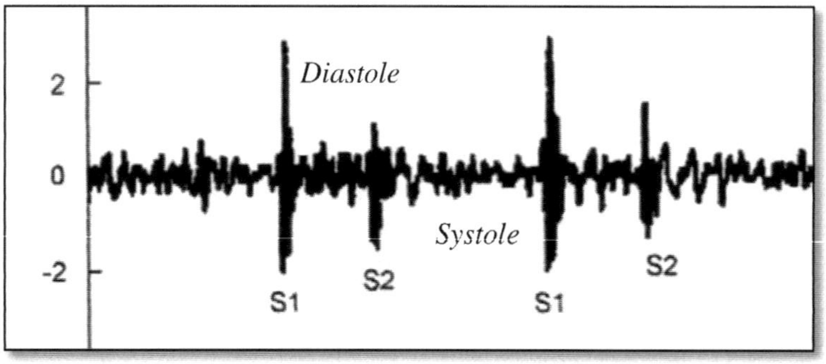

Figure 1: Representative heart sound

3. Methodology

Heart sounds from selected group of samples are recorded using simple circuit connecting the stethoscope to the laptop. Thinklabs Phonocardiography software powered by Audacity is used to record the heart sound. The comparison of the analysis of the diastole's frequency of the heart sound between before and after was made. The verse of the Quran played to the subjects is Al-Hasyr verses 18-24. Total duration of data taken is for 3 minutes. Each minute of the data taken is allocated for before, during and after listening to this verse of the Quran. The process flow on recording the heart sound is shown in Figure 2

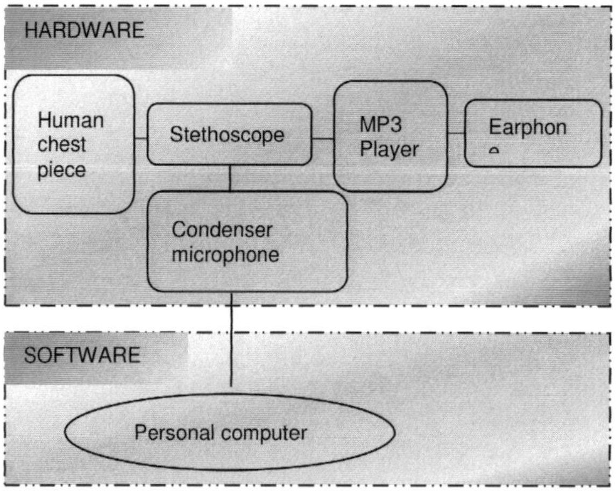

Figure 2: Flowchart of heart sound recording

The sentences of the Quran exposed to the subjects are chosen due to the contents it conveys. Since it is about the faith (imaan) of the muslims, we would like to see the effects of these verses to the listeners. Regarding the faith of the muslims, they are required to follow all commands from Allah and the ones taught by the Prophet Muhammad (Peace be upon him).

4. Fast Fourier Transform (FFT)

In determining the frequency contents of the heart sounds, the fundamental method; the Fourier transform is used. It gives the information on the frequency components that existed in the signal analysed. The Fourier transform relates the continuous-time representation x(t) to the continuous-frequency representation X(f). For processing on digital computers, it is necessary to have a similar transform that relates the discrete-time representation x(n) to the discrete-frequency representation X(k). The resulting transformation is known as the discrete Fourier transform (DFT). The Fast Fourier transform (FFT) on the other hand is essentially the high speed implementation of the DFT of a given signal. It transforms the signal in time or space into discrete frequency domain.

The FFT is given by;

$$X(k) = \sum_{n=0}^{N-1} x(n) e^{\frac{-j2\pi kn}{N}} \qquad 0 \le k \le N-1$$

= 0; elsewhere

Where k is the discrete frequency, n is the sample in time and N is the total number of samples.

5. Results and Discussions

Figure 3 illustrates the sample of the heart sound recorded using Thinklabs Phonocardiography software. This data is processed and then the frequency components of the heart sounds during diastolic are determined. Fast Fourier Transform from Matlab mathematical software is used for this purpose. The sampling frequency of audio used is 44.1 kHz and the duration of recording data is around 7-12seconds. The volunteer is asked to lie down as of to make their body static and minimize the movement and hence the interference during recordings. The heart sound's data is recorded from the pulmonary position of the human chest for three conditions; before during and after listening to the recitations of the Quran. The diastolic frequencies of the samples are recorded and analysed.

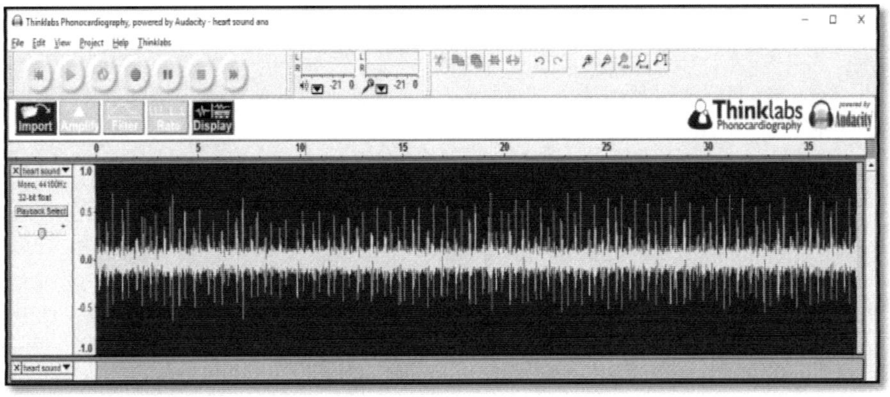

Figure 3: Screenshot one of recorded data of heart sound signal

Figure 4 shows the display representation of the heart sound (in both time and frequency domain) before listening to the Quran while Figure 5 shows these representations after listening to the recitations.

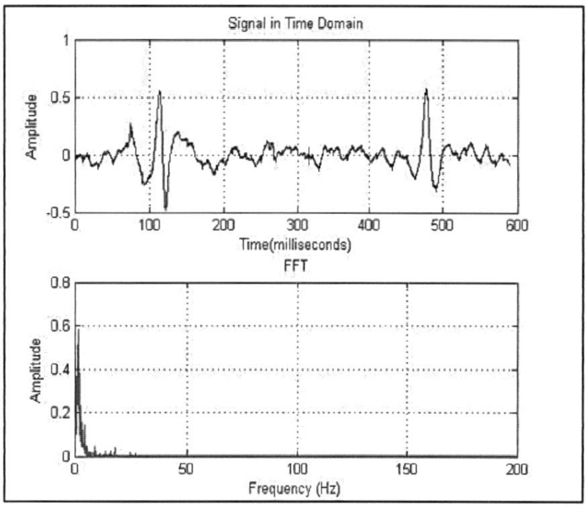

Figure 4: Sample of the heart sound before listening to Al-Quran

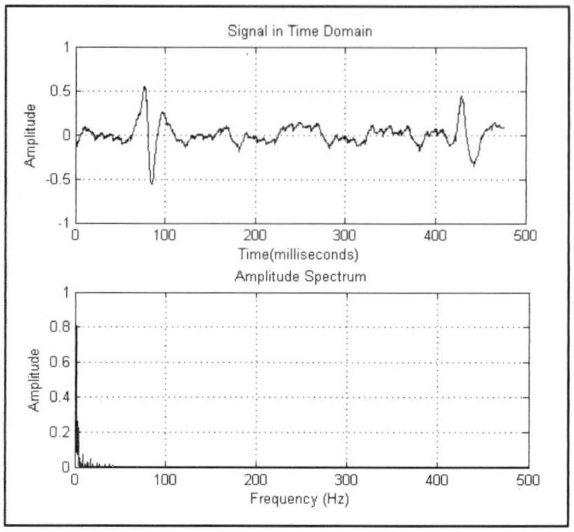

Figure 5: Time and frequency domain during listening to Al-Quran

Based on the analysis of the heart sounds for 20 volunteers, it is found that the frequency of the heart sound between S2 and S1 or diastolic frequency is in the range of 60Hz to 80Hz. During listening to Al-Quran, their respective frequency shows a decrease in number as shown in Figure 6. It indicates that the lower the frequency, the higher the level of relaxation. Thus, it is found that the frequency of the diastolic heart sound during listening to Al-Quran is lower than the frequency of before listening to Al-Quran. Meaning that, the state of calmness or relaxation can be accomplished by listening to Al-Quran. On average, the frequency of the heart sound has decrease by 8.6%. That indicates a significant decrease which leads to relaxation and calmness

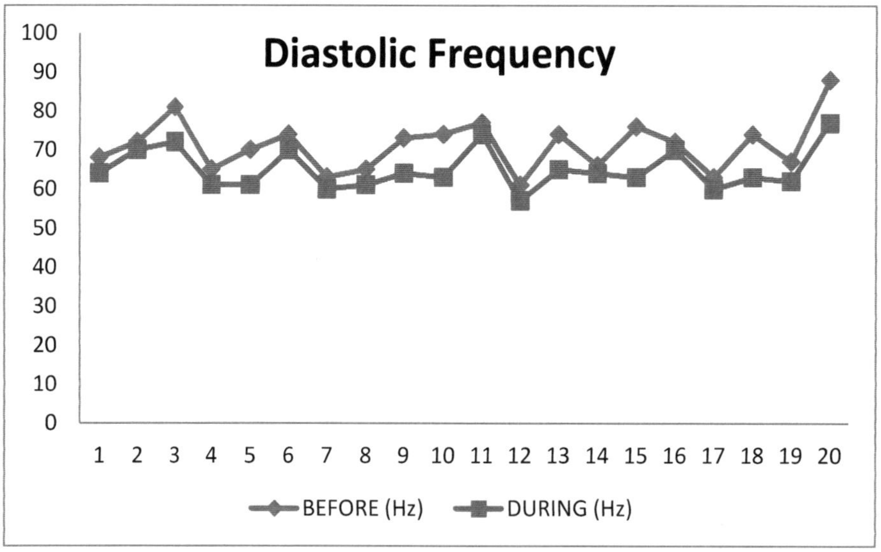

Figure 6: Diastolic frequency before and during listening to Al-Quran.

6. Conclusions

In conclusion, it is found out that by listening to Surah Al Hasyr, the state of calmness can be achieved. The frequency obtained between S2 and S1 frequency is found to be lower during listening to Al-Quran compared to the frequency before listening to Al-Quran. Lower frequency indicates a state of relaxation. Furthermore, improvements can be made on the procedure of taking the data by properly handling the stethoscope and the position of the volunteer so as there is no external factors affecting the data collections.

7. References

[1] Bakar, S.A.B.A. *Effects of Holy Quran Listening on Physiological Stress Response Among Muslim Patients in Intensive Care Unit.* in *Conference on Management and Muamalah.* 2014.

[2] Zulkurnaini, N.A., et al. *The comparison between listening to al-Quran and listening to classical music on the brainwave signal for the alpha band.* in *2012 Third International Conference on Intelligent Systems Modelling and Simulation.* 2012. IEEE.

[3] Khan, N., et al. *Mental and spiritual relaxation by recitation of the holy Quran.* in *Computer Research and Development, 2010 Second International Conference on.* 2010. IEEE.

[4] C. Cycle, V. Diastole, T. Ap, and V. Systole, "CV III : The Cardiac Cycle," vol. c, 2004.

[5] Z. Sharif, M. S. Zainal, A. Z. Sha'ameri, and S. H. S. Salleh, "Analysis and classification of heart sounds and murmurs based on the instantaneous energy and frequency estimations," *2000 TENCON Proceedings. Intell. Syst. Technol. New Millenn. (Cat. No.00CH37119)*, vol. 2, pp. 130–134, 2000.

[6] E. M. Cherry and F. H. Fenton, "Heart Structure , Function and Arrhythmias Heart Anatomy and Structure."

[7] Kaheel, A.A. *Healing by listening to Quran.* 2013.

[8] Oschman, J.L., *What is healing energy? Part 3: silent pulses.* Journal of Bodywork and Movement Therapies, 1997. **1**(3): p. 179-189.

[9]. Crowe, B.J. and M. Scovel, *An overview of sound healing practices: Implications for the profession of music therapy.* Music Therapy Perspectives, 1996. **14**(1): p. 21-29.

[10]. Goldman, J., *Healing sounds: The power of harmonics*2002: Inner Traditions/Bear & Co.

[11] A. Von Denffer, "An Introduction to the Sciences of the Qur'an Ahmad von Denffer Introduction," 1981.

[12] T. H. Qur, "Qur'anic Lessons Importance of The Holy Qur'an."

FPGA implementation of predictive degradation model for engine oil lifetime

M. F. M. Idros[1,2], A. H. A. Razak[1,2], S A M Al Junid[1,2], S.I. Suliman[2], A. K. Halim[1,2],

[1]Electronic Architecture and Application (EArA) Research Group
[2]Faculty of Electrical Engineering
Universiti Teknologi MARA Malaysia
40450 Shah Alam, Selangor

faizul_san287@salam.uitm.edu.my

Abstract—This paper presents the implementation of linear regression model for degradation prediction on Register Transfer Logic (RTL) using QuartusII. A stationary model had been identified in the degradation trend for the engine oil in a vehicle in time series method. As for RTL implementation, the degradation model is written in Verilog HDL and the data input are taken at a certain time. Clock divider had been designed to support the timing sequence of input data. At every five data, a regression analysis is adapted for slope variation determination and prediction calculation. Here, only the negative value are taken as the consideration for the prediction purposes for less number of logic gate. Least Square Method is adapted to get the best linear model based on the mean values of time series data. The coded algorithm has been implemented on FPGA for validation purposes. The result shows the prediction time to change the engine oil.

***Index Terms*—Engine oil degradation sensing, FPGA, Online Monitoring**

1. Introduction

An awareness of green environment and the need to reduce the yearly expenses, increase the requirement to have a new technique that can benefit the people to prolong their usage of engine oil and at the same time contribute to the green environment. There are a lot of parameters to be considered in determining the quality of engine oil. Viscosity, Total Acid Number (TAN), Total Base Number (TBN), oxidation and some of contamination including soot, water, glycol and other chemical contents are the type of parameters to determine the condition based on engine oil. Total Acid Number (TAN) and Total Base Number (TBN) also has been considered among the important parameter in term of quality indication due to the oxidation and the acidity content of the engine oil. The measurement method is based on the result of the carboxylic acids present after the oxidation process by using a titration method or optical spectroscopy [1]–[4].

Recent advances in optical technology have led to the more accurate in condition based monitoring. The parameters such as viscosity, oxidation, water and soot contamination can be monitored by exposed the sample into a certain optical wavelength. The optical wavelength can determine the type of material or acid contamination based on the light transmittance or absorbance. The degraded samples will reduce the percentage of light transmittance due to the increase of the absorbed light. At the receiver part, the amount of transmitted or absorbed light determined by the variation of voltage. From this optical signal, the warning limit of the used engine oil can be determined from the drastic decreased of transmittance percentage or gradual increased of absorbance [5], [6]. This technique can be implemented using Least Square Method (LSM) where the slope is monitored until it crosses the threshold value.

In recent years, efforts have been made to propose an online system to alert user the suitable time to change engine oil. But, the number of research which is focusing on online analysis technique is very few [7], [8]. Nevertheless the result of the research needs to be analysed according to the standard degradation parameters. In offline analysis, a deep knowledges about the lubricant technology are required in making decision on engine oil replacement [9]. In order to have online system, statistical analysis is has been largely applied in different purposes; utilization of particle filtering algorithm as Remaining Useful Lifetime (RUL) prediction tool [10], web-based decision support system (DSS) in managing waste lube oils collection and recycling operations [11], predictive models in different techniques including Projection Pursuit Regression (PPR), Partial Least Square (PLS), Support Vector Machines (SVM), Linear Models, Generalized Linear Models (GLM) and Random Forest (RF) [1], partial least squares algorithm in development of calibration, prediction and validation models [12] and predictive maintenance algorithm including classification analysis, factor analysis, and multiple regression analysis [13].

In this paper, the implementation LSM in predicting the engine oil lifetime based on optical properties is presented. The work focuses on the monitoring technique of the degradation based on the slope variation. The slope variation is based on the oxidation wavelength of $1710cm^{-1}$. The samples of engine oil SAE 10W40, 15W/40, CD-SE50 and SJ15W/40 (diesel) were used. Least Square Method were used as a method for prediction purposes which is coded using Quartus II software. Finally, the system design implemented on FPGA. The system is capable to predict the suitable time to change the engine oil and it relied to help the environment by reducing the amount of waste engine oils.

2. Experimental work

A) Sample collection and prediction algorithm

At this stage, the sample is collected for validation purposes of the developed prediction algorithm. CD-SE15W/40 is used as a first sample and second sample were taken from a type of CD-SE50 while the last sample is taken from SJ 15W/40. The samples have been analysed for oxidation effect to the optical properties where the virgin samples were oxidize in a temperature of $150^{o}c$. The oxidation process has been accelerated by the air pump and the copper wire catalyst. In this approach, the data resulted from FTIR analysis were extracted by using Perkin Elmer software by following the band location.

In this work, y-axis was denoted as percentage of transmittance %T while the x-axis denoted to running time (T). The main objective in this stage is to get the prediction time at x-axis for remaining oil condition, while the y-axis are the transmittance percentage (%T). The variation of %T are projected to the x-axis up to the warning and critical limit as shown in Figure 1.

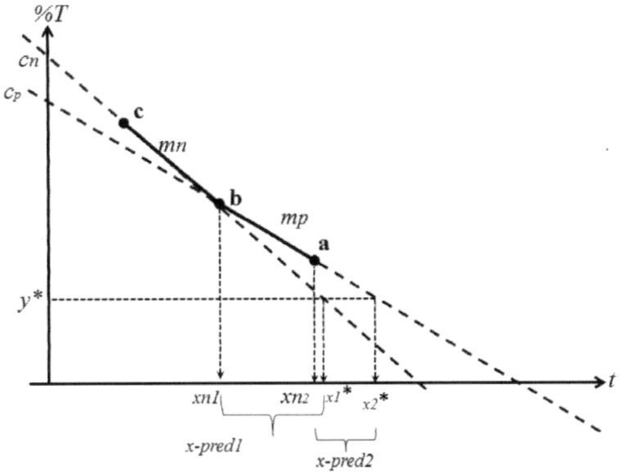

Figure 1 %T Projection and Engine Oil Lifetime Prediction

Line *mn* and *mp* created from the regression technique using LSM consisted of 10 data. As an example, the current condition of %T at point b. So, the line is projected to the x-axis by using the value of slope obtained by the following formula;

$$\hat{\beta}_1 = \frac{SS_{xy}}{SS_{xx}}. \quad (1)$$

y^* indicates half of *%T* from the original value. *mn* is a slope from data point *a* to data point *b*, whereas *mp* is the slope from data point *b* to data point *c*. For example, the *a–b* line is projected to point y^* and the value of time is recorded as time x_2^*. Thus, the remaining time to change the engine oil is determined by subtracting x_2^* from xn_2 at point *a* and expressed as *x-pred*$_2$. The *b–c* line also produced the predicted time of *x-pred*$_1$ by subtracting x_1^* from xn_1.

B) System design and architecture

This is the back end system design based the optical data input. The front end input is the optical sensor with the analog to digital converter. The back end system developments have been performed using Quartus II software environment where the algorithm is coded in Verilog language. The design is based on the LSM algorithm discussed in section 3.4.1. The LSM is the linear regression which is capable to be used as the prediction algorithm. The LSM is written in Verilog language to generate the Register Transfer Level (RTL) where the logic gates are generated. The system consists of clock divider, arithmetic and logic unit, register and input/output unit as shown in Figure 3.14. The system is composed of five blocks, namely, clock divider, data input processor, arithmetic unit, prediction and decision maker, and display unit. The input data is fed from *Data in* in 9-bit binary number representing the percentage of transmittance (*%T*) in the *y*-axis, whereas the hours in the *x*-axis is taken from the *Clock Divider* input. The 9-bit as an n-bit data are chosen to be counted at the maximum number 2^n of 512 that represent the counting hours. *Clock Divider* is used to generate the 1 hour pulse because the data are sampled at every 1 hour and have a capability to sample the data up to 512 hours due to the bit count of 9-bit. The maximum generating time is depending on the engine oil condition.

The engine oil degradation monitoring and engine oil lifetime prediction models discussed in sub-chapter 4.4 will be implemented in hardware. The main architecture of the system is shown in Figure 2. The system is composed of five blocks, namely, clock divider, data input processor, arithmetic unit, prediction and decision maker, and display unit. *Clock Divider* is used to generate the 1 hour pulse because the data are sampled at every 1 hour and continuously sampled up to 400 hour, depending on the engine oil condition. The input data is fed from *Data in* in 9-bit binary number representing the percentage of transmittance (*%T*) in the *y*-axis, whereas the hours in the *x*-axis are taken from the *Clock Divider* input. The 9-bit numbers are chosen to be counted at the maximum number of 512 that represent the counting hours.

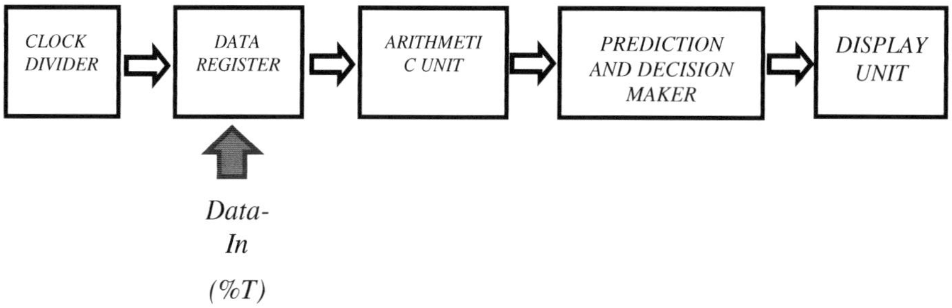

Figure 2 The design block of the real-time degradation monitoring system

3. Result and Discussion

A. *Simulation Analysis using MATLAB*

This simulation is based on the collected sample as discussed in sub-chapter 4.2.3. The degradation trend of optical transmittance for the three engine oil samples have been plotted in MATLAB as shown in Figure 4.10(a-b). Y-axis of the graph is referring to transmittance percentage (%T) while the x-axis is the running time (T).

In this work, the value of slope for the warning limit determined by the negative value of the slope. All three type of engine oil CD-SE50, CD-SE15W/40 and SJ 15W/40 which have been discussed in section 4.1.3 were analyzed using MATLAB for slope determination. As an example of the degradation trends for engine oil type CD-SE50 plotted in Figure 3(a), the warning condition occured after the flat line because of the drastic drop of the transmittance percentage. At this time, the slope for this area will be calculated for lifetime prediction in the monitoring system. Figure 3(b) shows the regression data from the sample type of CD-SE50 from Figure 3(a). From here, we can see that the value of slope is determined between the time of 90 hours and 160 hours. At this time, the data are at the flat value at 85 to 90 hours before gradually decreased to the negative value.

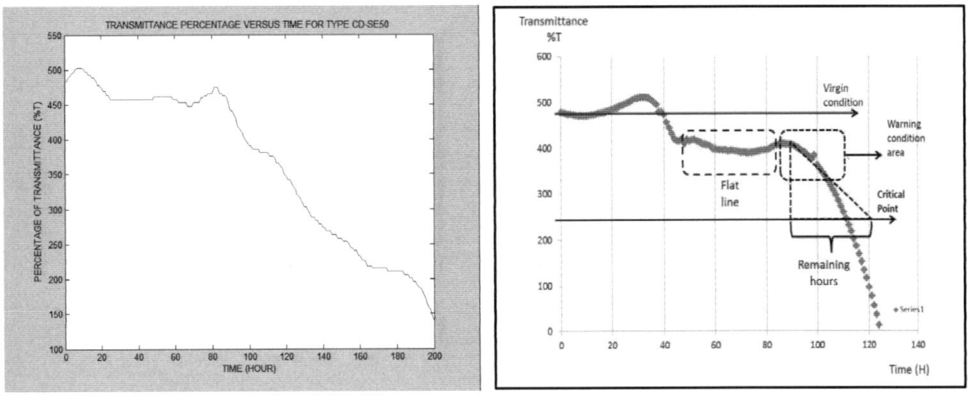

(a) (b)

Figure 3(a) Degradation trends for engine oil type CD-SE50

Figure 3 (b) Regression data from the sample type of CD-SE50

B. Functional Simulation using Quartus II Software

Slope is the main parameter in this development of monitoring system. It had been generated using LSM at every ten data. The first ten data were processed to get the value of the slope and stored in a Registers and named as *data_origin* as shown in Figure 4.29. The *data_origin* were taken from the first input data just after the engine oil was changed. It is also called as the virgin engine oil. This data will be used to compare the variation of degradation from the beginning of engine oil changes. According to the Figure 4, the origin data was 496 at '0' hour. The first value of slope were obtained at t=9. This is the accumulated values of data from t=0 to t=9.

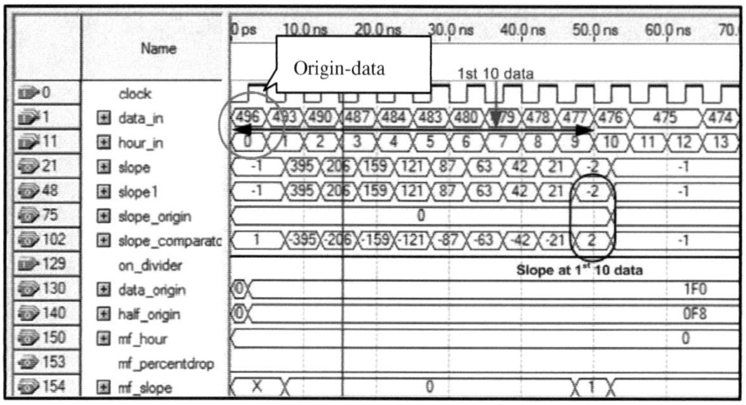

Figure 4 Accumulation of the first 10 data

The next slope will be obtained at t=10, 11, 12 ...... *n* as shown in Figure 5. These value of slopes are used as the input to the system which contain of a 9-bit data.

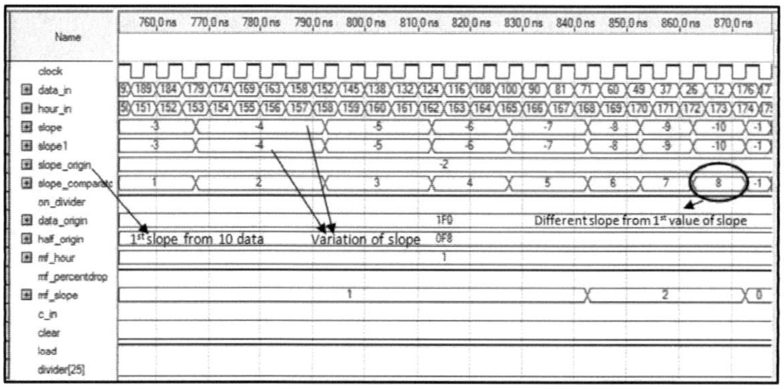

Figure 5 Slope variation for sample set 1

C. *FPGA Implementation for Data Validation*

In this work, the data was taken based on the oxidation wavelength and the prediction monitoring are based on the slope variation. LSM is used as an prediction algorithm and implemented on hardware as an online monitoring system by using Altera DE2-115 development board as shown in Figure 6.

Figure 6 Hardware implementation setup using Altera DE2-115 development board

The data from engine oil type CD-SE50 have been verified using DE-2 board. It has been found that the warning limit occurred at 52 hours as shown in Figure 7. Here, the remaining engine lifetime started at 21 hours and decreased until critical condition. The critical condition reached at 141 hours as shown in Figure 8. At this stage, we can see that the predicted lifetime decreased until 0.

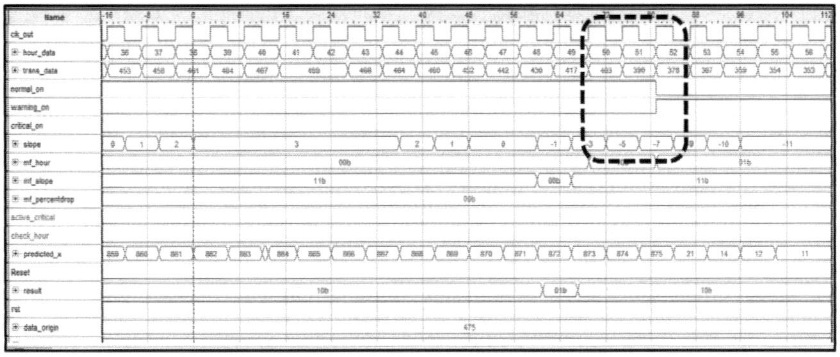

Figure 7 Warning Limit set 2 (CD-SE50)

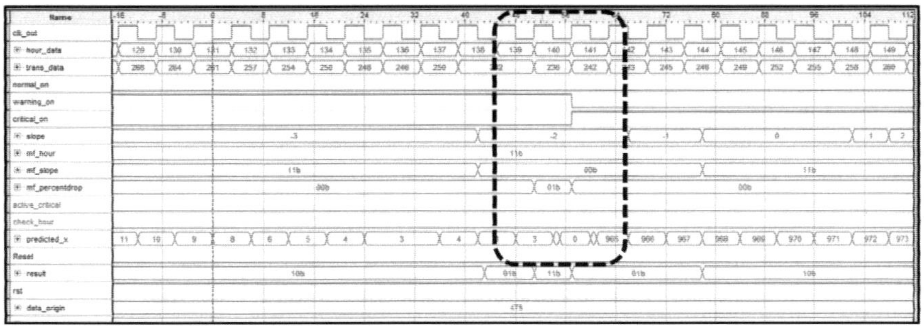

Figure 8 Critical Limit set 2 (CD-SE50)

The degradation monitoring based on the variation of slope was supported by several research which was stated that the warning limit of the engine oil occurred when there was a drastic drop of light transmittance [14]. These approaches has been implemented in a hardware where the condition limit will be displayed on LCD [5]. However, it was reported that the changes of optical data do not correlate well with actual oil degradation process [15]. The data was taken based on the variation of color which has no relation to the condition based parameter. In fact, the decision of engine oil changes are based on the offline analysis which is not suitable for online monitoring purposes.

Table 1 summarised the result of the condition based online monitoring system in Altera Board DE-2.

Table 1 Functional simulation result after debugging

Sample	Engine oil type	Warning limit	Prediction	Critical limit
1	CD-SE15W/40	84	54	136
2	CD-SE50	52	21	141
3	SJ 15W/40	52	22	114

From Table 4.10, it can be seen that, the sample 2 have the longest usage of engine oil because of the critical limit at 141 hours as compared to the sample 1 and sample 3 which the critical limit occurred at 136 hours and 114 hours respectively. For the overall functional test, the sample 1 gave the better result in term of the warning and prediction limit. The warning limit for sample occurred at 84 hours while the prediction shows the remaining 54 hours respectively. The degradation trend of this sample is more linear than the others.

4. Conclusion

The monitoring and prediction system was built based on the LSM. The LSM consisted of the combinational logic gate from adder, subtractor, multiplier and divider. Then, it have been combined with another sub-system including serial shift register, membership function, condition monitoring and lifetime prediction. The functional simulation have been tested in Quartus software and then implemented on FPGA for real hardware implementation. The result shows that the usage for engine oil in time based slope monitoring achieved the objective to prolong the usage of engine oil for more than 100 hours. The LSM was designed with the combination of other system and implemented in FPGA for hardware design. As a result, a user will have a system that will give alert about the exact time to change their engine oil based on the real parameter such as TAN, TBN, viscosity and oxidation rather than mileage based with the Fuzzy based decision making of time to change the engine oil. Hence, the second objective of this work is achieved successfully. A significant portion of the work presented in this thesis have been devoted to the complete implementation of the system on FPGA. The actual data have been used to validate the capability of the overall system in monitoring and predicting the condition of engine oil. This is to fulfill the third objective of this work.

5. Acknowledgement

Appreciation to the Universiti Teknologi MARA (UiTM) that provide the academic research training and also Ministry of education for supporting this research through Research Grant Lestari [File No: 600-RMI/DANA 5/3/LESTARI (45/2015)].

References

[1] B. Leal, D. Rivas, J. Vivancos, J. Ordieres-meré, and S. F. Capuz-rizo, "Chemometrics and Intelligent Laboratory Systems Determination of the total acid number (TAN) of used mineral oils in aviation engines by FTIR using regression models ☆," *Chemom. Intell. Lab. Syst.*, vol. 160, no. May 2016, pp. 32–39, 2017.

[2] J. Dong, F. R. van de Voort, V. Yaylayan, and and A. A. Ismail., "Determination of Total Base Number (TBN) in Lubricating Oils by MID-FTIR Spectroscopy," *J. Soc. Tribol. Lubr. Eng.*, pp. 1–24, 1999.

[3] N.Corporation, "A Comprehensive Look At the Acid Number Test," 2007.

[4] K. Qian, K. E. Edwards, G. J. Dechert, S. B. Jaffe, L. a Green, and W. N. Olmstead, "Measurement of total acid number (TAN) and TAN boiling point distribution in petroleum products by electrospray ionization mass spectrometry.," *Anal. Chem.*, vol. 80, no. 3, pp. 849–55, Feb. 2008.

[5] S. Kumar, P. S. Mukherjee, and N. M. Mishra, "Online condition monitoring of engine oil," *Ind. Lubr. Tribol.*, vol. 57, no. 6, pp. 260–267, 2005.

[6] S. Shirley and S. Donald, *Automotive Engine Oil Condition Monitoring*. 1997.

[7] A. Mujahid and F. L. Dickert, "Monitoring automotive oil degradation: analytical tools and onboard sensing technologies.," *Anal. Bioanal. Chem.*, vol. 404, no. 4, pp. 1197–209, Sep. 2012.

[8] P O Vähäoja and H V S Pikkarainen, "Trends in industrial oil analysis – a review," *Int. J. Cond. Monit.*, vol. 1, no. 1, 2011.

[9] S. Jamieson, "How Much Does Your Lube Tech Cost You ?," *Machinery Lubrication*, Nov-2006.

[10] J. Zhu, J. M. Yoon, D. He, Y. Qu, and E. Bechhoefer, "Lubrication Oil Condition Monitoring and Remaining Useful Life Prediction with Particle Filtering," no. 2008, pp. 1–15, 2013.

[11] P. P. Repoussis, D. C. Paraskevopoulos, G. Zobolas, C. D. Tarantilis, and G. Ioannou, "A web-based decision support system for waste lube oils collection and recycling q," *Eur. J. Oper. Res.*, vol. 195, no. 3, pp. 676–700, 2009.

[12] S. Marinović, A. Jukić, D. Doležal, B. Špehar, and M. Krištović, "Prediction of used Lubricating Oils Properties by Infrared Spectroscopy Using Multivariate Analysis," *Goriva I Maz. (Fuel Lubr.*, vol. 51, no. 3, pp. 205–215, 2012.

[13] H.-B. Jun, D. Kiritsis, M. Gambera, and P. Xirouchakis, "Predictive algorithm to determine the suitable time to change automotive engine oil," *Comput. Ind. Eng.*, vol. 51, no. 4, pp. 671–683, Dec. 2006.

[14] S. S. Wang and H. Lee, "The application of a.c. impedance technique for detecting glycol contamination in engine oil," vol. 40, pp. 193–197, 1996.

[15] J. Zhu, D. He, and E. Bechhoefer, "Survey of Lubrication Oil Condition Monitoring , Diagnostics , and Prognostics Techniques and Systems," vol. 2, pp. 100–115, 2013.

Implementation of digital equality comparator circuit on memristive memory crossbar array using material implication logic

Adib Haron, Fazren Mahdzair, Anas Luqman, Nazmie Osman and Syed Abdul Mutalib Al Junid

Electronic Architecture and Application Research Group (EArA)
Faculty of Electrical Engineering Universiti Teknologi MARA
40450 Shah Alam, Selangor, Malaysia

E-mail: adib@salam.uitm.edu.my

Abstract. One of the most significant constraints of Von Neumann architecture is the limited bandwidth between memory and processor. The cost to move data back and forth between memory and processor is considerably higher than the computation in the processor itself. This architecture significantly impacts the Big Data and data-intensive application such as DNA analysis comparison which spend most of the processing time to move data. Recently, the in-memory processing concept was proposed, which is based on the capability to perform the logic operation on the physical memory structure using a crossbar topology and non-volatile resistive-switching memristor technology. This paper proposes a scheme to map digital equality comparator circuit on memristive memory crossbar array. The 2-bit, 4-bit, 8-bit, 16-bit, 32-bit, and 64-bit of equality comparator circuit are mapped on memristive memory crossbar array by using material implication logic in a sequential and parallel method. The simulation results show that, for the 64-bit word size, the parallel mapping exhibits 2.8× better performance in total execution time than sequential mapping but has a trade-off in terms of energy consumption and area utilization. Meanwhile, the total crossbar area can be reduced by 1.2× for sequential mapping and 1.5× for parallel mapping both by using the overlapping technique.

1. Introduction

The classical CMOS-based Von Neumann architectures separate memory and processor. Fundamentally, the memory is used to store data, and the processor is used to compute data. The communication between memory and processor continues to be a bottleneck for many data-intensive applications due to the massive amount, and expensive cost of moving data back and forth between memory and processor [1, 2]. In addition, the Von Neumann architectures are developed based on the memory hierarchy that is used to overcome the problem of the growing gap between memory and processor speed. The memory that closer to the processor is faster due to the data locality that speeds up the computation but smaller in size and capacity that limits the amount of data can be handled and make it difficult to process Big Data applications [3, 4]. Meanwhile, the CMOS technology gradually scales down and intrinsically reach to its physical limit by about 2024 [5, 6]. Consequently, CMOS technology faces challenges such as reduce reliability, increase leakage power consumption and saturated performance.

The concept of in-memory processing is one of the most promising solutions that use processing within the memory architecture to alleviate the communication bottleneck in Von Neumann architecture [7–11]. The processing within memory architecture is a memristive memory crossbar architecture that can perform as memory, processor and interconnect. Thus, the memristive memory crossbar array is not only used to store data but also to compute and move data along the crossbar nanowire architecture [12, 13]. The data can be programmed as resistance, meanwhile, the logic operation and data movement are driven by the voltage driver. Therefore, there is no specific load and store instructions are required for the computation. The data can be directly computed where ever they reside within the crossbar architecture. Physically, the memristive memory crossbar array can be seen as a full memory devices without any computing element surrounded or integrated. The logic operations are performed by the voltage driver. The technology enabler for this architecture is memristor devices. [14]. Memristor is one of the promising devices that has features such as non-volatility, higher scalability, high integration density, and CMOS compatible [15].

The memristive memory crossbar array provides a huge amount of parallelism that can be exploited. The vast number of the horizontal and vertical nanowire enables massive parallelism for computation and communication. However, the parallelism is limited due to the nature of architecture that shares a single horizontal line to many vertical lines or vice versa. Therefore, either horizontal or vertical lines only can be used for parallelism at one time. In terms of logic gates operation, material implication logic is a suitable method to execute the nature of logic gates within the memristive memory crossbar array. However, most studies on implication logic operations show that the implication logic executed in sequential [7, 16]. Therefore, the total steps of computation and communication by using implication logic contributes to a long execution time and high energy consumption.

The concept of in-memory processing allows flexibility for the designer to store data on any best location within the crossbar, perform computation in parallel and manage the communication between task efficiently to achieve the best performance. The implementation of material implication logic on memristive memory crossbar array depends on the logic design of the circuit, i.e., digital equality comparator circuit. To exploit the massive parallelism on the crossbar array either on horizontal or vertical lines, the steps of the imply logic must be executed in a sequence that is equal to the number of switching energy for computation. The first question arises as to how can we parallelize the computation steps of the equality comparator circuit on memristive memory crossbar array so that we can reduce the number of steps for computation? Secondly, can we mapped the equality comparator circuit on memristive memory crossbar array so that we can decrease the area of the crossbar array used to compute the logic gates of the equality comparator circuit?

This paper proposes four design methodologies to implement the equality comparator circuit on memristive memory crossbar array. The logic gates of the equality comparator circuit are mapped on the memristive memory crossbar array in such a way the research question aforementioned can be solved. The contributions of this paper are:

- Reduced the number of steps for computation of equality comparator circuit on memristive memory crossbar array by exploiting gate-level parallelism.
- Reduced the number of memristors used to implement equality comparator circuit on memristive memory crossbar array by using material implication logic.

The rest of this paper is structured as follows. Section 2 describes the background and related work of digital equality comparator circuit, memristive memory crossbar array and material implication logic. Section 3 proposes the design methodology to implement equality comparator circuit on memristive memory crossbar array. Section 4 evaluates the performance of the proposed implementation. Finally, section 5 concludes the results.

2. Background and Related Work

This section presents first the digital equality comparator circuit and its conversion to NAND gate. Subsequently, the architecture of memristive memory crossbar array and lastly the material implication logic.

The equality comparator logic circuit is part of the computation in the pairwise DNA sequence alignment [17] and is used to compare two DNA sequence. The purpose is to find the regions of similarity that may indicate relationships between two biological sequences. Figure 1 shows the 2-bit digital equality comparator circuit using the classical XNOR and AND gates. The variety of the different types of logic gates can be converted into a NAND gate only as shown in figure 2. The conversion is important due to the simplicity of its implementation on memristive memory crossbar array using the material implication logic.

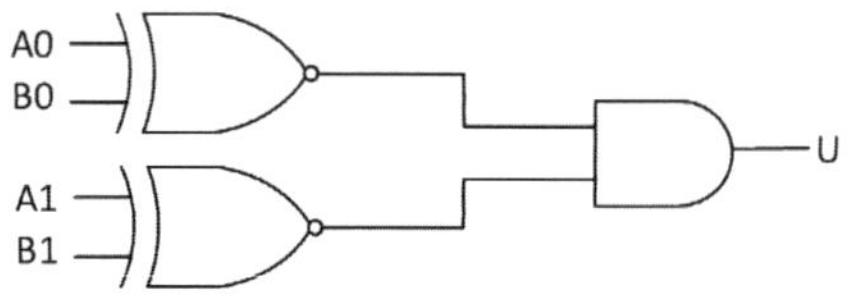

Figure 1. 2-bit equality comparator logic circuit using XNOR and AND gate.

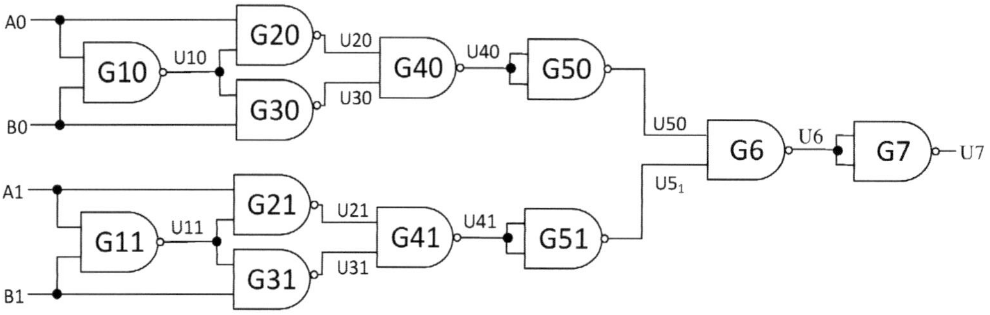

Figure 2. 2-bit equality comparator logic circuit using NAND gate.

The memristive devices are interconnected in the junction of the crossbar array architecture consists of horizontal and vertical nanowires. The crossbar array allows the resistance value of the memristive devices to be read and written at any junction in the crossbar. Therefore, each memristive device at the junction of the crossbar can perform as a memory. Section 3 illustrates several concepts of the memristive memory crossbar array architecture.

Besides, the memristive devices can behave as logic circuits in which the computation can be executed on the memory itself. One of the basic and deterministic logical element that can be used to perform computation on memristive memory crossbar array is material implication logic [9]. Figure 3 and 4 show the circuit of material implication logic. Each memristive device is used to store input, output and latch the intermediate output for the next logic operation. The memristor P is used to store input 1, and memristor Q is used to store input 2. The output is stored in the memristor S. The output position of the implication logic can be configured on the right side and the left side of the memristors following the principle of the voltage divider. The implication logic can perform only one operation at one time due to the fact that only one memristor can be switched in a horizontal or vertical line for every single voltage control applied.

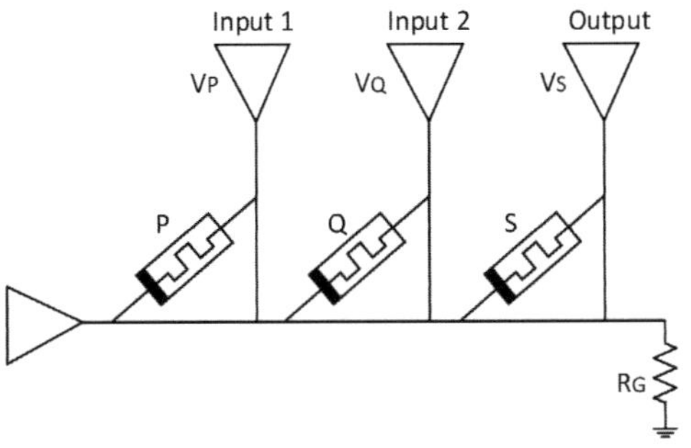

Figure 3. Right side output configuration.

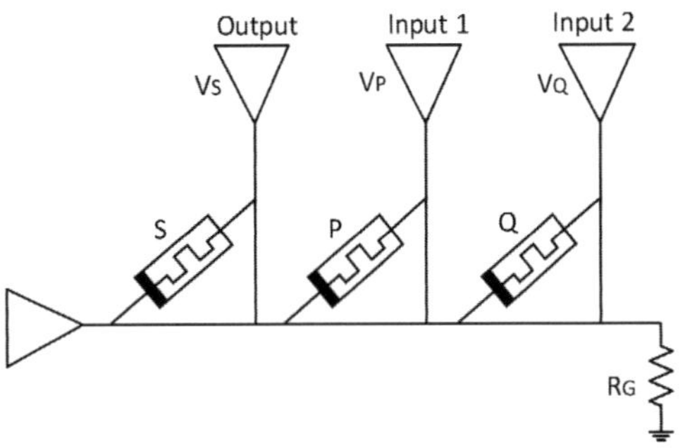

Figure 4. Left side output configuration.

3. Methodology

Currently, there is no standard way to map a set of combinational logic gates on the memristive memory crossbar array. A designer has to predetermine the location of input, intermediate output and the final output before the execution. In this section, we propose a design methodology to map the equality comparator circuit on memristive memory crossbar array architecture. The mapping scheme can be divided into four types:

- Sequential Mapping Non-Overlapping (SMNO)
- Sequential Mapping Overlapping (SMO)
- Parallel Mapping Non-Overlapping (PMNO)
- Parallel Mapping Overlapping (PMO)

The purpose of parallel over sequential technique is to reduce the number of computation steps. The parallelism exploited is gate-level parallelism. The purpose of overlapping over non-overlapping technique is to reduce the number of memristors used to compute a set of logic gates. The Non-Overlapping technique uses the right side output configuration (see figure 3) and the overlapping technique uses both the right side output configuration and left side output configuration (see figure 3 and 4) in order to reduce the number of working memristor.

Sequential Mapping Non-Overlapping (SMNO) is a naive method to execute the equality comparator circuit on the memristive memory crossbar array. Each logic gates in equality comparator are executed in sequential, and therefore each logic gates need to be labeled with a number (see figure 2). The sequential method is slow due to the fact that only one implication logic step operation can be executed at one time for a vertical or horizontal line. As a result, the number of steps for the execution time is usually high and equal to the number of memristor switching for the energy consumption. Table 1 shows the step-by-step of 2-bit sequential mapping non-overlapping with the sequence of logic gates and number of steps needed for each gate operation. The symbol $\uparrow$ and $\sim$ indicate the NAND gate and NOT gate operation respectively. Figure 5 shows the arrangement of memristors that are responsible for storing input data, output data, and the intermediate data.

Table 1. 2-bit sequential mapping non-overlapping.

Sequence	Logic Gate	Memristor Mapping	Number of steps
1	G10	$A0 \uparrow B0 \rightarrow U10$	3
2	G20	$A0 \uparrow U10 \rightarrow U20$	3
3	G30	$U10 \uparrow B0 \rightarrow U30$	3
4	G40	$U20 \uparrow U30 \rightarrow U40$	3
5	G50	$\sim U40 \rightarrow U50$	2
6	G11	$A1 \uparrow B1 \rightarrow U11$	3
7	G21	$A1 \uparrow U11 \rightarrow U21$	3
8	G31	$U11 \uparrow B1 \rightarrow U31$	3
9	G41	$U21 \uparrow U31 \rightarrow U41$	3
10	G51	$\sim U41 \rightarrow U51$	2
11	G6	$U50 \uparrow U51 \rightarrow U6$	3
12	G7	$\sim U6 \rightarrow U7$	2

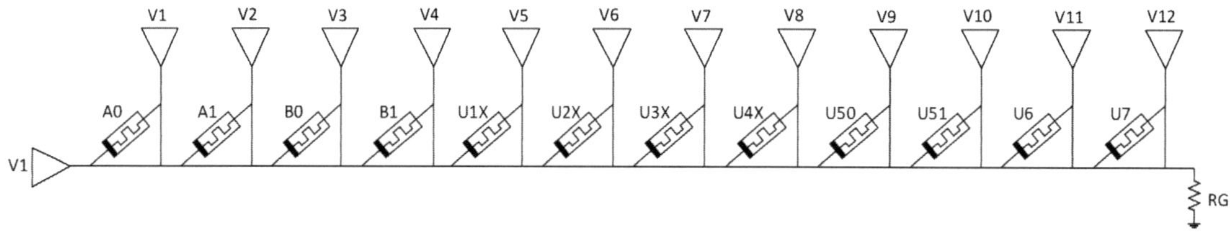

Figure 5. 2-bit sequential mapping non-overlapping on memristor-based crossbar array.

Sequential Mapping Overlapping (SMO) is a method to improve the Sequential Mapping Non-Overlapping in terms of the utilization of the number of memristors. This method is capable of reducing the number of memristors whereby the same physical memristor can be reused to store the intermediate output for the next operation so that the intermediate output of the previous operation is overwritten. The process requires the relocation of the output position as shown in figure 3 and 4. Table 2 shows the step-by-step of 2-bit Sequential Mapping Overlapping and the arrangement of memristors are illustrated in figure 6. The arrangement and location of data input and output of SMO are similar to SMNO except that the memristor U6 and U4X can be eliminated. Therefore, the overall number of memristors of SMO can be reduced when compared with SMNO.

IOP Conf. Series: Materials Science and Engineering **341** (2018) 012025 doi:10.1088/1757-899X/341/1/012025

Table 2. 2-bit sequential mapping overlapping.

Sequence	Logic Gate	Memristor Mapping	Number of steps
1	G10	$A0 \uparrow B0 \rightarrow U10$	3
2	G20	$A0 \uparrow U10 \rightarrow U20$	3
3	G30	$U10 \uparrow B0 \rightarrow U30$	3
4	G40	$U20 \uparrow U30 \rightarrow U10$	3
5	G50	$\sim U10 \rightarrow U50$	2
6	G11	$A1 \uparrow B1 \rightarrow U11$	3
7	G21	$A1 \uparrow U1 \rightarrow U21$	3
8	G31	$U11 \uparrow B1 \rightarrow U31$	3
9	G41	$U21 \uparrow U31 \rightarrow U11$	3
10	G51	$\sim U11 \rightarrow U51$	2
11	G6	$U50 \uparrow U51 \rightarrow U11$	3
12	G7	$\sim U11 \rightarrow U7$	2

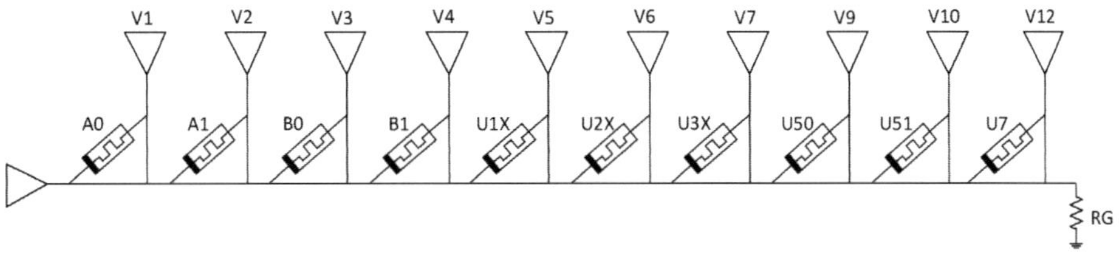

Figure 6. 2-bit sequential mapping overlapping on memristor-based crossbar array.

Parallel Mapping Non-Overlapping (PMNO) is a mapping method that developed to improve the execution time of Sequential Mapping Non-Overlapping. The parallelism is achieved due to the ability to switch multiple memristors by a single voltage control in horizontal or vertical line. Table 3 shows the step-by-step of 2-bit parallel mapping non-overlapping. The total number of steps of sequences for PMNO are eight which is less than SMNO that required twelve steps of sequences. In sequence 6, we need to execute the communication operation or moving the particular data downward, and the rest are computation operations. As shown in figure 7, the horizontal line is used for computation, and the vertical line is used for communication. We arrange the input data in the vertical line instead of the horizontal line, i.e., A0, A1 and B0, B1. The voltage driver V1 and V2 are able to perform the logic operation for both A0, A1 and B0, B1 in parallel. The final result from the first vertical line, i.e., U50 is transferred to the memristor U50'. Then, the second vertical line performs the computation until the final result in U7.

Table 3. 2-bit parallel mapping non-overlapping.

Sequence	Logic Gate	Memristor Maping	Logic Gate	Memristor Mapping	Steps
1	G10	$A0 \uparrow B0 \rightarrow U10$	G11	$A1 \uparrow B1 \rightarrow U11$	3
2	G20	$A0 \uparrow U10 \rightarrow U20$	G21	$A1 \uparrow U11 \rightarrow U21$	3
3	G30	$U10 \uparrow B0 \rightarrow U30$	G31	$U11 \uparrow B1 \rightarrow U31$	3
4	G40	$U20 \uparrow U30 \rightarrow U40$	G41	$U21 \uparrow U31 \rightarrow U41$	3
5	G50	$\sim U40 \rightarrow U50$	G51	$\sim U41 \rightarrow U51$	2
6	—Communication Steps : U50 move downward inline with U51—				3
7	G6	$U50 \uparrow U51 \rightarrow U6$			3
8	G7	$\sim U6 \rightarrow U7$			2

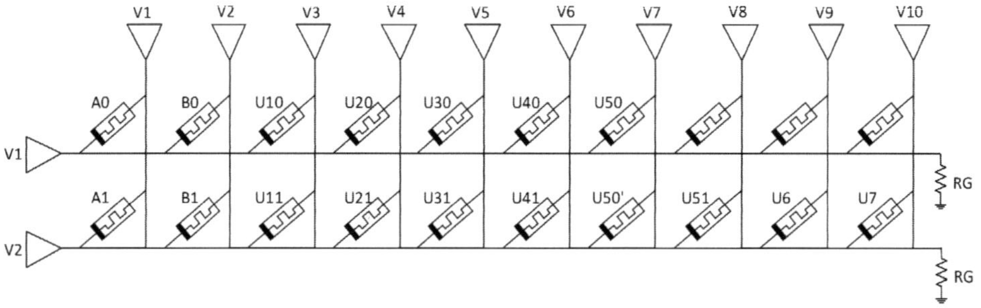

Figure 7. 2-bit parallel mapping non-overlapping on memristor-based crossbar array.

Parallel Mapping Overlapping (PMO) is a method that used to achieve both; reduce the number of steps computation and reduce the number of memristors by using the parallel and overlapping techniques. The PMO method is the most efficient method than all other methods previously. Table 4 shows the step-by-step of sequences of 2-bit parallel mapping overlapping and the memristors arrangement and data locality are shown in figure 8. All of the mapping techniques applied to 2-bit can be extended to 4-bit, 8-bit, 16-bit, 32-bit and 64-bit word size.

Table 4. 2-bit parallel mapping overlapping.

Sequence	Logic Gate	Memristor Maping	Logic Gate	Memristor Mapping	Steps
1	G10	$A0 \uparrow B0 \rightarrow U10$	G11	$A1 \uparrow B1 \rightarrow U11$	3
2	G20	$A0 \uparrow U10 \rightarrow U20$	G21	$A1 \uparrow U11 \rightarrow U21$	3
3	G30	$U10 \uparrow B0 \rightarrow U30$	G31	$U11 \uparrow B1 \rightarrow U31$	3
4	G40	$U20 \uparrow U30 \rightarrow U10$	G41	$U21 \uparrow U31 \rightarrow U11$	3
5	G50	$\sim U10 \rightarrow U50$	G51	$\sim U11 \rightarrow U51$	2
6	—Communication Steps : U50 move downward inline with U51—				3
7	G6	$U50 \uparrow U51 \rightarrow U11$			3
8	G7	$\sim U11 \rightarrow U7$			2

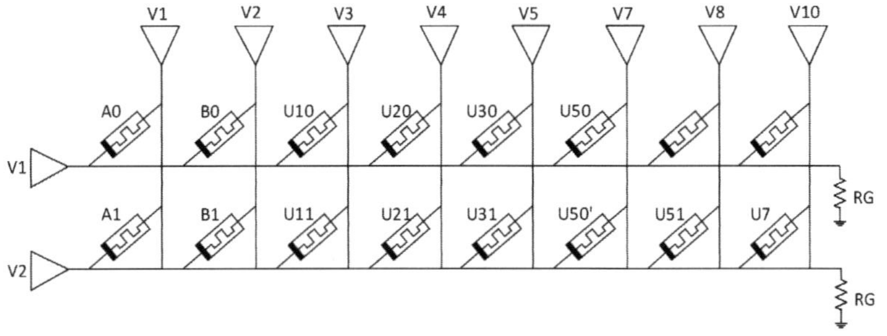

Figure 8. 2-bit parallel mapping overlapping on memristor-based crossbar array.

4. Simulation Result and Evaluation

In this section, we evaluate all the mapping schemes in section 3. We quantify the number of steps and number of switching for both computation and communication operations, as well as the number of memristors. The number of steps is used to evaluate the execution time (second), the number of switching is used to evaluate the energy consumption (joule), and lastly, the number of memristors is used to evaluate the area utilization (meter square). The number of steps and switching for both computation and communication are shown in equation 1 and 2 for sequential and parallel mapping respectively. Equations 3,4,5,6 shows the number of memristor used in SMNO, SMO, PMNO, and PMO respectively. The n in all equation is referred to 2-bit, 4-bit, 8-bit, 16-bit, 32-bit and 64-bit and tabulated in table 5, 6, 7 and 8. The cost of the delay of a single switching memristor is 200 *picosecond*, the energy of a single switching memristor is 0.25 *femtojoule* and finally, the area of one memristor is $1{\times}10^{-4}$ um^2. Subsequently, the performances are plotted as a graph in figure 9, 10, and 11.

$$sm = 14 * n + 5 * (n - 1) \tag{1}$$

$$pm = 14 + 5 * (n - 1) + 3 * n/2 \tag{2}$$

$$smno = 4 + 3 * n + 2 * (n - 1) \tag{3}$$

$$smo = 3 + 3 * n + (n - 1) \tag{4}$$

$$pmno = n * (4 + n + 2 * (n - 1)) \tag{5}$$

$$pmo = n * (3 + n + (n - 1)) \tag{6}$$

Table 5 shows the sequential mapping non-overlapping performance metric from 2-bit to 64-bit. The number of steps and switching for computation and the number of memristors increase in parallel with the number of bits or word size. The number of steps and switching for computation are equal due to the sequential steps executed by the implication logic. Meanwhile, the number of steps and switching for communication is zero because the data movement operations in sequential mapping are not required.

Table 5. Sequential mapping non-overlapping.

Number of bits	2-bit	4-bit	8-bit	16-bit	32-bit	64-bit
Number of steps computation	33	71	147	299	603	1211
Number of steps communication	0	0	0	0	0	0
Number of switching computation	33	71	147	299	603	1211
Number of switching communication	0	0	0	0	0	0
Number of memristors	12	22	42	82	162	322

Table 6 tabulates the sequential mapping overlapping performance metrics from 2-bit to 64-bit. The number of steps and switching both for computation and communication SMO are equal to SMNO. However, the number of memristors used in SMO are less than SMNO due to the overlapping technique.

Table 6. Sequential mapping overlapping.

Number of bits	2-bit	4-bit	8-bit	16-bit	32-bit	64-bit
Number of steps computation	33	71	147	299	603	1211
Number of steps communication	0	0	0	0	0	0
Number of switching computation	33	71	147	299	603	1211
Number of switching communication	0	0	0	0	0	0
Number of memristors	10	18	34	66	130	258

Table 7 shows the parallel mapping non-overlapping performance metric from 2-bit to 64-bit. The number of steps for computation in PMNO is less than the number of steps for computation in SMNO due to the exploitation of parallelism at gate-level. However, the number of switching for computation in PMNO is equal to the SMNO due to the similar design of logic gate in the equal comparator. Moreover, in PMNO, the data movement in vertical line contribute to the number of steps and switching for communication.

Table 7. Parallel mapping non-overlapping.

Number of bits	2-bit	4-bit	8-bit	16-bit	32-bit	64-bit
Number of steps computation	19	29	49	89	169	329
Number of steps communication	3	6	12	24	48	96
Number of switching computation	33	71	147	299	603	1211
Number of switching communication	3	6	12	24	48	96
Number of memristors	16	56	208	800	3136	12416

Table 8 shows the parallel mapping overlapping performance metric from 2-bit to 64-bit.The number of steps for computation and communication in PMO are equal to PMNO. Similarly,

IOP Conf. Series: Materials Science and Engineering **341** (2018) 012025 doi:10.1088/1757-899X/341/1/012025

the number of switching for computation and communication has the same value for both for PMO and PMNO due to the sequential steps in communication. Note that the communication or data movement for PMNO and PMO are executed in sequential. However, the number of memristors in PMO is less than PMNO due to the overlapping technique applied.

Table 8. Parallel mapping overlapping.

Number of bits	2-bit	4-bit	8-bit	16-bit	32-bit	64-bit
Number of steps computation	19	29	49	89	169	329
Number of steps communication	3	6	12	24	48	96
Number of switching computation	33	71	147	299	603	1211
Number of switching communication	3	6	12	24	48	96
Number of memristors	12	40	144	544	2112	8320

Figure 9 illustrates the performance of execution time for all mapping schemes. The sequential mapping, i.e., SMNO and SMO both has the same execution time due to the sequential mapping technique. The parallel mapping, i.e., PMNO and PMO are equal in execution time due to the parallel mapping technique. On average, from 2-bit to 64-bit, the PMNO exhibits 2.4× better execution time than SMNO. The result is also similar for PMO over SMO.

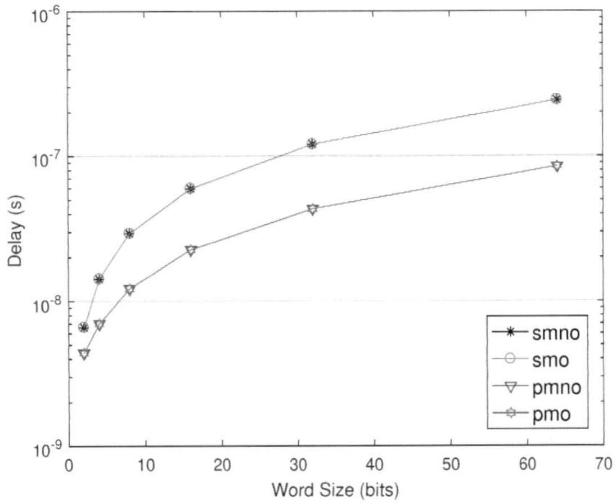

Figure 9. Execution time.

Figure 10 illustrates the energy consumption for all mapping schemes. The parallel mapping exhibits slightly higher energy consumption than the sequential mapping due to the communication or data movement cost. On average, parallel mapping consumes 1.08× higher energy than sequential mapping.

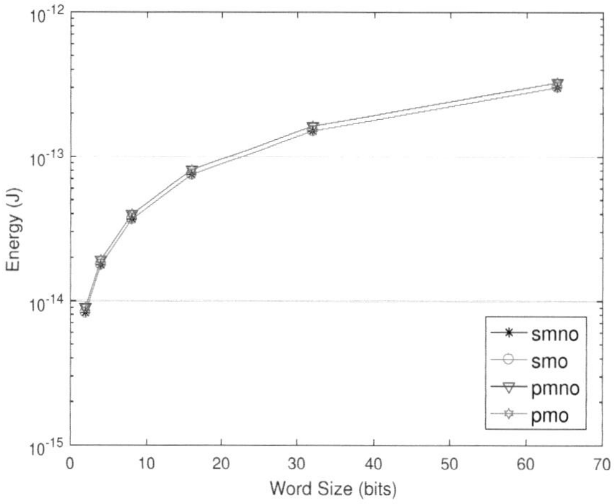

Figure 10. Energy consumption.

Figure 11 illustrates the area utilization for all mapping schemes. The parallel mapping schemes perform better in execution time than sequential mapping schemes. As a trade-off, the parallel mapping non-overlapping consumes 12.8× larger area than the sequential mapping non-overlapping in average. Meanwhile, the parallel mapping overlapping consumes 10.7× larger area than sequential mapping overlapping in average.

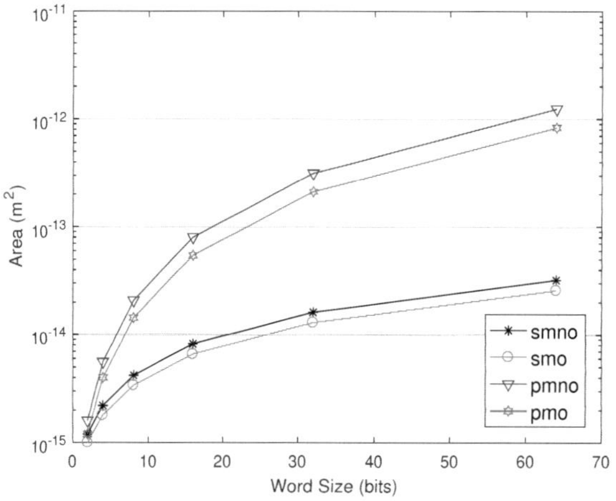

Figure 11. Area Utilization.

5. Conclusion

We show that we can reduce the amount of execution time by using the gate-level parallelism technique and the amount of area utilization by using the overlapping technique for all 2-bit, 4-bit, 6-bit, 8-bit, 16-bit, 32-bit, 64-bit of digital equality comparator circuit on memristive memory crossbar arrray architecture.

Acknowledgments

This work was funded by a research grant from the LESTARI 2017 under the file NO 600-IRMI/DANA KCM 5/3/LESTARI (128/2017). The authors would like to thank to the Institute of Research Management & Innovation (IRMI), Universiti Teknologi MARA for the funding of this research.

References

[1] Al Junid S, Tahir N, Haron M, Abd Majid Z, Idros M and Osman F 2010 *International Journal of Simulation–Systems, Science & Technology* **11**

[2] Saliman N F A, Sabri N D A, Al Junid S A M, Tahir N M and Majid Z A 2013 *Systems, Process & Control (ICSPC), 2013 IEEE Conference on* (IEEE) pp 268–274

[3] Idros M, Ali S and Islam M S 2012 *Intelligent Systems, Modelling and Simulation (ISMS), 2012 Third International Conference on* (IEEE) pp 278–282

[4] Idros M, Ali S and Islam M S 2011 *Research and Development (SCOReD), 2011 IEEE Student Conference on* (IEEE) pp 225–228

[5] Borkar S 2009 *Proceedings of the 46th Annual Design Automation Conference* (ACM) pp 93–94

[6] Author et al 2016 MORE MOORE Tech. rep. International Roadmap for Devices and Systems

[7] Kvatinsky S, Satat G, Wald N, Friedman E G, Kolodny A and Weiser U C 2014 *IEEE Transactions on Very Large Scale Integration (VLSI) Systems* **22** 2054–2066

[8] Levy Y, Bruck J, Cassuto Y, Friedman E G, Kolodny A, Yaakobi E and Kvatinsky S 2014 *Microelectronics Journal* **45** 1429–1437

[9] Borghetti J, Snider G S, Kuekes P J, Yang J J, Stewart D R and Williams R S 2010 *Nature* **464** 873–876

[10] Kvatinsky S, Belousov D, Liman S, Satat G, Wald N, Friedman E G, Kolodny A and Weiser U C 2014 *IEEE Transactions on Circuits and Systems II: Express Briefs* **61** 895–899

[11] Linn E, Rosezin R, Tappertzhofen S, Böttger U and Waser R 2012 *Nanotechnology* **23** 305205

[12] Xie L, Du Nguyen H A, Taouil M, Hamdioui S and Bertels K 2015 *Nanoscale Architectures (NANOARCH), 2015 IEEE/ACM International Symposium on* (IEEE) pp 124–129

[13] Du Nguyen H, Xie L, Yu J, Taouil M and Hamdioui S 2017 *Design & Technology of Integrated Systems In Nanoscale Era (DTIS), 2017 12th International Conference on* (IEEE) pp 1–6

[14] Hamdioui S, Taouil M, Du Nguyen H A, Haron A, Xie L and Bertels K 2015 *Memristive Systems (MEMRISYS) 2015 International Conference on* (IEEE) pp 1–3

[15] ITRS Emerging Research Devices (ERD) report URL `http://www.itrs.net`

[16] Kvatinsky S, Kolodny A, Weiser U C and Friedman E G 2011 *Computer Design (ICCD), 2011 IEEE 29th International Conference on* (IEEE) pp 142–147

[17] Al Junid S A M, Tahir N M, Majid Z A and Idros M F M 2012 *Intelligent Systems, Modelling and Simulation (ISMS), 2012 Third International Conference on* (IEEE) pp 187–190

Locally Linear Embedding of Local Orthogonal Least Squares Images for Face Recognition

Fadhlan Hafizhelmi Kamaru Zaman

Faculty of Electrical Engineering, Universiti Teknologi MARA, 40450 Shah Alam, Selangor, Malaysia

fadhlan@salam.uitm.edu.my

Abstract. Dimensionality reduction is very important in face recognition since it ensures that high-dimensionality data can be mapped to lower dimensional space without losing salient and integral facial information. Locally Linear Embedding (LLE) has been previously used to serve this purpose, however, the process of acquiring LLE features requires high computation and resources. To overcome this limitation, we propose a locally-applied Local Orthogonal Least Squares (LOLS) model can be used as initial feature extraction before the application of LLE. By construction of least squares regression under orthogonal constraints we can preserve more discriminant information in the local subspace of facial features while reducing the overall features into a more compact form that we called LOLS images. LLE can then be applied on the LOLS images to maps its representation into a global coordinate system of much lower dimensionality. Several experiments carried out using publicly available face datasets such as AR, ORL, YaleB, and FERET under Single Sample Per Person (SSPP) constraint demonstrates that our proposed method can reduce the time required to compute LLE features while delivering better accuracy when compared to when either LLE or OLS alone is used. Comparison against several other feature extraction methods and more recent feature-learning method such as state-of-the-art Convolutional Neural Networks (CNN) also reveal the superiority of the proposed method under SSPP constraint.

1. Introduction

Since over two decades ago, face recognition has received unprecedented attentions from across numerous research fields such as pattern recognition, computer science, neuroscience, artificial intelligence and many more. The reasons for its popularity is two-fold – largely because of (1) the intricate, unsolved mystery behind how human recognize faces as well as (2) the wide range of practical applications that can typically benefit from face recognition. Previous researchers reported that in face recognition there are numbers of facial features that represent the spatial and configural information of the face which help human brain recognize faces [1-5]. Applications such as access control, security, image database investigations, and surveillance [6] can highly benefit from face recognition due to its non-intrusiveness and high accuracy under constrained condition [7]. However, reliability of face recognition in unconstrained environment is somehow inferior to other intrusive biometric features such as fingerprints and iris. This is because face recognition always suffers under unconstrained variations in face image, for instance change in lighting, pose, facial expressions and age. These are non-trivial problems and images collected from unconstrained environments such as images from CCTV and Webs have a very profound effect of these variations. Over the years, many previous works have focused on how to deal with such variations to increase the reliability of face recognition in unconstrained environments.

Facial appearances can commonly be represented in two ways, either by component-based (local) or holistic (global) representations which have their own advantages and weaknesses. Component-based method is more suitable to be used when the face variations are local, such as variations in expressions, while holistic method is more suited to when the variations are global, such as variations in pose. By holistic representation, a vector of pixel intensities is used to represent the whole image while component-based representation uses multiple group of vectors of pixel intensities to represent various components or regions on face. These components can be evenly divided, or segregated per several common pre-defined regions, and it is either unweighted or linearly/non-linearly weighted at classifier level, for better classification results especially when used with ensembles of classifiers. In both types of representations, dimensionality reduction, or sometimes referred to subspace projection, are usually applied to extract the facial feature from the pixel intensities and to reduce the curse of dimensionality effect. The dimensionality reduction can be linear, non-linear or manifold-learning methods, and can be applied holistically for simpler implementation but with higher processing cost, or locally for better computation performance but slightly complex implementation [8].

Some of popular linear dimensionality reduction methods that has been used in extracting facial features are Eigenfaces [9], Fisherfaces [10], and Local Binary Patterns (LBP) [11]. Some current variants of these methods are discussed in [12, 13]. Non-linear methods such as Kernel Principal Component Analysis (KPCA) [14] which extends PCA to nonlinearity by projecting the data into a higher-dimensional feature space via the kernel trick and Independent Component Analysis (ICA) have also been used in face recognition. On the other hand, manifold-based techniques, such as local linear embedding (LLE) [15], ISOMAP [16] and curvilinear component analysis (CCA) [17], and their linear variants (e.g., locality preserving projection (LPP) [18] and preserving projection (OLPP) [19]) has been used extensively. Investigation on the performance of several linear, non-linear and manifold methods in face recognition have been reported extensively in [20].

Based on findings reported in [20] the applications of these dimensionality projection methods particularly LLE has improved recognition performance, and LLE has been successfully used previously in face recognition [20-22]. However as pointed in [8], the computation cost has increased with increasing complexity of each projection method. We propose in this paper to use Orthogonal Least Square (OLS) [23] to overcome this problem. OLS is known as orthogonal constraint which can preserve more discriminant information in the subspace and it is very popular in other fields and applications especially in system identification [24]. Recently, a unified two-stage orthogonal least squares methods instead of the fast recursive-based methods has been proposed for neural network construction [25]. OLS also has been shown to be able to produce considerable improvement in neural networks training compared to the randomly initialized networks [26]. Similarly, decomposition of the recursive OLS procedure into Sequential Partial Orthogonalization (SPO) has been used for extreme machine learning in [27]. Other recent use of OLS includes to approximate sparse solution [28] and recovery of sparse signals [29]. The amount of work focusing on the use of OLS in feature extraction specifically used for face recognition is limited. Only just recently, Zhao et al. proposed an orthogonal least squares regression model which is used for the extraction of facial features in face recognition [30]. However, in this paper only OLS is used for feature extraction, without fusion with any other subspace method.

In this paper, we specifically focus on increasing computation performance of LLE while maintaining its face classification accuracy. We use locally-applied OLS as pre-processing step prior the application of LLE to reduce the number of features and in doing so we propose a new compact projected face images called Local Orthogonal Least Square (LOLS) images. The feature extraction approach used in this paper is component-based approach since most of the face datasets used for experiments encompasses local variations and component-based method delivers better performance under this constraint. Another constraint implemented in this paper is Single Sample Per Person

(SSPP) [31, 32] constraint – this is where only single image is used for training (gallery), imitating real-world condition where in most applications only single image is available as reference (e.g. identification card, passport, and license). This paper is arranged as follows: Section 2 will describe our proposed method in detail. Section 3 presents experimental results which is followed by discussions in Section 4, and we conclude the paper in Section 5.

2. Methodology

In this section, we discuss the process of computing LOLS images by locally applying OLS on face images. Then we use component-based application of LLE on LOLS images to acquire LLE-LOLS features for face classification. Finally, we classify the face using ensembles of classifiers based on ensembles of Cosine Similarity Metric Learning [33].

2.1. Local Orthogonal Least Squares (LOLS) Images

Orthogonal least square algorithm (OLS) is an algorithm which implements the forward selection method for subset model selection and capable of estimating the parameter estimators. The OLS algorithm transforms the set of regressors P_i into orthogonal basis vectors [23]. Here the OLS is used as discriminant analysis method, such that it improves the model sparsity and control the model complexity [30]. As described in [23], the orthogonal polynomials can be calculated by applying Classical Gram-Schmidt (CGS) (this method is susceptible to round-off error) and Modified Gram-Schmidt (MGS) procedures. The MGS procedure calculates A one row at a time and orthogonalizes P as follows: at the kth stage make the columns subscripted $k+1,\dots,M$ orthogonal to the kth column and repeat the operations for $k=1,\dots,M-1$. Specifically, denoting $p_i^{(0)}=p_i, i=1,\dots,M$, then

$$\left.\begin{array}{r}\left.\begin{array}{r} w_k = p_k^{(k-1)} \\ \alpha_{ki} = \dfrac{\langle w_k, p_i^{(k-1)}\rangle}{w_k, w_k}, i = k+1,\dots,M \\ p_i^{(k)} = p_i^{(k-1)} - \alpha_{ki} w_k, i = k+1,\dots,M \end{array}\right\} k = 1,2\dots,M-1 \\ w_M = p_M^{(M-1)} \end{array}\right\} \quad (1)$$

The elements of g are computing by transforming $z^{(0)}=z$ in a similar way

$$\left.\begin{array}{r} g_k = \dfrac{\langle w_k, z^{(k-1)}\rangle}{w_k, w_k} \\ z^{(k)} = z^{(k-1)} - g_k w_k \end{array}\right\} k = 1,2,\dots,M \quad (2)$$

We locally apply OLS on each of facial component that is extracted using a sliding window of size W that traverse across the surface of the facial image by certain amount of stride S. These facial components are projected into smaller component called LOLS component. LOLS component has smaller dimension than the original facial component which would result into much smaller final LOLS images – a combination of all LOLS component that would yield a reduced representation of original image. This process is illustrated in Figure 1.

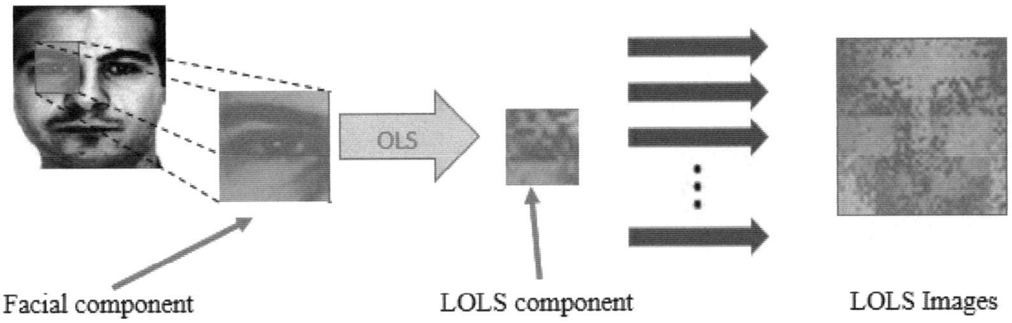

Figure 1. The process of extraction of OLS features from each facial component. The combination of all LOLS components will result into a complete facial image with reduced dimension.

Based on Figure 1, the resulting LOLS image still resembles the original face, however with some details missing. The missing details are those feature that has been eliminated by the OLS, which may not contribute much to the discrimination of the facial identity. This is actually similar to the implementation of Linear Discrimination Analysis (LDA) [34] on face images. The LOLS images acquired in this step is further reduced using LLE, as described in the following section.

2.2. Locally Linear Embedding of LOLS Images(LLE-LOLS)

LLE [15] is a nonlinear manifold method which maps nonlinear high dimensional multivariate data onto lower dimensional subspace having single global coordinate scheme without local minima optimization. The dimension of LOLS facial components acquired from previous steps are further reduced by projection into LLE subspace as illustrated in Figure 2. First we define the number of LLE neighborhood K_{LLE} for each data point x_i , and then by minimizing the cost function $\phi(w)$ as constraint least-square problem, the weight of w_{ij} of x_i from its neighbor x_j is computed by (3). After that, by minimizing the cost function $\phi(y)$, we calculate the projected vector y_j, which is solved as eigenvalue problem in (4).

$$\phi(w) = \min_w \sum_i \left\| x_i - \sum_j w_{ij} x_j \right\|^2 \tag{3}$$

$$\phi(y) = \min_y \sum_i \left\| y_i - \sum_j w_{ij} y_j \right\|^2 \tag{4}$$

where in the case of data x_j is not the neighborhood of x_i, $\sum_j w_{ij} = 1$ and $w_{ij} = 0$. The reduction of LOLS components into LLE vectors take into consideration the lateral LOLS components that belong in the training data. This is similar to the Locally Lateral Subspace (LLS) strategy adopted in [21] which delivers better performance with considerably lower memory requirements. The final features acquired from this step is LLE-LOLS vectors, which is practically has significantly lower dimension than the original image (around more than 80%-dimensional reduction in feature, as pointed out in later sections) while maintaining the salient facial features of the original image.

2.3. Face Classification using Ensembles of Cosine Similarity Metric Learning

We use an ensemble of classifier strategy similar to the one adopted in [21] to classify the acquired LLE-LOLS features. This includes finding the contribution of each LLE-LOLS components on final classification termed as component's confidence, *c*. This confidence is calculated based on the distance vector d_m between component *m* and all lateral components from *P* gallery (training) images. Given that h is the minimum values in d_m, such that $h = \arg\min \{d_m\}$, vector of component's confidence c_m can be computed from modification of softmax normalization in (5).

$$c_m = \log\,(h+2)\,/\log\,(d_m+2) \qquad (5)$$

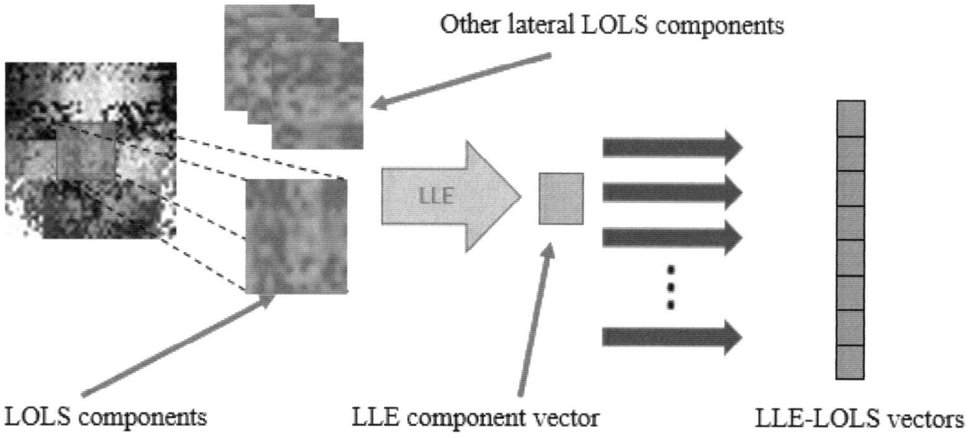

Figure 2. The process of extraction of LLE features from each LOLS component. The combination of all LLE components will result into LLE-LOLS features for face classification. This strategy used in this paper follows LLS strategy proposed in [21].

As shown in (5), maximum confidence is assigned to the nearest feature and others are assigned exponentially lesser confidence per their distance. Then, the component confidence vector, c is calculated using (5). The process is recurring for all available component features and the class label is assigned based on the summation of component confidence $C = \sum_{m=1}^{M} c_m$. For example, if linearly weighted summation of confidence is used, the final classification label of image i would be given to class with $\arg max\{C\}$. Given two vectors of attributes A and B where A_i and B_i are the components of vector A and B respectively, the distance metrics d used in this paper can be calculated using equations (6) which is based on the Cosine Similarity Metric Learning [33].

$$d = 1 - \frac{\sum_{i=1}^{n} A_i B_i}{\sqrt{\sum_{i=1}^{n} A_i^2}\sqrt{\sum_{i=1}^{j} B_i^2}} \qquad (6)$$

3. Results and Discussions

In this section, we present the results and discussions on our experiments on the performance of LLE-LOLS in face classification. For benchmarking purpose, we use publicly available datasets such as ORL, AR, YALE B, and FERET face datasets. Details on datasets used in this paper is summarized in Table 1.

Table 1. Datasets distribution of gallery and probe images for experiments conducted in this paper

Dataset	SSPP Gallery	Probes
AR	100 images of 100 persons	25 images per person
YALE B	38 images with $0^{\circ} \leq (\varphi, \theta) < 12^{\circ}$	2376 images with $(\varphi, \theta) \geq 12^{\circ}$
ORL	40 images of 40 persons	5 to 6 images per person
FERET	1196 images from *fa*	2111 images from *fb, fc, dup I* and *dup II*

The AR dataset [20] contains images taken in two sessions with 13 images per session where 14 images per subject depict expressions and illumination variation, 12 images per subject depict the subject as wearing sunglasses and scarf. We use 100 subjects of 50 males and 50 females for the experiment following procedures adopted in [21]. YALE B dataset contains images having different illumination conditions with light direction with respect to camera axis defined by azimuth angle φ and elevation θ. The cropped version is used which contains 2414 full-frontal images from 38 subjects [18]. ORL dataset [17] consists of 40 different subjects with 10 different images each. The images were taken at different times, varying the lighting, facial expressions and facial details in an upright, frontal position with some side movement. For the experiment on FERET dataset, five subsets used namely *fa* (gallery), *fb* (1,195 expression-variant images), *fc* (194 illumination-variant images), *dup I* (722 images taken later in time), and *dup II* (234 images, which is a subset of *dup I*), following the standard FERET evaluation protocol [19].

All images used in this paper are normalized to 63x63 pixels and they are histogram-equalized before LOLS feature extraction is applied on the images. The size of sliding window is set constant for all images that is W = 7x7 with stride size S=7. Samples of LOLS images obtained from our proposed method is illustrated in Figure 3. In this illustration, these images are reduced to 45x45 LOLS images after we apply the proposed LOLS algorithm. Other parameters used in experiments are as follows: Retained features for each facial component and neighbourhood for LLE are $f_{LLE} = 8$ and $K_{LLE} = 100$ respectively, while retained features for PCA and ICA are also 8. Parameters for Convolutional Neural Network (CNN) are *number of filter* = 1000, *filter size* = 7, *max pooling* = 2, *stride* = 2, *initial learning rate* = 0.01. The learning method used is stochastic gradient descent while the iterations are set to be maximum of 1000 iterations. All experiments are in accordance to SSPP constraint where only 1 image per subject is used for training. We analyse the results of experiments from two different aspects which is classification accuracy and computation time. These are discussed in detail below.

3.1. Face classification performance (classification accuracy).

Face classification accuracy results from the experiment are tabulated in Table 2, where the best classification accuracy of our proposed method is compared against several other popular face extraction methods. Baseline method denotes classification using only pixel intensities as features. Based on results shown, for AR dataset, LLE-LOLS produces the best result of 83.08% classification accuracy which is slightly better than LLE method, which is at 82.23%. This is also significantly higher than Baseline accuracy, which is at a mere 65.08% accuracy. The number of dimensions reduced are also significant when compared to Baseline – the reduction is as much as 83.67%, which means that 83.67% of the original features are 'removed' from the final LLE-LOLS features. Based on the result in Table 2, it is also shown that OLS managed to reduce 38.77% of dimension from original features before further processing by LLE. Based on the result, the best number of OLS retained feature for AR dataset is 26 for each facial component.

Furthermore, results on YaleB face classification demonstrates the superiority of LLE-LOLS method compared to other methods. It delivers 94.75% classification accuracy which the highest out of all tested methods. It produces even better results than the recent Orthogonal Least Squares Regression (OLSR) [30] method which delivers only 89.14% accuracy. Besides, the CNN method fails to deliver good result since the SSPP constraint affect CNN's ability to generalize from the limited sample. For both AR and YaleB datasets, CNN delivers 66.39% and 54.42% classification accuracy respectively. On two other datasets, the ORL and FERET datasets, the proposed LLE-LOLS method also delivers the best result – 94.03% and 56.47% classification accuracy respectively. LLE-OLS also produces better results when compared against OLSR method, which shows that LLE-OLS is more than 3% better than OLSR in terms of accuracy for ORL dataset. The result for ORL and FERET also demonstrates that CNN once again fall short of expectation where the results were severely affected by limitation of samples. In all tested datasets, we observed that OLS alone cannot produce good results on face recognition, while LLE alone can be good but LLE-OLS manages to improve LLE further. LLE-LOLS produced best results and shows improvement in classification accuracy when compared to results obtained from just using either OLS or LLE in feature extraction stage. The difference is as much as 3% (LLE-LOLS compared against LLE in ORL dataset) and 19% (LLE-LOLS compared against OLS in AR dataset) respectively. For YaleB, ORL and FERET datasets, the best number of OLS features used in LLE-LOLS method is 34, 26 and 20 features per facial component respectively.

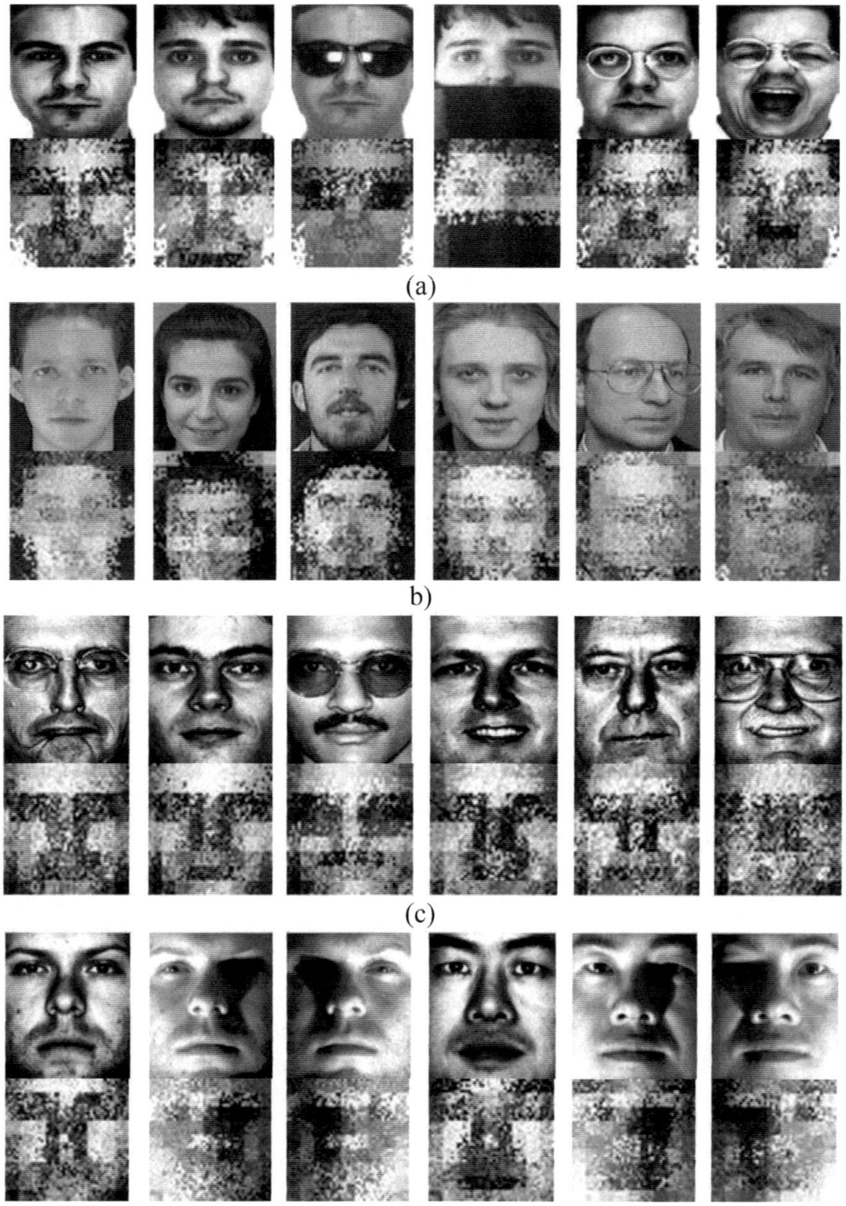

(a)

b)

(c)

(d)

Figure 3: Sample of images from datasets used in this paper, namely (a) AR, (b) ORL, (c) FERET and (d) YaleB. These images are normalized to 63x63 and the subsequent images shown after each row are the equivalent 45x45 LOLS images.

Table 2. Face recognition results for several methods as compared to our proposed method. All experiments are conducted in adherence to SSPP constraint

Dataset	Method	Number of Retained Local Features	Reduced Dimension (%)	Classification Accuracy (%)
AR	Baseline	49	0	65.08
	PCA	8	83.67	76.31
	ICA	12	75.51	76.77
	CNN [35]	49	0	66.39
	OLS	36	26.53	64.62
	LLE	8	83.67	82.23
	Weighted LLE [21]	8	83.67	80.27
	LLE-LOLS	26 8	38.77 (OLS) 83.67 (LLE)	**83.08**
YaleB	Baseline	49	0	93.81
	PCA	8	83.67	79.30
	ICA	12	75.51	92.69
	CNN [35]	49	0	54.42
	OLS	30	38.77	93.39
	OLSR [36]	NA	NA	89.14
	LLE	8	83.67	92.87
	LLE-LOLS	34 8	30.61 (OLS) 83.67 (LLE)	**94.75**
ORL	Baseline	49	0	78.11
	PCA	8	83.67	91.54
	ICA	12	75.51	86.07
	CNN [35]	49	0	86.07
	OLS	12	75.51	79.61
	OLSR [36]	NA	NA	90.43
	LLE	8	83.67	91.04
	LLE-LOLS	26 8	46.94 (OLS) 83.67 (LLE)	**94.03**
FERET	Baseline	49	0	51.87
	PCA	8	83.67	51.25
	ICA	12	75.51	52.44
	CNN [35]	49	0	43.15
	OLS	18	63.26	52.34
	LLE	8	83.67	56.08
	LLE-LOLS	20 8	59.18 (OLS) 83.67 (LLE)	**56.47**

Figure 4 further illustrates the comparison between the classification accuracy of LLE-LOLS, LLE and LOLS. It shows the classification accuracy when different number OLS features are used during feature extraction. For AR dataset, the classification accuracy of LLE-LOLS is not too sensitive towards different number of OLS features used. The LLE-LOLS accuracy is significantly higher than OLS, but just slightly better than LLE. For YaleB dataset, LLE-LOLS is sensitive towards different number of OLS features used where large fluctuations can be observed when different number of features are used. Its classification accuracy is also much better than both LLE and OLS method. In both ORL and FERET datasets, the LLE-LOLS method is just slightly affected by number of OLS features, while the best accuracy for LLE-OLS is better than LLE and OLS accuracy.

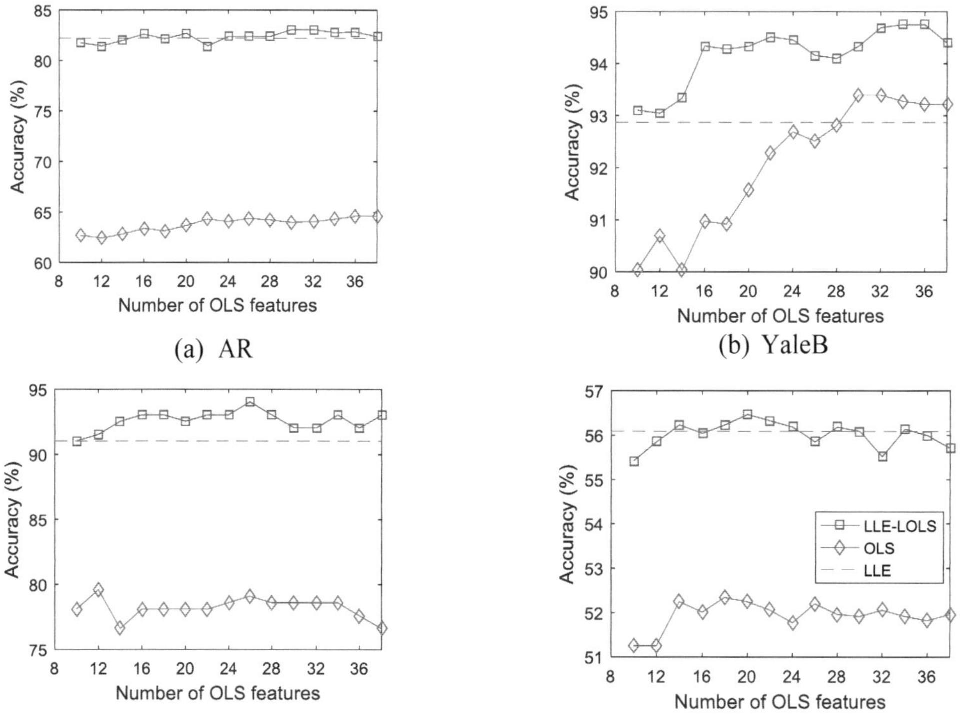

(a) AR (b) YaleB

(c) ORL (d) FERET

Figure 4. The performance of face classification accuracy (in percentage) for (a) AR, (b) YaleB, (c) ORL and (d) FERET datasets when different number of OLS feature are used in feature extraction stage. Shown together in this figure is the accuracy of LLE face classification.

3.2 Face classification performance (processing time).

In terms of processing time, we measure the reduction in processing time achieved by LLE-LOLS method when compared against when only LLE is used for feature extraction. Figure 4 shows the results in detail. In this figure, we plot the processing time of LLE-LOLS which is normalized against the processing time taken by LLE. In the same figure, we plot the classification accuracy (normalized to 1.0 – where 1.0 is the highest accuracy) to further emphasize the relationship between the processing time and face classification accuracy. According to Figure 5, for AR and YaleB and FERET datasets, the processing time of LLE-LOLS are significantly less than the processing time of LLE by considerable amount. In fact, the reduction of processing time is as much as 10% when compared against LLE processing time. Even though smaller number of OLS features consumes lower processing time, the classification accuracy is not severely affected by the small number of features. This is highly desirable since if the classification accuracy is not sensitive to number of features used, we could opt for better processing time should such requirement arise. However, for ORL dataset, OLS features higher than 18 would increase the processing time of LLE-LOLS, making it to be higher than LLE processing time. In fact, higher number of features does not guarantee better performance, however it does increase the processing time required. It is thus observed that the proposed method works best when large datasets are used (i.e. AR, YaleB and FERET) while smaller datasets such as ORL does not benefit much in terms of processing time.

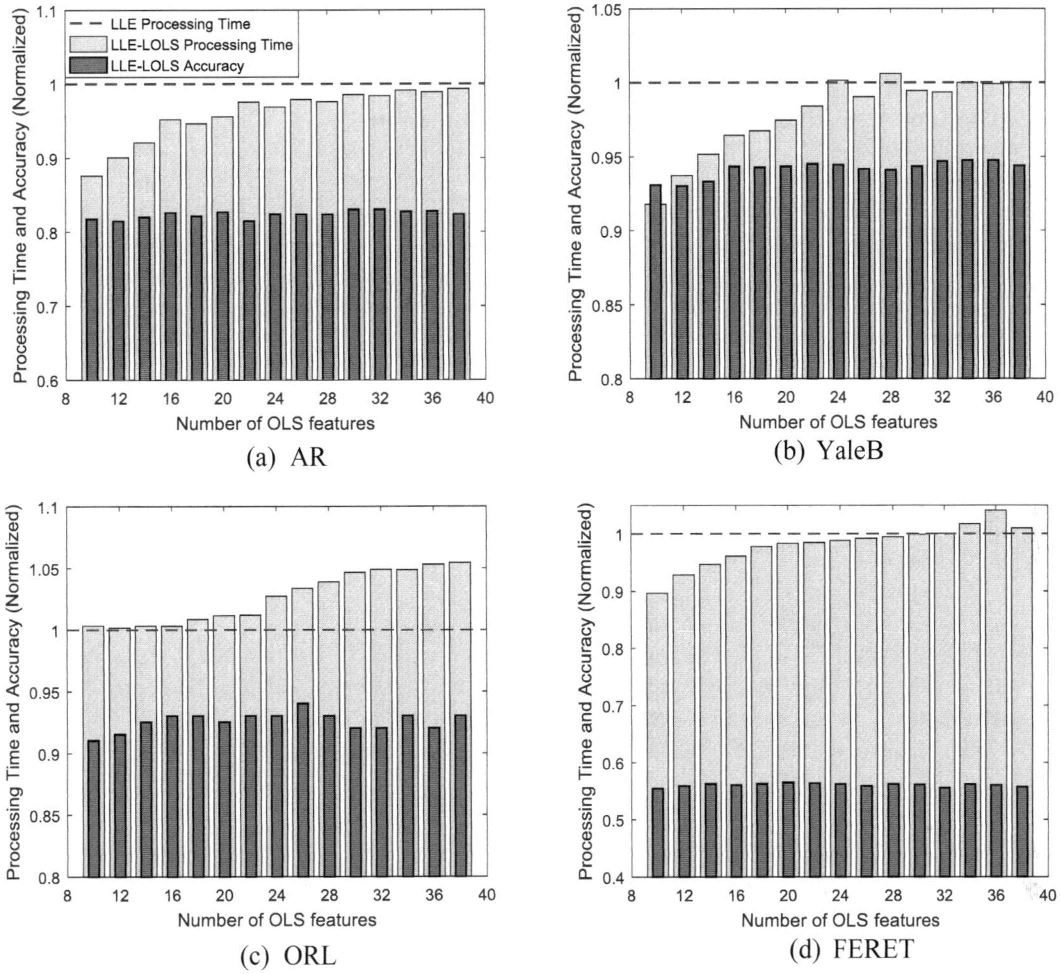

Figure 5: The normalized processing time of LLE-LOLS (normalized against LLE processing time) for (a) AR, (b) YaleB, (c) ORL and (d) FERET datasets when different number of OLS feature are used in feature extraction stage. Shown together in this figure is the LLE-LOLS face classification accuracy.

4. Conclusions

In this paper, we present a method to improve LLE processing time in face extraction stage while improving or at least maintaining the face classification accuracy. We propose a method called LLE-LOLS method where components of face images are dimensionally reduced using OLS which results into LOLS images. These LOLS images are further reduced in dimension using conventional LLE method. The LLE-LOLS features practically has significantly lower dimension than the original image (around more than 80%-dimensional reduction in feature) while maintaining the salient facial features of the original image, as proven by the classification accuracy. In all tested datasets, LLE-LOLS produced best results when tested against several popular face recognition methods, and shows improvement in classification accuracy when compared to results obtained from just using either OLS or LLE in feature extraction stage. The difference is as much as 3% (LLE-LOLS compared against

LLE in ORL dataset) and 19% (LLE-LOLS compared against OLS in AR dataset) respectively. Comparisons against state-of-the-art CNN also reveals the robustness of our proposed method against limitation in number of available samples for training. In terms of processing time, LLE-LOLS manages to improve the processing time in almost all tested datasets, where improvement of as much as 10% in processing time is achieved. To further improve this method, it is recommended to investigate the aptness of this method to be used with other popular feature extraction methods such as Gabor Wavelets or Scale Invariant Feature Transform (SIFT) features.

Acknowledgements
The authors would like to thank Dr. Aleix M. Martinez for providing the AR Face Dataset. Some portions of the research in this paper also use Extended Yale Face Database B. The authors would like to thank the FERET Technical Agent, the US National Institute of Standards and Technology (NIST) for providing the FERET dataset. The work presented here is sponsored by Universiti Teknologi MARA, Malaysia under grant 600-RMI/DANA 5/3/LESTARI (20/2015).

References

[1] A. W. Young, D. Hellawell, and D. C. Hay, "Configural information in face perception," *Perception,* vol. 16, pp. 747-759, 1987.

[2] A. W. Young, F. Newcombe, E. H. F. de Haan, M. Small, and D. C. Hay, "Face perception after brain injury: Selective impairments affecting identity and expression," *Brain,* vol. 116, pp. 941-959, 1993.

[3] J. W. Tanaka and J. A. Sengco, "Features and their configuration in face recognition," *Memory and Cognition,* vol. 25, no. 5, pp. 583-592, 1997.

[4] A. Schwaninger, C. Wallraven, D. W. Cunningham, and S. D. Chiller-Glaus, "Processing of facial identity and expression: a psychophysical, physiological, and computational perspective.," *Progress in Brain Research,* vol. 156, pp. 321-343, 2006.

[5] J. Sadr, I. Jarudi, and P. Sinha, "The role of eyebrows in face recognition,," *Perception,* vol. 32, pp. 285-293, 2003.

[6] D. N. Parmar and B. B. Mehta, "Face Recognition Methods & Applications," *Computer Technology & Applications,* vol. 4, no. 1, pp. 84-86, 2013.

[7] Y.-Z. Lu, "A Novel Face Recognition Algorithm for Distinguishing Faces with Various Angles," *International Journal of Automation and Computing,* vol. 5, no. 2, pp. 193-197, 2008.

[8] F. Kamaruzaman, A. A. Shafie, and Y. M. Mustafah, "Robust Face Recognition Against Expressions and Partial Occlusions," *International Journal of Automation and Computing,* 2015.

[9] M. Turk and A. Pentland, "Eigenfaces for recognition," *Journal of Cognitive Neuroscience,* vol. 3, no. 1, pp. 71-86, 1991.

[10] P. Belhumeur, J. Hespanha, and D. Kriegman, "Eigenfaces vs. Fisherfaces: recognition using class specific linear projection," *IEEE Trans. Pattern Analysis and Machine Intelligence,* vol. 19, no. 7, pp. 711-720, 1997.

[11] T. Ahonen, A. Hadid, and M. Pietikainen, "Face Description with Local Binary Patterns: Application to Face Recognition," *Pattern Analysis and Machine Intelligence, IEEE Transactions on,* vol. 28, no. 12, pp. 2037-2041, 2006.

[12] H. V. Nguyen, L. Bai, and L. Shen, "Local Gabor binary pattern whitened pca: a novel approach for face recognition from single image per person," in *Lecture Notes in Computer Science, Proceedings of the Third International Conference on Advances in Biometrics*, Alghero, Italy, 2009, vol. 5558, pp. 269-278.

[13] F. Hafiz, A. A. Shafie, and Y. M. Mustafah, "Face Recognition From Single Sample Per Person by Learning of Generic Discriminant Vectors," *Procedia Engineering,* vol. 41, pp. 465-472, 2012.

[14] H. Hoffmann, "Kernel PCA for novelty detection," *Pattern Recognition,* vol. 40, no. 3, pp. 863-874, 2007.

[15] S. T. Roweis and L. K. Saul, "Nonlinear dimensionality reduction by locally linear embedding," *Science,* vol. 290, no. 5500, pp. 2323-2326, 2000.

[16] J. B. Tenenbaum, V. De Silva, and J. C. Langford, "A global geometric framework for nonlinear dimensionality reduction," *Science,* vol. 290, no. 5500, pp. 2319-2323, 2000.

[17] P. Demartines and J. Hérault, "Curvilinear component analysis: a self-organizing neural network for nonlinear mapping of data sets," *IEEE Trans. Neural Networks,* vol. 8, pp. 148-154, 1997.

[18] X. Niyogi, "Locality preserving projections," *Neural information processing systems,* vol. 16, p. 153, 2004.

[19] D. Cai, X. He, J. Han, and H. J. Zhang, "Orthogonal Laplacianfaces for Face Recognition," *IEEE Transactions on Image Processing,* vol. 15, no. 11, pp. 3608-3614, 2006.

[20] W. Huang and H. Yin, "On nonlinear dimensionality reduction for face recognition," *Image and Vision Computing,* vol. 30, pp. 355-366, 2012.

[21] F. K. Zaman, A. A. Shafie, and Y. M. Mustafah, "Robust face recognition against expressions and partial occlusions," *International Journal of Automation and Computing,* journal article vol. 13, no. 4, pp. 319-337, 2016.

[22] N. Mekuz, C. Bauckhage, and J. K. Tsotsos, "Face recognition with weighted locally linear embedding," in *Computer and Robot Vision, 2005. Proceedings. The 2nd Canadian Conference on*, 2005, pp. 290-296.

[23] S. Chen, S. A. Billings, and W. Luo, "Orthogonal least squares methods and their application to non-linear system identification," *International Journal of Control,* vol. 50, pp. 1873-1896, 1989.

[24] L. Mirea, "Dynamic multivariate B-spline neural network design using orthogonal least squares algorithm for non-linear system identification," in *2014 18th International Conference on System Theory, Control and Computing (ICSTCC)*, 2014, pp. 720-725.

[25] L. Zhang, K. Li, E. W. Bai, and G. W. Irwin, "Two-Stage Orthogonal Least Squares Methods for Neural Network Construction," *IEEE Transactions on Neural Networks and Learning Systems,* vol. 26, no. 8, pp. 1608-1621, 2015.

[26] M. Lehtokangas, J. Saarinen, K. Kaski, and P. Huuhtanen, "Initializing Weights of a Multilayer Perceptron Network by Using the Orthogonal Least Squares Algorithm," *Neural Computation,* vol. 7, no. 5, pp. 982-999, 1995.

[27] N. Wang, M. J. Er, and M. Han, "Parsimonious Extreme Learning Machine Using Recursive Orthogonal Least Squares," *IEEE Transactions on Neural Networks and Learning Systems,* vol. 25, no. 10, pp. 1828-1841, 2014.

[28] A. Hashemi and H. Vikalo, "Sparse linear regression via generalized orthogonal least-squares," in *2016 IEEE Global Conference on Signal and Information Processing (GlobalSIP)*, 2016, pp. 1305-1309.

[29] J. Wang and P. Li, "Recovery of Sparse Signals Using Multiple Orthogonal Least Squares," *IEEE Transactions on Signal Processing,* vol. 65, no. 8, pp. 2049-2062, 2017.

[30] H. Zhao, Z. Wang, and F. Nie, "Orthogonal least squares regression for feature extraction," *Neurocomputing,* vol. 216, pp. 200-207, 12/5/ 2016.

[31] X. Tan, S. Chen, Z.-H. Zhou, and F. Zhang, "Face recognition from a single image per person: A survey," *Pattern Recognition,* vol. 39, no. 9, pp. 1725-1745, 2006.

[32] F. H. K. Zaman, "Single Sample Face Recognition Using a Network of Spiking Neurons," PhD Thesis, Kulliyyah of Engineering, International islamic University Malaysia, Kuala Lumpur, 2015.

[33] H. V. Nguyen and L. Bai, "Cosine Similarity Metric Learning for Face Verification," in *Computer Vision – ACCV 2010: 10th Asian Conference on Computer Vision, Queenstown,*

New Zealand, November 8-12, 2010, Revised Selected Papers, Part II, R. Kimmel, R. Klette, and A. Sugimoto, Eds. Berlin, Heidelberg: Springer Berlin Heidelberg, 2011, pp. 709-720.

[34] W. Zuo, K. Wang, D. Zhang, and H. Zhang, "Combination of two novel LDA-based methods for face recognition," *Neurocomputing,* vol. 70, no. 4–6, pp. 735-742, 2007.

[35] A. Krizhevsky, I. Sutskever, and G. E. Hinton, "ImageNet classification with deep convolutional neural networks," *Advances in Neural Information Processing Systems,* 2012.

[36] H. Zhao, Z. Wang, and F. Nie, "Orthogonal least squares regression for feature extraction," *Neurocomput.,* vol. 216, no. C, pp. 200-207, 2016.

IOP Conf. Series: Materials Science and Engineering **341** (2018) 012027 doi:10.1088/1757-899X/341/1/012027

Problems and Issues of High Rise Low Cost Housing in Malaysia

Noraziah Wahi[1], Rosli Mohamad Zin[2], Vikneswaran Munikanan[3],Ismail Mohamad[4],Syahrizan Junaini[5]

[1]Faculty of Architectural,Planning and Surveying,Universiti Teknologi MARA, Malaysia

[2]Faculty of Civil Engineering, University of Technology Malaysia

[3]Faculty of Engineering, National Defence University of Malaysia

[4]Science Faculty, University of Technology Malaysia

[5]Brooke Dockyard and Engineering Works Corporation, Malaysia

Email: belle_rusd@yahoo.com

Abstract. Major cities in developing countries are undergoing an enormous migration of peoples from countryside regions. This migration from the countryside regions were mostly to develop carrier and expecting for higher salary for their living survival. Consequently, the large amount of immigrants from countryside to the cities each year had created a great demand for urban housing. The impact from that, Kuala Lumpur, Selangor and its surrounding area now is crowded by the low-income group who cannot afford to own an affordable house. The government of Malaysia had aware of this situation and therefore had created the low cost housing especially for urban poor. However, there are many issues and problems arise regarding the low cost housing in Malaysia especially in urban area. The research is regarding a study on problems and issues of high rise low-cost housing in Malaysia. The need to examine the problems associated with the high rise low cost housing is to ensure the success of future low cost housing development in Malaysia.

1. Introduction

Housing becomes an important part of daily life, not only because of its cost, but also because of the access that housing can afford to other attributes of a viable urban life. Specifically, housing can help residents address the struggle to maintain economic livelihood, the threats of a changing climate, the challenges of urban violence, and the inequities of governance [1].

In Malaysia, housing requirement between the years of 2006 to 2010 is 709,400 houses. From that solid figure, twenty thousand units are allocated for hardcore poor, people living under poverty line, their need for housing, too urgent. 165,400 units belong to low-income earner as low-cost housing and 85,505 units are for low-medium-cost housing [2].

The total number of housing units in Malaysia targeted was 800,000 units under Seventh Malaysia Plan and 782,300 units of housing were targeted to be constructed under Eighth Malaysia Plan. During the Ninth Malaysia Plan, requirement for new houses is expected to be about 709,400 units of which

19.2% will be in Selangor followed by Johor at 12.9%, Sarawak 9.4% and Perak 8.2% [3].Therefore, the Ninth Malaysia Plan (9MP) itself gave special attention to low-cost housing. The overall performance of houses built under the low-cost housing category is encouraging with 200,513 units completed or 86.4% of the set target. Out of this number, 103,219 units or 51.5% were constructed by the public sector including state economic development corporations. To ensure an adequate supply of low-cost houses, any mixed-development projects undertaken by private developers are required to allocate a minimum of 30% to low-cost housing [2].

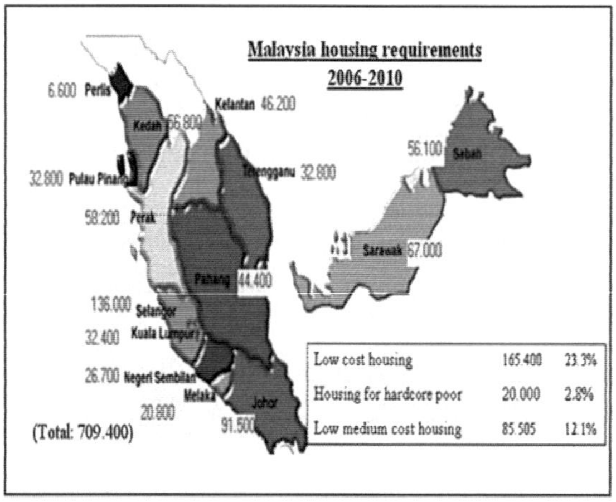

Figure 1.0: Malaysian Housing Requirements 2006-2010 [2]

2. Overview of low cost housing policy in Malaysia

Public house schemes are perceived by the government as the way for low income groups to own their houses and to live under better housing conditions by ensuring the provisions of minimum acceptable standards, amenities, and facilities within and outsides the dwelling units. This will contribute to an improvement in the quality of life for residents [4].

The Malaysia government always tries to identify clearly the target groups entitled to low-income housing. The target group has continued to expand in accordance with the higher aspirations of the people, often matched by a corresponding increase in the capacity for delivery. The low-cost housing has relatively low selling prices or rentals so as to maintain high levels of affordability by the low income group [5]. The government is actively promoting policies to ensure quality and affordability of houses to all nations. Thus, this scenario shows that sustainability is becoming prominent in Malaysian housing industry [6].

Federal government in Malaysia involved directly in housing production and allocation system with the cooperation of the State governments for low cost housing. Funding for public low cost housing programmes usually comes from Federal Government, the State Governments are only responsible to identify the available state's land for housing construction and selection of eligible low cost house buyers. The local government generally did not involve directly in housing production, except approving the planning permission and building plans for housing construction by both public and private developers. However in major cities such as Kuala Lumpur and Petaling Jaya, local governments are also involved in administration and maintenance of public low cost housing with the Federal and State funding [7].

Table 1.0: Total strata public and private low cost housing according to state, 2013

State	Stratified/High Rise Low Cost Housing				
	Public Scheme	Private Scheme	Total	Total Unit	Estimated No. of Occupants
Federal Territory KL	59	2317	2376	130,566	652,830
Penang	29	340	369	73,541	367,705
Sabah	24	-	24	16,167	80,835
Melaka	23	80	103	14,218	71,090
Pahang	18	64	82	7326	36,180
Johor	18	205	223	57,234	286,180
Sarawak	18	-	18	12,832	64,160
Perak	15	74	89	11,430	57,150
Kedah	13	76	89	7754	38,770
Terengganu	12	14	26	3706	1853
Selangor	11	1063	1074	221,703	1,108,515
Negeri Sembilan	8	76	84	13,586	6793
Kelantan	3	6	9	1353	6765
Perlis	3	0	3	1378	6890
Total	**254**	**4315**	**4569**	**572,704**	**2,785,706**

Source:[7]

3. Problems and issues of low cost housing in Malaysia

The success of housing programmes does not only depend on merely provision of housing units, but also on other factors that affect the needs of residents. The failure of many housing projects may be attributed to the lack of knowledge on the determinants of residential satisfaction. Residential satisfaction reflects the degree to which individuals' housing needs are fulfilled. Therefore, it acts as a guide to policy makers to monitor the implementation of housing policies [8]. Several issues were also recognized related to low cost housing based on Kuala Lumpur Structure Plan 2020 which includes low space standards, under the provision of community facilities, high maintenance costs and poor quality of construction and materials in low cost housing projects [9].

The provision of strata low cost housing units in Malaysia contribute significantly in ensuring more houses were built to relocate the slums, provide decent and adequate home for the low income people since 1990s. However development of high rise housing in urban area for the low income people are not without any problems especially in public housing. Housing maintenance, management and anti-social behaviours problems are still the major issues that need to be addressed by the government [7].

Nowadays, there is a crisis in affordable housing where prices have risen steadily and construction does not meet standards that guarantee good quality over the life of a project [10]. Qualitatively the type of low cost housing in Malaysia has not been satisfactory to the family housing needs, comfort, social, cultural and religious needs [11]. Zainal et al. [12] stated that most of low cost houses in Malaysia have limited space and recreation areas such as multipurpose hall, and playground to be used for community and recreation activities. Therefore, local participation between policymakers and residents is increasingly necessary to make environmental elements and services available, make those neighborhoods liveable and improve the quality of urban life [13].

Urban housing problems normally arise when the developer in urban areas have less interested to build medium and low cost housing due to small profit gained from the development which consequently affected in terms of affordability. In addition, with shortage of land supply and increased value of land in urban areas may also affect in the provision of open space. This is much concerned with the compliance of development plan and standard guidelines provided for planning approval where at some point, developers refuse to comply with the provision as there is lack of enforcement carried out by the authority [14]. Providing good quality housing units should promote more equal social, educational, and economic opportunities and help create more equitable environment for all [10].

4. Methodology

A self-administered survey was conducted to collect data regarding issues and problems of the low cost housing residents. The respondents eligible to participate in the survey were those who stay in the high-rise low cost housing. The data were collected by face-to-face interviews and distribution of questionnaires survey.

For the purpose of analysis of quantitative data collection through questionnaires survey, the Statistical Package for Social Science (SPSS) software version 20.0 was adopted. The objectives of using quantitative approach is to summarizes statistically large information such as questionnaires responses feedback translated into figures that can be understood and interpreted.

5. Results and Findings

The research study was conducted in the state of Selangor in peninsular Malaysia. Census 2010 revealed that the total population of Malaysia was 28.3 million. Selangor state has a geographical area of 8,104 square kilometers and a population of 5,462,141 as the most populated state in Malaysia, census 2010 [9].

There were 700 sets of questionnaires distributed to all occupants in the high rise low cost housing in Selangor. However, there were only 510 sets of questionnaires had been returned to the researcher. The questionnaires survey contained three section. These were Demographic information; Problems/Issues of Current High Rise Low Cost Housing on Occupied Unit (Internal unit to perimeter drain area).

5.1. Section A of the questionnaires: Demographic Information

Section A of the questionnaire survey provided data on the age, gender, marital status, race, education level, total household income levels and employment status of the respondents, Table 2.0 illustrate the demographic information of the respondents in Selangor.

Table 2.0: Demographic information of the respondents

No	Item	Frequency	Percentage
1.	***Age (years):***		
	20-30	75	14.7
	31-40	170	33.3
	41-50	170	33.3
	Above 50	95	18.6
2.	***Gender:***		
	Female	280	54.9
	Male	230	45.1
3.	***Marital status:***		
	Married	387	75.9
	Single	67	13.1
	Single mother	39	7.6
	Single father	17	3.3

4.	***Race:***		
	Malay	199	39.0
	Indian	153	30.0
	Chinese	158	31.0
5.	***Home ownership:***		
	Owner	179	35.1
	Rent from landlord/owner	167	32.7
	Rent from State Government	164	32.2
6.	***Length of occupancy:***		
	Below 1 year	2	0.4
	2-4 years	72	14.1
	5-7 years	138	27.1
	8-10 years	150	29.4
	More than 10 years	148	29.0
	TOTAL	**510**	**100%**

Based on the Table 2.0, 14.7% of the respondents were aged 20-30 years old. A further 33.3% of the respondents fell into 31-40 and 41-50 years old bracket respectively. Aged above 50 years old contributes to 18.6% of the total respondents. Majority respondents residing in low cost housing in Selangor are aged range between 31-50 years old. The results turned out to have similarities such as a study conducted by Sulaiman et al., 2016 that stated the highest of respondent age is 31 to 50 years old are commonly living in low-cost housing area in Malaysia

There were slightly more female (54.9%) than male (45.1%) among the 510 respondents involved. This might be because the distribution of questionnaires took place during weekday working hours; many male dwellers would have been at work and unable to respond to the questionnaires. Based on the Table 2.0, it is stated that 75.9% of the respondents are married; 13.1% of the respondents are single; 7.6% of the respondents are single mother; and 3.3% of the respondents are single father. As for ethnicity, there were Malay (39%), Chinese (31%) and Indian (30%). The proportion of race is vital to know to make sure that there is multicultural race exist in the high rise low cost housing area community.

It is also showed that 35.1% of the respondents owned their house during the case study period. Another 64.9% of the respondents rented their house, either rented from landlord or the state government. People usually rented a house as they feel convenient to travel to their working place and can save more time. They also can avoid traffic congestion during the peak hours.

There were 29.0% of the respondents had residency lengths of more than 10 years, 29.0% had habitation periods of 8–10 years, 27.1% had habitation periods of 5–7 years, 14.1% had a length of residency of 2–4 years and a further of 0.4% had lived in the low cost house below than a year. This high level prevalence of occupancy for periods over four years as described indicates that the respondents were well-placed to convey their experiences of living in the high rise low cost housing neighbourhoods. These demographic results have; thus, confirmed that the respondents in this case study are reasonably representative of a low cost housing in Malaysia. The duration of occupancy equal to five years or more, means that these respondents have, potentially, developed a sense of community and neighbourliness in the high rise low cost housing in Malaysia.

5.2. Problems/Issues of Current High Rise Low Cost Housing on Occupied Unit (Internal unit to perimeter drain area)

Section B of the questionnaire requires the respondents to rate their satisfaction level regarding the issue and problems faced by them during their stay in the high rise low cost housing. The respondents were need to rate their satisfaction level based on five numerical Likert-scale; "1" (Strongly satisfied), "2"

(Satisfied), "3" (Neutral), "4" (Dissatisfied), and "5" (Strongly dissatisfied). Table 3.0 illustrates the occupants' satisfaction level towards the listed issues.

Table 3.0: Percentage of 'strongly satisfied and satisfied' level on high rise low cost housing issues

No	Issue	Strongly Satisfied	Satisfied	Strongly Satisfied & Satisfied	Percentage
1	Roof leakage	6	187	193	**37.8%**
2	Criminal cases	0	192	192	**37.6%**
3	Corridor lighting	0	76	76	**14.9%**
4	Noise	0	70	70	**13.7%**
5	Drainage system	0	58	58	**11.4%**
6	Fire door	0	24	24	**4.7%**
7	Corridor spaces	0	20	20	**3.9%**
8	Safety walking under uncovered corridor	0	16	16	**3.1%**
9	Staircase condition	0	15	15	**2.9%**
10	House design	0	7	7	**1.4%**
11	Plumbing system	0	5	5	**1.0%**
12	House workmanship	0	4	4	**0.8%**
13	Lift numbers	0	3	3	**0.6%**
14	Lift size	0	3	3	**0.6%**
15	Handicap facilities	0	3	3	**0.6%**
16	Material quality	0	2	2	**0.4%**
17	Lift maintenance	0	1	1	**0.2%**
18	Internal ventilation	0	0	0	**0.0%**

Table 3.0 above shows the questionnaires results in relation to occupants' satisfactory and strongly satisfactory level to the issues and problems in their the low cost housing area. As depicted in the table above, there are 38.7% respondents satisfied with the issue of roof leakage. This is because most of the respondents do not experience such issue as their housing unit is in the mid or low levels. Respondents are quite satisfied with the low criminal cases within their low cost housing vicinity. This is due to the high density populated low cost house living in close community. Thus everyone takes the responsibility and awareness for crime prevention. 14.9% respondents are satisfied with corridor lighting and 13.7% shows satisfaction for noise issue at low cost housing complex. A total of 7 issues and problems having less than 1% respondents for satisfied and strongly satisfied. These are namely house workmanship, numbers of lifts, size of lifts, availability of facilities for handicap, quality of building materials, issues of lift maintenance and internal ventilation.

Table 4.0: Percentage of 'Strongly dissatisfied and dissatisfied' level on high rise low cost housing issues

No	Issue	Strongly Dissatisfied	Dissatisfied	Strongly Dissatisfied & Dissatisfied	Percentage
1	Lift maintenance	386	112	498	**97.65%**
2	Lift size	355	130	485	**95.10%**
3	Lift numbers	357	125	482	**94.51%**
4	Handicap facilities	295	170	465	**91.18%**
5	House workmanship	217	224	441	**86.47%**
6	Plumbing system	231	207	438	**85.88%**
7	House design	185	233	418	**81.96%**
8	Material quality	147	258	405	**79.41%**
9	Safetywalking under uncovered corridor	199	197	396	**77.65%**
10	Staircase condition	141	214	355	**69.61%**
11	Corridor spaces	134	214	348	**68.24%**
12	Fire door	120	221	341	**66.86%**
13	Internal ventilation	68	219	287	**56.27%**
14	Drainage system	85	181	266	**52.16%**
15	Corridor lighting	47	193	240	**47.06%**
16	Noise	71	122	193	**37.84%**
17	Criminal cases	89	49	138	**27.06%**
18	Roof leakage	19	59	78	**15.29%**

Table 4.0 above shows the dissatisfaction and strongly dissatisfaction level of the low cost house residents' towards the problems and issues faced. The top three issues faced are related to the lift, namely lift maintenance (97.65%), lift size (95.10%) and numbers of lift (94.51%). The numbers of lift are insufficient and to make it worse most are having malfunctioning lift. This is then followed by the availability of facilities for disabled. The next four problems are related to the quality of the low cost house which are house workmanship, plumbing system, design of the house and construction material quality used. In between 79% to 84% of respondents agreed to these common quality problems. Subsequent group of problems are associated with the common area within the low cost housing complex. The issues of safety while walking under the uncovered corridor, condition of the staircases, corridor spaces, condition of the fire door, internal ventilation and drainage system are recorded as dissatisfaction and strongly dissatisfaction among 52% to 77% of the respondents. Finally four issues recorded below than 50% response are corridor lighting issue, noise problem, numbers of criminal cases and roof leakage. Low criminal cases shows that the level of safety and security at low cost housing area are good as many dwellers are house wives and senior citizens which normally occupy the house during day time. Close community at low cost housing complex also contributes to low crime case. Problem having lowest response is roof leakage as this is only experienced by several homeowners residing at the topmost level in each block.

Table 5.0: Percentage of 'Neutral' level on high rise low cost housing issues

No	Issue	Neutral	Percentage
1	Noise	247	**48.4%**
2	Roof leakage	239	**46.9%**
3	Internal ventilation	223	**43.7%**
4	Corridor lighting	194	**38.0%**
5	Drainage system	186	**36.5%**
6	Criminal cases	180	**35.3%**
7	Fire door	145	**28.4%**
8	Corridor spaces	142	**27.8%**
9	Staircase condition	140	**27.5%**
10	Material quality	103	**20.2%**
11	Safety walking under uncovered corridor	98	**19.2%**
12	House design	85	**16.7%**
13	Plumbing system	67	**13.1%**
14	House workmanship	65	**12.7%**
15	Handicap facilities	42	**8.2%**
16	Lift numbers	25	**4.9%**
17	Lift size	22	**4.3%**
18	Lift maintenance	11	**2.2%**

Out of 510 respondents, it is also important to point out the statistic of those in neutral position, neither satisfied and strongly satisfied nor dissatisfied and strongly dissatisfied. Noise issue is less of a concern among 48.4% of respondents though low cost housing complex is normally densely populated. Secondly followed by the problem of roof leakage which is only experienced by dwellers at the topmost floor. This is followed by three physical properties of the low cost house which is internal ventilation (43.7%), corridor lighting (38.0%) and drainage system (36.5%). 35.3% respondents are neutral on social issue such as criminal cases that happened within their low cost housing area. Issues related to safety during emergency such as fire door condition, space on the corridor and condition of staircase is responded as neutral by 28.4%, 27.8% and 27.5% respectively. The lesser the percentages means respondents regard these issues either as satisfaction or dissatisfaction. This is shown by the bottom three issues that is marked as neutral namely lift numbers, lift size and lift maintenance in which is the top three dissatisfaction issues as depicted in Table 4.0 above.

6. Discussion

Studies suggest that low-income households are somewhat less likely to move than higher income groups. The studies income household have more imply the fact that higher choices in the housing market as they are less deterred by the high transaction cost. Thus, higher income group are more likely to move as compared to low income households [15]. Due to this reason, there is a need to improve on all the problems faced by all residents in the high rise low cost housing in Malaysia. Governments has put up continuous efforts in the provision of low cost housing for the benefits of low income earners to own a house and thus provide better living for their families. However the issues and problems faced by the dwellers need to be emphasized and rectified. The most common problem at high rise low cost house is the availability and sufficiency of lifts which affects all residents. Numbers of lift is insufficient thus increases the frequency of use. Lifts are not being regularly maintained thus resulting in frequent breakdown. The quality of low cost house units in terms of construction material, internal design, layout and plumbing system also need to be taken seriously.

IOP Conf. Series: Materials Science and Engineering **341** (2018) 012027 doi:10.1088/1757-899X/341/1/012027

References

[1] Vale, L.J. et al., 2014. What Affordable Housing Should Afford : Housing for Resilient Cities. *A Journal of Policy Development and Research*, 16(2), pp.21–50.

[2] Bakhtyar, B. et al., 2013. Housing for poor people: A review on low cost housing process in Malaysia. *WSEAS Transactions on Environment and Development*, 9(2), pp.126–136.

[3] Zainun, N.Y., Rahman, I.A. & Eftekhari, M., 2010. Forecasting Low-Cost Housing Demand in Johor Bahru, Malaysia Using Artificial Neural Networks (ANN). *Journal of Mathematics Research*, 2(1), pp.14– 19.

[4] Al-Abed, R.D.A., 2000. Satisfaction level with neighbourhoods in low-income public housing in Yemen. *Property Management*, 18(4), pp.230–242.

[5] Abdullahi, B.C. & Aziz, W.N.A.W.A., 2011. Pragmatic housing policy in the quest for low-income group housing delivery in Malaysia. *Journal of Design and Built ...*, 8(1), pp.1–18.

[6] Goh, K.C., Seow, T.W. & Goh, H.H., 2013. *Challenges of implementing Sustainability in Malaysian Housing Industry*,

[7] Shuid, S., 2016. The housing provision system in Malaysia. *Habitat International*, 54, pp.210–223.

[8] Abdul Ghani Salleh, 2008. Neighbourhood factors in private low-cost housing in Malaysia. *Habitat International*, 32, pp.485–493.

[9] Hashim, A.E. et al., 2012. Assessing Factors Influencing Performance of Malaysian Low-Cost Public Housing in Sustainable Environment. *Procedia - Social and Behavioral Sciences*, 50(July), pp.920– 927. Available at: http://dx.doi.org/10.1016/j.sbspro.2012.08.093.

[10] Abed, A.R., Awada, E. & Sen, L., 2013. The Impact of Affordable Sustainable Housing Neighborhoods on Housing Cost Efficiency. *Sustainable Development*, 6(9), pp.62–72.

[11] Mohit, M.A., Ibrahim, M. & Rashid, Y.R., 2010. Assessment of residential satisfaction in newly designed public low-cost housing in Kuala Lumpur , Malaysia. *Habitat International*, 34(1), pp.18–27. Available at: http://dx.doi.org/10.1016/j.habitatint.2009.04.002.

[12] Zainal, N.R. et al., 2012. Housing Conditions and Quality of Life of the Urban Poor in Malaysia. In *Procedia - Social and Behavioral Sciences*. pp. 827–838.

[13] Ardeshiri, A. et al., 2016. The values and benefits of environmental elements on housing rents. *Habitat International*, 55, pp.67–78. Available at: http://dx.doi.org/10.1016/j.habitatint.2016.02.004.

[14] Yakob, H., Yusof, F. & Hamdan, H., 2012. Land Use Regulations towards a Sustainable Urban Housing: Klang Valley conurbation. *Procedia - Social and Behavioral Sciences*, 68, pp.578 – 589.

[15] Wahi, N. & Junaini, S., 2012. Residential satisfaction and the propensity to stay: An analysis of potential movers from low cost house. *ISBEIA 2012 - IEEE Symposium on Business, Engineering and Industrial Applications*, pp.585–590.

[16] Ariff, K. et al., 2016. Reviewing Youth Facility Requirements for Low-Cost Housing in Malaysia. *Procedia Social and Behavioral Sciences*, 222, pp.702–709. Available at: http://dx.doi.org/10.1016/j.sbspro.2016.05.231.

[17] Bajunid, A.F.I. & Ghazali, M., 2012. Affordable Mosaic Housing: Rethinking Low-Cost Housing. *Procedia - Social and Behavioral Sciences*, 49, pp.245–256.

[18] Department of Statistics, 2011. *Population distribution and basic demographic characteristics*,

[19] Economic Planning Unit, 2015. Strategy Paper 6: Providing Adequate and Quality Affordable House. *The Economic Planning Unit, Prime Minister's Department, Malaysia*, pp.1–14. Available at: http://rmk11.epu.gov.my/pdf/strategy-paper/Strategy Paper 06.pdf.

[20] Sulaiman, F.C., Hasan, R. & Jamaluddin, E.R., 2016. Users Perception of Public Low Income Housing Management in Kuala Lumpur. *Procedia - Social and Behavioral Sciences*, 234, pp.326–335. Available at: http://dx.doi.org/10.1016/j.sbspro.2016.10.249.

The High Rise Low Cost Housing : *Sustainable Neighbourhood Elements (Green Elements) in Malaysia*

Noraziah Wahi[1], Ismail Mohamad[2], Rosli Mohamad Zin[3], Vikneswaran Munikanan[4],Syahrizan Junaini[5]

[1] Faculty of Architectural,Planning and Surveying,Universiti Teknologi MARA, Malaysia
[2] Science Faculty, University of Technology Malaysia
[3]Faculty of Engineering, National Defence University of Malaysia
[4] Faculty of Civil Engineering, University of Technology Malaysia
[5]Brooke Dockyard and Engineering Works Corporation, Malaysia

Email: belle_rusd@yahoo.com

Abstract. The sustainable development is a vital measure to alleviate the greenhouse gas effect, global warming and any other environment issues. The sustainable neighbourhood concept is not new in Malaysia, However, the concept still needs attention and awareness from the stakeholders. This paper discusses on the sustainable neighbourhood elements specifically green elements application on the high rise low cost housing in Malaysia. Malaysia should have focused sustainable neighbourhood planning and design especially on the high rise low cost housing therefore the future generation can be benefited from this type development.

1. Sustainable Development

In recent years, as a result of developments in the globalisation era, the world has witnessed environmental, social, economic and urban planning changes. These changes have affected the living conditions of the population, natural resources, environment and economy [1]. However, the environmental principle requires that society protect its environmental resources. It assumes that ecosystems have limited regenerative capability and that the earth's land, air, water, and biodiversity will be compromised by irresponsible actions. Population growth combined with excessive consumption, pollution, and depletion of natural resources has pushed the limits of the earth's carrying capacity [2].

According to the Merriam-Webster Online Dictionary the meaning is, 'capable of being sustained', and 'of, relating to, or being a method of harvesting or using a resource so that the resource is not depleted or permanently damaged', and from the Collins English Dictionary, '(of economic development, energy sources, etc) capable of being maintained at a steady level without exhausting natural resources or causing severe ecological damage'[3]. While according to Ibrahim et al. 2015, the word sustainability derived from Latin word where it means to hold. Hence, sustainable development can be seen as a development that meets the needs of the present without compromising the ability of future generations to meet their needs.

IOP Conf. Series: Materials Science and Engineering **341** (2018) 012028 doi:10.1088/1757-899X/341/1/012028

Sustainable development requires a balance between economic growth, social expansion and environmental protection. In order to pursue sustainable development, the construction industry itself has to be sustainable and give emphasis to environmental matter in addition to economic gains and social obligations [6]. It also seeks to meet the needs and aspirations of the present without compromising the ability to meet those of the future. Far from requiring the cessation of economic growth, it recognizes that the problems of poverty and underdevelopment cannot be solved unless we have a new era of growth in which developing countries play a large role and reap large benefits [13]. During the past decade, the concept of sustainable development attracted significant attention from researchers, governments of different countries and international environmental organizations. The reason is clear, that is, the increased rate of environmental pollution and degradation [8].

2. Sustainable Development in Malaysia

Ever since the United Nations Conference on the Human Environment in 1972, Malaysia has been serious in dealing with environmental issues. In 1992, Malaysia showed its commitment on the Rio Summit after which the Malaysian National Environmental Policy was established. The policy became the basis for the country to give attention to environmental issues. The figure below represents Malaysia's sequence of actions and involvements since 1972 [9].

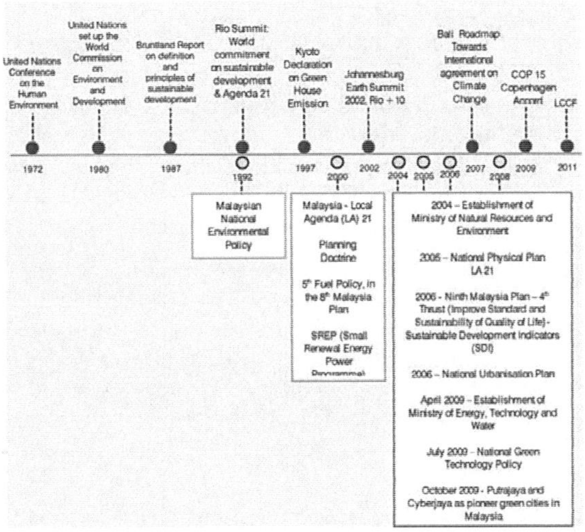

Figure 1.0: Malaysia's Involvements in Sustainability Development Agenda since 1972 [9]

Malaysia, like many countries across the world, is grappling with the challenge of balancing a growing population and demand, with a natural environment that is increasingly under stress. In a global context of increasing intensity and frequency of extreme weather events, adopting green growth has now become an imperative for Malaysia. It represents Malaysia's commitment to renew and, indeed, increase its commitment to the environment and long-term sustainability [5].

Efforts to achieve green and sustainable development in Malaysia implemented at the state, agency and organization levels through various programs, such as Local Agenda 21, formulation of Low Carbon Cities : Framework and Assessment System' by the Ministry of Energy, Green Technology and Water (KETTHA), Green Building Index' and `Green Building Index For Township' by GBI Sdn.

Bhd., Low Carbon Green City by Putrajaya Corporation, Low Carbon Society Blueprint for Iskandar Malaysia' by IRDA, implementation of 3R program by the National Solid Waste Management Department, implementation of rainwater harvesting, 1Malaysia Green 1Malaysia Clean Campaign and My Beautiful Neighbourhood' and my Beautiful Malaysia programs [10].

On 24th July 2009, the government of Malaysia unveiled the National Green Technology Policy (NGTP). This was a turning point in the history of initiatives on sustainable development in Malaysia where a policy focusing on technology, solution and road map to minimising impacts of development on the environment is formulated. The policy, built upon four pillars, namely energy, environment, economy and social aspects, underlines the following five main objectives [9]:-

i. Decreasing growth of energy consumption while enhancing economic development;
ii. Facilitating growth of the green technology industry and enhancing its contribution to the national economy;
iii. Increasing national capability and capacity for innovation in green technology development and enhancing Malaysia's green technology competitiveness in the global arena;
iv. Ensuring sustainable development and conserving the environment for future generations; and
v. Enhancing public education and awareness of green technology and encouraging its widespread use.

The policy also outlines the five (5) strategic thrusts on which the road map for implementation will be concentrated. The success of this policy and its initiatives will be further measured through three sets of sectoral indicators, namely environment, economy and social indicators. One of the indicators under the social perspective is that there "should be more cities, townships and communities in Malaysia embracing green technology and which are classified as green townships" [9].

Whilst green cities or townships have varied definitions and characteristics, more often than not, they have resulted in a definition equivalent to a 'sustainable city'. It makes more sense that a green city would offer long-term sustainability in a holistic manner. Thus the general definition of a green city can be considered to be the same as sustainable city where the characteristics are made up of the three tenets of sustainable development, namely environment, economy and the social perspective [9].

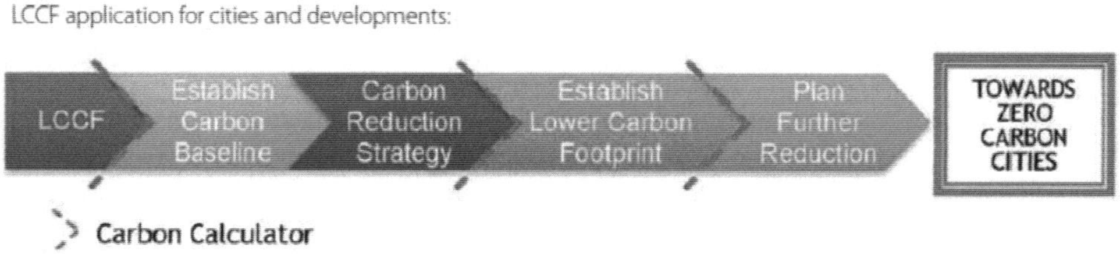

Figure 2.0: LCCF application for cities and developments [9]

The key environmental issues including climate change, improving air quality, and protecting the reserve forest and natural resources become the key principle of sustainable development. Now Malaysia is one of the few countries which are on track to meet their targets for reducing greenhouse gas emissions. Concepts like green building, sustainable living indicators and environmental regulation have a central role to play in this process. They can provide long term protection and improves the environment, and deter actions that are damaging to the environment. But there need much attention on the neighbourhood scale as it impacted the society until the grass- root level [12].

3. Sustainable Neighbourhood

Neighborhoods are building blocks of our cities, and, from the early years of twentieth century, planners have experimented with programs for improving the quality of life in urban neighborhoods (Sharif & Murayama 2013). To improve neighbourhood sustainability, a consideration of their buildings, public spaces, infrastructure, and cooperation between their parts is necessary (Reith & Orova 2015).

Green neighborhood can be defined as a neighborhood that is planned to be integrated with priority to the protection and use of natural resources, application of green technology, green practices and recycling, with the aim of preserving the environment, improving public health, safety and general welfare of city residents (JPBD 2012).

A growing interest in sustainable neighborhoods development has been evolved in the last few years (Abed et al. 2013). Neighbourhood concept was first introduced by Clarence Perry in 1910 to solve the problem of transportation in most of the urban centre and housing area. His concept evolved from Ebenezer Howard Garden City theory and social reformed aimed at growing urban population (Azmi & Karim 2012). Neighbourhood quality plays an important role for development, city officials, planners, realtors and researchers. In order for their construction projects to succeed, developers must understand the types of locations and amenities most valued by buyers (Ardeshiri et al. 2016).

The neighbourhood as a basic module in developing green cities becomes the front line to incorporate efforts in designs and activities to reduce greenhouse gas emissions while meeting a host of other community goals. As the quality of people's homes is influenced by the spaces around them, there is an increasing recognition that well-designed, well-managed green spaces by and in between housing are crucial to making neighbourhoods liveable, and contribute to people's quality of life (Rosly & Puzi 2010).

3.1 Sustainable Neighbourhood in Malaysia

Rapid urban development has contributed to the degradation of environmental quality especially the quality of water, air and noise. The unsustainable urban development has impact vastly in the urban neighbourhood infrastructures which lead to negative impact on environment and urban quality of living (Zakaria et al. 2012).

In December 2009, during the United Nations Conference on Climate Change Conference 2009 (COP15) in Copenhagen, Denmark, Malaysia has expressed its commitment to reduce carbon emissions up to 40% by 2020 compared to 2005. In order to achieve these objectives, various initiatives have been implemented immediately, especially by state and local authorities, developers and so on as the executor (JPBD 2012).

Principles of green neighbourhood set by (JPBD 2012)are as follows:

(i) emphasis on sustainability and conservation of the environment;
(ii) provide a community area that emphasizes the beauty and comfort of the population;
(iii) has the characteristics of affordable ie for people able to carry out economic activities, have a home, work and recreation in the neighborhood
(iv) creating a harmonious, prosperous, healthy society;
(v) creating local economic activity that is competitive;
(vi) create a highly accessible to the residents of connecting public transport and walking and cycling;
(vii) offers a variety of facilities and services for the daily activities of the population;
(viii) using renewable energy and energy efficiency;
(ix) provide green infrastructure such as rainwater harvesting and district cooling system; and
(x) designing human-scale neighborhood development as in the design of the streetscape, buildings, public spaces and so to ensure a green neighborhood can work well, comfortable and safe;

Federal Department of Town and Country Planning had formulated policies, plans, guidelines and assessment system for urban sustainability, namely [10]:

(i) Malaysian Urban Indicators Network (MURNInet –since 2002) and Malaysia Urban-Rural National Indicator Network for Sustainable Development (MURNInets – 2012). MURNInet and MURNInets were aimed to assess and improve the sustainability level of towns in Malaysia through the performance of a set of sustainable development indicators identified. It is

recognised nationally as one of the most important programs ever undertaken in the country to measure urban sustainability;
(ii) Planning Guidelines for Green Neighbourhood;
(iii) Implementation of Five Green Neighbourhood Initiatives;
(iv) Planning Guidelines for Livable Compact City;
(v) Planning Guidelines for `Ubiquitous City'; and
(vi) Planning Guidelines for the Elderly.

3.2 Green Building in Malaysia

The earth's surface needs to retain some of the sun's heat in order to regulate mean global temperatures, and naturally occurring gases in our atmosphere such as the ozone, water vapour, methane, nitrous oxide and carbon dioxide serve this purpose by trapping the required amount of heat from the sun so that conditions are conducive to the survival of all living creatures. Global temperature records show that since the start of the industrial age, the temperature of the earth has not risen so rapidly compared to any given 100-year block of recorded history, since the start of recorded history era to date. Carbon dating of the ice columns drilled out of the polar caps also gives support to this evidence by tracing history back to a few million years [9].

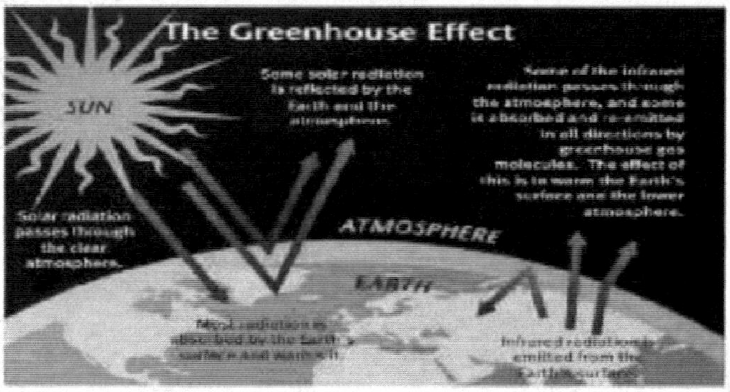

Figure 3.0: The Greenhouse Effect [9]

The building is one of the key elements symbolize a green neighborhood. The main goal of the construction of buildings in the neighborhood are green buildings that meet green building features, namely (JPBD 2012),

i. use energy efficiently;
ii. using recycled materials for the construction of buildings;
iii. planning and management of sustainable building
iv. using green technology innovation;
v. produce quality interior ambience and comfortable for the user; and
vi. reflect the identity and culture of the local community.

The promotion of green strategy in housing development has significant contribution to the implementation of sustainable development principles (Zhang et al. 2011). It is believed that 21st century cities must be greener and smarter; hence, promoting sustainable issue for many developing countries. The concept of sustainability is a broad global issue comprising various interrelated studies about people, the environment and society (Ghaffarianhoseini et al. 2013).

4. Methodology

A self-administered survey was conducted to collect data regarding issues and problems of the low cost housing residents. The respondents eligible to participate in the survey were those who stay in the high-rise low cost housing. The data were collected by face-to-face interviews and distribution of questionnaires survey.

For the purpose of analysis of quantitative data collection through questionnaires survey, the Statistical Package for Social Science (SPSS) software version 20.0 was adopted. The objectives of using quantitative approach is to summarizes statistically large information such as questionnaires responses feedback translated into figures that can be understood and interpreted.

5. Results and Findings

The research study was conducted in the state of Selangor in peninsular Malaysia. Census 2010 revealed that the total population of Malaysia was 28.3 million. Selangor state has a geographical area of 8,104 square kilometers and a population of 5,462,141 as the most populated state in Malaysia, census 2010 [4].

There were 700 sets of questionnaires distributed to all occupants in the high rise low cost housing in Selangor. However, there were only 510 sets of questionnaires had been returned to the researcher. The questionnaires survey contained two section. These were Demographic information and Sustainable Neighbourhood Elements (Green Elements).

5.1. Section A of the questionnaires: Demographic Information

Section A of the questionnaire survey provided data on the age, gender, marital status, race, education level, total household income levels and employment status of the respondents, Table 1.0 illustrate the demographic information of the respondents in Selangor.

Table 1.0: Demographic information of the respondents

No	Item	Frequency	Percentage
1.	***Age (years):***		
	20-30	75	14.7
	31-40	170	33.3
	41-50	170	33.3
	Above 50	95	18.6
2.	***Gender:***		
	Female	280	54.9
	Male	230	45.1
3.	***Marital status:***		
	Married	387	75.9
	Single	67	13.1
	Single mother	39	7.6
	Single father	17	3.3
4.	***Race:***		
	Malay	199	39.0
	Indian	153	30.0
	Chinese	158	31.0
5.	***Home ownership:***		
	Owner	179	35.1
	Rent from landlord/owner	167	32.7
	Rent from State Government	164	32.2
6.	***Length of occupancy:***		
	Below 1 year	2	0.4
	2-4 years	72	14.1
	5-7 years	138	27.1
	8-10 years	150	29.4
	More than 10 years	148	29.0
	TOTAL	**510**	**100%**

Based on the Table 1.0, 14.7% of the respondents were aged 20-30 years old. A further 33.3% of the respondents fell into 31-40 and 41-50 years old bracket respectively. Aged above 50 years old contributes to 18.6% of the total respondents. Majority respondents residing in low cost housing in Selangor are aged range between 31-50 years old. The results turned out to have similarities such as a study conducted by Sulaiman et al., 2016 that stated the highest of respondent age is 31 to 50 years old are commonly living in low-cost housing area in Malaysia

There were slightly more female (54.9%) than male (45.1%) among the 510 respondents involved. This might be because the distribution of questionnaires took place during weekday working hours; many male dwellers would have been at work and unable to respond to the questionnaires. Based on the Table 2.0, it is stated that 75.9% of the respondents are married; 13.1% of the respondents are single; 7.6% of the respondents are single mother; and 3.3% of the respondents are single father. As for ethnicity, there were Malay (39%), Chinese (31%) and Indian (30%). The proportion of race is vital to know to make sure that there is multicultural race exist in the high rise low cost housing area community.

It is also showed that 35.1% of the respondents owned their house during the case study period. Another 64.9% of the respondents rented their house, either rented from landlord or the state government. People usually rented a house as they feel convenient to travel to their working place and can save more time. They also can avoid traffic congestion during the peak hours.

There were 29.0% of the respondents had residency lengths of more than 10 years, 29.0% had habitation periods of 8–10 years, 27.1% had habitation periods of 5–7 years, 14.1% had a length of residency of 2–4 years and a further of 0.4% had lived in the low cost house below than a year. This high level prevalence of occupancy for periods over four years as described indicates that the respondents were well-placed to convey their experiences of living in the high rise low cost housing neighbourhoods. These demographic results have; thus, confirmed that the respondents in this case study are reasonably representative of a low cost housing in Malaysia. The duration of occupancy equal to five years or more, means that these respondents have, potentially, developed a sense of community and neighbourliness in the high rise low cost housing in Malaysia.

5.2. Sustainable Neighbourhood Elements (Green Elements)

Section B of the questionnaire requires the respondents to rate their level of agreement regarding proposed green elements in the high rise low cost housing. The respondents were need to rate their agreement level based on five numerical Likert-scale; "1" (Strongly agree), "2" (Agree), "3" (Neutral), "4" (Disagree), and "5" (Strongly disagree). Table 2.0 illustrates the occupants' agreement level towards the listed elements.

Table 2.0: Percentage of 'strongly agree and agree' level on the high rise low cost housing green elements

No	Green elements	Strongly agree	Agree	Strongly Agree & Agree	Percentage
1	Shady trees	344	123	467	**91.6%**
2	Diversifying landscaping	334	86	420	**82.4%**
3	Preserved greenery	165	166	331	**64.9%**
4	Green network	188	131	319	**62.5%**
5	Garden	266	43	309	**60.6%**
6	Green roof	186	120	306	**60.0%**
7	Building orientation	195	111	306	**60.0%**
8	Community agriculture	177	122	299	**58.6%**
9	Tall trees	89	97	186	**36.5%**
10	Natural ponds	71	114	185	**36.3%**

As shown in Table 2.0 above, 8 out of 10 elements were having more than 50% respondents that strongly agree and agree to the sustainable green element. Top of the list is provision of shaded trees at 91.6%. Diversifying landscape followed next with 82.4% as this gives no only the comfort of having greenery but provide the aesthetic look to the neighbourhood area. 64.9% respondents both strongly agree and agree that natural greenery at the low cost housing area must be preserved. Secondary or planted greenery may takes time to grow and mature. Availability of garden is agreed and strongly agreed by 60.6% of respondents. Subsequently, 60.0% respondents in in support of garden roof. The next green element is related to construction of the building orientating to the direction of the sun. This element is agreed by 60.0% of the respondents. Tall trees and natural ponds both recorded 36.5% and 36.3% respectively.

Table 3.0: Percentage of 'strongly disagree and disagree' level on the high rise low cost housing green elements

No	Green elements	Strongly disagree	Disagree	Strongly disagree & disagree	Percentage
1	Natural ponds	14	156	170	**33.3%**
2	Community agriculture	53	83	136	**26.7%**
3	Tall trees	33	80	113	**22.2%**
4	Garden	71	38	109	**21.4%**
5	Building orientation	38	68	106	**20.8%**
6	Green network	33	32	65	**12.7%**
7	Preserved greenery	13	48	61	**12.0%**
8	Diversifying landscaping	17	31	48	**9.4%**
9	Green roof	17	24	41	**8.0%**
10	Shady trees	20	3	23	**4.5%**

Green elements are often related to the environmental factor. It is shown in the above Table 3.0 that respondents' response to disagree or strongly disagree with green element is low. It indicates that respondents agree that sustainable green elements are important to be implemented at their low cost housing area. Nonetheless there are 33.3% respondents disagree or strongly disagree with the provision of natural pond at their housing area. Natural pond acts as climate control to its surrounding area. Community agriculture element is ranked second at 26.7% and followed by tall trees element with 22.2% respondents either disagree or strongly disagree. Allocation of an area within or neighboring to the low cost area for the purpose of small agricultural activities also serves as a common socializing place for the residents. Besides, residents may sell the crops for additional income. Another 21.4% respondents disagree with the provision of garden. Merely 20.8% respondents disagree with the building orientation as they do not foresee any significant impact on their daily lives. Three elements having less than 10% respondents that disagree and strong disagree respectively are diversifying landscaping (9.4%), green roof (8.0%) and shady trees (4.5%). These elements are regarded as one of the important green elements for implementation at low cost housing area. The second last green elements having only 8.0% respondents disagree that green roof is to be made available at their low cost house. This could be due to the understanding on the actual meaning and the effect of implementation of green roof. Only 4.5% respondents disagree that shady trees must be provided at low cost housing area.

Table 4.0: Percentage of 'neutral' level on the high rise low cost housing green elements

No	Green elements	Neutral	Percentage

1	Tall trees	211	**41.4%**
2	Green roof	163	**32.0%**
3	Natural ponds	155	**30.4%**
4	Green network	126	**24.7%**
5	Preserved greenery	118	**23.1%**
6	Building orientation	98	**19.2%**
7	Garden	92	**18.0%**
8	Community agriculture	75	**14.7%**
9	Diversifying landscaping	42	**8.2%**
10	Shady trees	20	**3.9%**

Table 4.0 above shows percentage of respondents in neutral position, neither they agree nor disagree with sustainable green elements. Top three elements are tall trees (41.4%), green roof (32.0%) and natural ponds (30.4%). Respondents approach these elements in view of associated risk it carry or lack understanding on the benefits of these elements.

Chart 1.0: Comparison of agreement level on the sustainable green elements

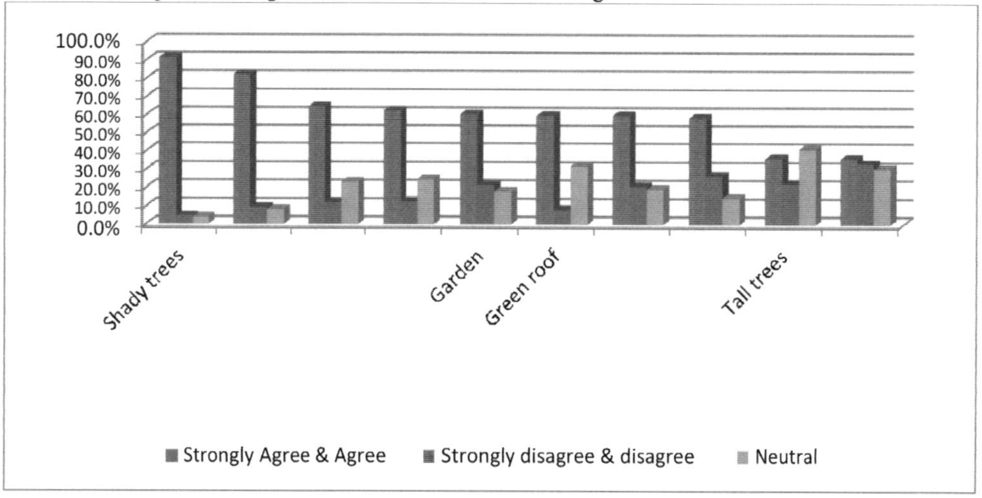

Chart 1.0 above shows the comparison of respondents that are strongly agree and agree, strongly disagree and disagree, and neutral on the sustainable green elements at the high rise low cost housing area. In general, respondents strongly agree and agree with all green elements except for tall trees and natural pond which respondents for strongly disagree, disagree and neutral surpasses those who strongly agree and agree. It is understood that both of these green elements carry risks for example parents are concern with the presence of natural pond within the vicinity of their housing area which may lead to issues of children playing in the pond or even drowning. Nonetheless, from the sustainable green element perspective, these elements are necessary to be made available. Shady trees are more preferred by respondents as opposed to tall trees as there were many cases of tall trees uprooted due to strong wind during bad weather. It is apparent that safety factor plays important role from the mindset of respondents as opposed to the benefits brought by these essential green elements.

6. Discussion

Globalization era has reshaped the needs for the provision of sustainable development in order to cater for the demand in providing better living conditions, preserving the environment and natural resources and at the same time continue on with the development. Malaysia is not left behind in terms of efforts and commitment in dealing with environmental issues. Policies, guidelines and targets had been set by public and private sector to ensure environmental issues is addressed in every development phase of construction.

The research focuses on implementation of sustainable green elements at high rise low cost housing is Selangor. It is apparent from the result that majority low cost dwellers agree that sustainable green element must be implemented at their low cost housing area. This shows that the level of understanding on the requirement and the needs for such elements to be present is acceptably high. Other stakeholders such as the developers and local authorities are at the position to put in place such environmental elements. The cost of implementation of sustainable green elements at the high end buildings might be acceptable but to implement such initiatives at low cost housing is still a challenge due to the associated cost to provide for such facilities. However this must not be the reason which leads to lack of implementation albeit requirements and standards set by the Government. Nonetheless, some of the green element can also be provided by the dwellers for example by planting shaded tress and creating specific area for a garden and landscaping.

References

[1] Alqahtany, A., Rezgui, Y. & Li, H., 2013. A proposed model for sustainable urban planning development for environmentally friendly communities. *Architectural Engineering and Design Management*, 9(3), pp.176–194.

[2] Bansal, P., 2002. The corporate challenges of sustainable development. *Academy of Management Executive*, 16(2), pp.122–131.

[3] Burnett, J., 2007. Sustainability and Sustainable Buildings. *The Hong Kong Institution of Engineers Transactions*, 14(3), pp.1–9.

[4] Department of Statistics, 2011. *Population distribution and basic demographic characteristics*,

[5] Economic Planning Unit, 2015. *Eleventh Malaysia Plan 2016-2020 (Anchoring Growth on People)*, Available at: http://rmk11.epu.gov.my/book/eng/Elevent-Malaysia-Plan/RMKe-11 Book.pdf.

[6] Hussein, T.S.D.I.J. et al., 2010. Towards Sustainable Construction: Malaysian Construction Industry Initiatives. *The Ingenieur*, 44(DEC 2009-FEB 2010), pp.8–15.

[7] Ibrahim, F.I., Omar, D. & Mahamad, N.H.N., 2015. Theoretical Review on Sustainable City Indicators in Malaysia. *Procedia - Social and Behavioral Sciences*, 202(December 2014), pp.322–329. Available at: http://www.sciencedirect.com/science/article/pii/S1877042815048843.

[8] Jalil, D.M.A., 2010. Sustainable Development in Malaysia: A Case Study on Household Waste Management. *Journal of Sustainable Development*, 3(3), pp.91–102.

[9] Kementerian Tenaga Teknologi Hijau dan Air, 2011. *Low Carbon Cities Framework and Assessment System*,

[10] Khir, M.F.H.M., 2013. Town Planning Expert Talk 2013 Exploring the `Future Cities'. *Federal Department of Town and Country Planning,Peninsular Malaysia*, (May).

[11] Sulaiman, F.C., Hasan, R. & Jamaluddin, E.R., 2016. Users Perception of Public Low Income Housing Management in Kuala Lumpur. *Procedia - Social and Behavioral Sciences*, 234, pp.326–335. Available at: http://dx.doi.org/10.1016/j.sbspro.2016.10.249.

[12] Vikneswaran, M., Said, M.I.M. & Zakaria, R., 2010. Implementation of sustainable neighbourhood elements (SNEs) model in Malaysian neighbourhood planning system. In *CRIOCM 2010 - International Symposium on Advancement of Construction Management and Real Estate "Towards Sustainable Development of International Metropolis,Chinese Research Institute of Construction Management.* pp. 96–105.

[13] World Commission on Environment and Development, 1987. Report of the World Commission on Environment and Development: Our Common Future (The Brundtland Report). *Medicine, Conflict and Survival*, 4, p.300.

Evaluation of Embedded System Component Utilized in Delivery Integrated Design Project Course

Syed Abdul Mutalib Al Junid*[1,2], Yusnira Hussaini[2], Fairul Nazmie Osman[1,2], Abdul Hadi Abdul Razak[1,2], Mohd Faizul Md Idros[1,2], Abdul Karimi Halim[1,2]

[1] Electronic Architecture and Application Research Interest Group (EArA), Faculty of Electrical Engineering, Universiti Teknologi MARA, Shah Alam, Selangor, Malaysia

[2] Centre for Electronic Engineering Studies, Faculty of Electrical Engineering, Universiti Teknologi MARA, Shah Alam, Selangor, Malaysia

Email: samaljunid@salam.uitm.edu.my,

Abstract. This paper reports the evaluation of the embedded system component utilized in delivering the integrated electronic engineering design project course. The evaluation is conducted based on the report project submitted as to fulfil the assessment criteria for the integrated electronic engineering design project course named; engineering system design. Six projects were assessed in this evaluation. The evaluation covers the type of controller, programming language and the number of embedded component utilization as well. From the evaluation, the C-programming based language is the best solution preferred by the students which provide them flexibility in the programming. Moreover, the Analog to Digital converter is intensively used in the projects which include sensors in their proposed design. As a conclusion, in delivering the integrated design project course, the knowledge over the embedded system solution is very important since the high density of the knowledge acquired in accomplishing the project assigned.

1. Introduction

Nowadays, the revolution of technology is moving towards the Industrial Revolution (IR) 4.0 which includes the cyber-physical systems, the Internet of things, cloud computing and cognitive computing as reported in [1-2]. Undoubtedly, all the area listed in the IR 4.0 required the integration of the embedded system in solving the problem towards creating the new solution. In more specific, the integration of the system required expert to detail out the requirement and analysis over the sub-system integration of embedded system before the deployment taking part especially for the embedded system controller.

On the other hand, in embedded system technology, there are enormous numbers of embedded solution available on the shelf which allowed the system integrator and designer to choose the best as to accomplish their idea on the solution development according to their project and design constraints [3]. The selection can be in the form of lower end of the embedded solution up to the high-end without neglect that the cost and level of complexity of the solution are the others influencing factors as well to be considered.

Despite from the demand in the industry, the electronic engineering based degree program will be exposed to the embedded system course as to prepare the students with the current technology used in the industry. Although the implementation of the course is differing from one university to the others, the important fundamental element to expose the student with the embedded system controller is the main idea of the course implementation. This includes the basic knowledge of the embedded system exposure and the utilization of the embedded system towards completion of the specific engineering problem given.

In Faculty of Electrical Engineering, Universiti Teknologi MARA, Shah Alam Selangor, Malaysia, the fundamental of embedded system is teach in semester 6 under the Embedded System Design and Interfacing course. As to apply the knowledge acquired, an Engineering System Design was introduced at semester 8 which required the student to use the fundamental knowledge acquired from the subject to solve the engineering problem and come out with a complete prototype of the project. As to evaluate the density of knowledge used in the project, this paper will assess the component of embedded system utilized in the project based on the main important components taught at the embedded system course.

2. Methodology of Evaluation

The evaluation was conducted at Faculty of Electrical Engineering Universiti Teknologi MARA Shah Alam, Selangor, Malaysia. A group of Bachelor of Electronics Engineering degree program student taking Engineering System Design course from March 2017 to July 2017 session were assessed in this studies. There are six projects submitted by the student with the main theme to design new equipment and tools for material and devices fabrication process laboratory. The evaluation is conducted by assessing the component of embedded system used in the project from the final report and prototype submitted by the student at the end of the semester.

3. Result and Discussion

There are six completed projects have been assessed in this embedded system component assessment including the Low Cost Spin Coater Machine as shown in **Figure 1**.

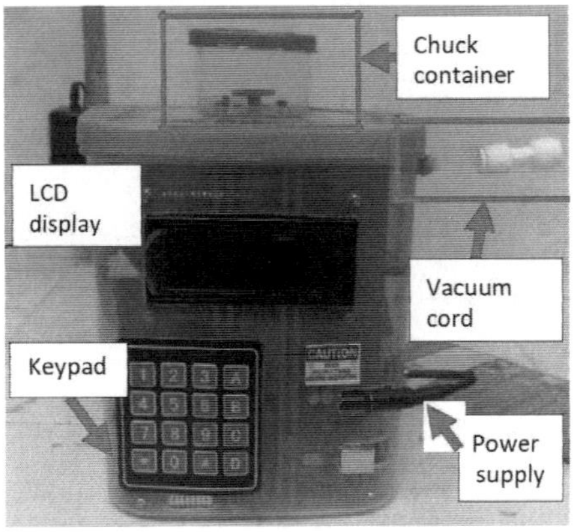

Figure 1. Low cost spin coater machine

As depicted in the **Figure 1**, the Low Cost Spin Coater Machine is using ATmega2560 development board as main controller, keypad to control the speed level and timing for coating, motor as a spinner, vacuum pump to hold the slide and LCD as display and indicator. Moreover, the detail projects submitted and assessed in this embedded system component assessment are listed in **Table 1**.

Table 1. List of the project submitted for Engineering System Design course

No	Project Title	Description of Project
1	Hot Plate Stirrer Machine	The project is to design and develop a complete hot plate stirrer at low cost which function to heat and stir the sample at the same time.
2	Low Cost Spin Coater Machine	The project is to design and develop a reliable and low cost spin coating machine which function to deposit uniformly thin film to the substrate
3	Low Cost Dry Cabinet	The project is to design and develop a low cost dry cabinet which function to provide a dry and low level of humidity container
4	Dip Coating Machine	The project is to design and develop a low cost dip coating machine which function to dip and formation of thin film
5	Ultrasonic Cleaner	The project is to design and develop ultrasonic cleaner which function is to remove the contamination from the object with high frequency of sound waves
6	Glove Box	The project is to design and develop glove box which function is to provide the sealed container for allow one to manipulate objects

As recorded in **Table 1**, the main objective of the project is to design and develop a new solution for the material and fabrication research laboratory with 66.7% focusing on producing low cost design solution as their main focus or element in their design. In addition, the type of controller is one of the main influencing factors for the cost of the project. Therefore, the type of microcontroller and its main board type were assessed in this evaluation and the result is summarized in Table 2.

Table 2. List of the controller type and main board used in the project

No	Project Title	Type of Microcontroller	Main Board
1	Hot Plate Stirrer Machine	ATmega 2560	Development board
2	Low Cost Spin Coater Machine	ATmega 2560	Development board
3	Low Cost Dry Cabinet	ATmega 2560	Development board
4	Dip Coating Machine	ATmega 2560	Development board
5	Ultrasonic Cleaner	ATmega 2560	Development board
6	Glove Box	ATmega 2560	Development board

As summarized in Table 2, all of the projects are using Arduino platform which one of the famous cost effective open-source electronic prototyping platform been used before in many of the prototyping project as reported in [4]-[10]. All of the projects are using ATmega 2560 development board in their project.

There two main components of embedded system assessed from the project; hardware and software. In the hardware components part, the assessment include printed circuit board (PCB) fabrication, system input and output, hardware components and controller type. On the other hand, the software part focusing on the type of programming language used and embedded system function/library. The result for the embedded system hardware components assessment is recorded in **Table 3**.

Table 3. Embedded system hardware based component assessment

No	Hardware components assessment	List of Project					
		1	2	3	4	5	6
	Input and Output (I/O)						
1.1	Number of input pin used	10	8	7	8	7	8
1.2	Number of output pin used	12	12	10	12	14	10
1.3	Number of Analogue to Digital Converter (ADC) pin	1				1	2
	Mounting type technology						
2.1	Through hole technology only						
2.2	Surface mount technology (SMT) only						
2.3	Mix of both	√	√	√	√	√	√
	Circuit board type technology for I/O						
3.1	PCB						
3.2	Custom PCB	√	√	√	√	√	√
3.3	Donut board						
3.4	Straight board						
3.3	Breadboard						
	Type of common component used						
4.1	Liquid Crystal Display (LCD)	√	√	√	√	√	√
4.2	Push button	√					√
4.3	Relay	√	√	√		√	√
4.4	Keypad	√	√	√	√	√	
4.5	Solenoid	√	√	√			√
4.6	Heater	√				√	
4.7	Transducer					√	
4.8	Valve		√				√

As recorded in **Table 3**, all of the projects are using more than 16 inputs and outputs to support the sensor and others components connected to the microcontroller. Therefore, the ATmega2560 are been selected to support the inputs and outputs requirement. On the other hand, the mix of mounting type of technology are recorded in the projects; SMT and through hole. All of the projects are developing their own PCB as their application board to be connected to the ATmega2560 prototyping board. From the list of components used in the project, all of the projects are using LCD to display the status, setting and configuration of their project.

Table 4. Embedded system software based component assessment

No	Software components assessment	List of Project 1	2	3	4	5	6
	System function / library						
1.1	Liquid crystal display	√	√	√	√	√	√
1.2	Secure Digital (SD) memory function						
1.3	Serial Peripheral Interface (SPI)	√	√	√	√	√	√
1.4	Basic digital input/output function	√	√	√	√	√	√
1.5	Analogue to digital converter (ADC)	√				√	√
1.6	Keypad function	√	√	√	√	√	
1.7	Interrupt function	√	√		√		
	Type of programming language for microcontroller						
2.1	Assembly language						
2.2	C/C++ programming language	√	√	√	√	√	√
2.3	Mix of both						

As recorded in **Table 4**, all of the projects are using C/C++ programming as their programming language for the project. Moreover, In Serial Programming (ISP) method was used to program the microcontroller for all of the projects. All of the project On the other hand, all of the projects are using basic digital input/output function in their project as to control the switches, LED and buzzer. Although the ADC function is significantly related to the interfacing of the analogue sensor, three projects are using this ADC function to read their sensors value to the microcontroller. The interrupt function has been used by (1), (2) and (4) as to control the operation of the machine.

4. Conclusion

Evidently, the embedded system components are intensively utilised in delivery the engineering system design project proposed. These assessments clearly show that adoption of the embedded system components is very high in delivery and successful implementing the engineering system design project with almost all of the components learn are covered in completion of the project. Finally, this is shows that, the implementation of integrated design project is the best platform for students to apply their embedded system knowledge intensively.

5. References

[1] Yongxin Liao, Fernando Deschamps, Eduardo de Freitas Rocha Loures & Luiz Felipe Pierin Ramos, 2017 Past, present and future of Industry 4.0 - a systematic literature review and research agenda proposal *International Journal of Production Research* **55**:12, p3609-3629 DOI: 10.1080/00207543.2017.1308576

[2] R. Drath and A. Horch, 2014 Industrie 4.0: Hit or Hype? [Industry Forum] *IEEE Industrial Electronics Magazine* vol.**8** no.2 p56-58 doi: 10.1109/MIE.2014.2312079

[3] N. Parimala & S.R.N. Reddy 2015 Processor Selection for Embedded System Design, *International Journal of Computers and Applications* **30**:4 p348-353 DOI: 10.1080/1206212X.2008.11441916

[4] Rashid Z.A., Thani S.K.S.O., Al Junid S.A.M. 2014 Trees' cooling effect on surrounding air temperature monitoring system: Implementation and observation International *Journal of Simulation: Systems, Science and Technology* **15** 70-77

[5] Husin R., Al Junid S. A. M., Abd Majid Z., Othman Z., Md Shariff K. K., Hashim H. and Saari M. F. 2012 Automatic Street Lighting System for Energy Efficiency Based on Low Cost Microcontroller *International Journal of Simulation: Systems, Science and Technology* **13** 29-34

[6] Othman N., Manan M. I. A., Othman Z., Al Junid S.A.M., 2013 Performance analysis of dual-axis solar tracking system *Proc. IEEE Int. Conf. on Control System, Computing and Engineering (ICCSCE 2013)* p370-375

[7] Tahir N. A. M., Al Junid S. A. M., Othman Z., Abd Majid Z. and Thani S. K. S. O. 2012 Design Automatic Meter Reading (AMR) Data Logger with Xbee *International Journal of Simulation: Systems, Science and Technology* **13** 67-73

[8] Ussdek M. E. M., Al Junid S. A. M., Majid Z. A., Osman F. N. and Othman Z. 2013 High-Sensitivity Gas Detection and Monitoring System for High-Risk Welding Activity *Proc. 2013 IEEE Conf. on Systems, Process & Control (ICSPC)* p256-261

[9] Bakri, N. A. M.; Al Junid, S. A. M.; Razak, A. H. A.; Idros, M. F. M.; Halim, A. K., 2015 Mobile carbon monoxibe monitoring system based on Arduino-Matlab for environmental monitoring application *IOP Conference Series: Materials Sciences and Engineering* **99** p1-10

[10] Ahmad Faizal Zainal Abidin, Mohammad Huzaimy Jusoh, Elster James, Syed Abdul Mutalib Al Junid, Ahmad Ihsan Mohd Yassin, 2015 Real-Time Remote Monitoring with Data Acquisition System *IOP Conference Series: Materials Sciences and Engineering* **99** p1-7

Acknowledgments
The authors wishing to acknowledge the Universiti Teknologi MARA, Malaysia for the financial support under the ARAS Research Grant Scheme (ARAS): ARAS (0040/2016).